D0202429

Darwin to DNA,
Molecules to Humanity

Darwin to DNA,

Molecules to Humanity

G. Ledyard Stebbins

Professor Emeritus
University of California, Davis

W. H. Freeman and Company
San Francisco

Project Editor: *Larry Olsen*
Copy Editor: *Nancy Segal*
Designer: *Sharon H. Smith*
Production Coordinator: *Linda Jupiter*
Illustration Coordinator: *Cheryl Nufer*
Artists: *Darwen and Vally Hennings, John and Judy Waller*
Compositor: *Graphic Typesetting Service*
Printer and Binder: *The Maple–Vail Book Manufacturing Group*

Library of Congress Cataloging in Publication Data

Stebbins, G. Ledyard (George Ledyard), 1906–
    Darwin to DNA, molecules to humanity.

    Bibliography: p.
    Includes index.
    1. Evolution.   I. Title.
QH366.2.S718   575   81-15152
ISBN 0-7167-1331-4   AACR2
ISBN 0-7167-1332-2 (pbk.)

Copyright © 1982 by W. H. Freeman and Company

No part of this book may be reproduced by any mechanical, photographic, or electronic process, or in the form of a phonographic recording, nor may it be stored in a retrieval system, transmitted, or otherwise copied for public or private use, without written permission from the publisher.

Printed in the United States of America

1 2 3 4 5 6 7 8 9 0   MP   0 8 9 8 7 6 5 4 3 2

QH
366.2
.S718
1982

Contents

308223

Preface

As a scientist, I have devoted my career to the study of the evolution of plants, animals, and microorganisms from their earliest beginnings. I am often asked, by my friends and academic colleagues who are not trained biologists, what one should read in order to be well informed about modern concepts of evolution. Until about ten years ago, the answer was easy: For animals and plants, read *The Meaning of Evolution* by George Gaylord Simpson, and for human beings, read *Mankind Evolving* by Theodosius Dobzhansky. During the past decade, however, knowledge about evolution has increased in a spectacular fashion, and the central focus of evolutionary theory has undergone a radical shift.

In the wake of the molecular revolution in biology, information provided by biochemists about both the processes and the directions of evolution is in some ways more precise than that which paleontologists, morphologists, and systematists had previously gathered by looking at organisms with the naked eye. The tremendous increase of interest in ecology, shared by both biologists and laymen, focused attention on the evolution of biotic communities and the interrelationships among the various organisms in a community. Charles Darwin had explored this aspect of evolution, but now scientists could express their hypotheses in quantitative terms and test certain theories by controlled experiments. Finally, the rise of the discipline of ethology—the study of animal behavior—and the succeeding attention given to the new discipline of sociobiology have made possible a new evaluation of the relationship of human beings to other kinds of organisms, particularly the apes and the still poorly known ancestors that apes share with us.

All these recent scientific discoveries call for a new, nontechnical book on evolution. Because I taught heredity and evolution to nonbiologists at the University of California at Davis for six years before my retirement in 1973 and lectured on the subject on other campuses for four years thereafter, I felt qualified to accomplish this task. The result is the present volume. Particularly in those fields with which I was previously less familiar, such as vertebrate paleontology, biochemistry, anthropology, and history, I have read extensively, discussed these topics with many other scientists, and prevailed upon my colleagues to review the manuscript for scientific accuracy. I believe, therefore, that the facts presented here are correct and that the interpretations are in accord with opinions of the best-informed experts in these fields. Nearly all biologists accept evolution as the only valid explanation for the origin of the millions of animals, plants, and microorganisms that populate the earth, and those who know the facts best agree with Dobzhansky that "nothing in biology makes sense without evolution."

The book is organized according to a definite plan, but it differs in this respect from general surveys of physics, chemistry, mathematics, and other sciences that study the inanimate universe in that it is not based on a framework of general laws. The living world is so diverse and has varied so much through time that all broad generalizations about biological phenomena must be qualified by numerous complications and exceptions. An oft-repeated aphorism is that the only generalization or law that holds without exception in biology is that exceptions exist to every law. Generalizations about evolution are unquestionably subject to this aphorism. For instance, natural selection is often advanced as a primary law that governs evolutionary change. Nevertheless, if an environment remains constant with respect to the adaptive needs of a population, natural selection retards or even inhibits, rather than promotes, evolutionary change.

What, then, is the most appropriate framework for a general discussion of evolution? Because of my fondness for classical music, I have chosen the framework of theme and variations. The first movement (Part One) begins with a brief introductory chapter, which raises the main questions we shall attempt to answer and states the main theme of the entire work, that evolutionary change depends chiefly on the ability of populations to respond, over shorter or longer periods of time, to challenges posed by changing environments. This theme is then developed, first separately and then as a whole. The time scale receives first attention, followed by the genetic structure of populations, which permits some responses and inhibits others (Chapter Two). The medium of interaction between populations and their environments, natural selection, is briefly developed in Chapter Three. Chapter Four is a contrapuntal three-part invention in which time, population structure, and the nature of environments are interwoven in such a way as to make their interrelationships as clear as possible. The movement closes (Chapter Five) with a contemplative section, andante con moto, that reflects upon the nature of novelty and its relationship to quantitative changes.

The second movement (Part Two) is a series of variations on the main theme, based on successive changes that populations and the biotic factors of their environments have undergone through time. The variations have been selected to illustrate both the origin of biological diversity and the succession of changes that led to the origin of humanity.

The third movement (Part Three) deals particularly with our own species, humanity. In it, the main theme of genetic heredity and natural selection is interwoven with a similar but in some ways contrasting theme, emphasizing cultural selection of traits that originate via invention and are transmitted by imitation, teaching, and learning. This theme begins softly but gains in power until, at the end of the movement, in Chapters Thirteen and Fourteen, it becomes the dominant motif of human history. The movement ends with another contemplative section in Chapter Fifteen. Since this movement represents part of an ongoing process, its coda cannot be a series of chords that denote finality. Instead, in addressing the question of human evolution in the future, the motif of future uncertainty colored with hope is the only one that appears acceptable to me.

Following Chapter Fifteen I have provided Notes for each chapter, which include suggestions for further reading. The complete citations are listed in the References following the Notes. Key terms have been set in italic typeface in the text, and definitions for most of these terms appear in the Glossary.

I have been greatly aided in this project by many friends, both scientists and laymen, who have read all or parts of the manuscript. They are Austin Armer, Francisco Ayala, Robert Boyd, Charles Brown, Curtis Clark, William Clement, John Curtin, Thomas Dietz, Stephen Jay Gould, William J. Hamilton, Richard Lewontin, Ernst Mayr, Timothy Prout, Peter Richerson, Donald Savage, Lyman Donald Smith, Steven Stanley, James W. Valentine, Michael Vasey, Peter Ward, Ernest Williams, Edward O. Wilson, and Steel Wotkyns. In addition, students in two classes—one taught in collaboration with John Curtin at San Francisco State University in 1978 and the other at The Ohio State University in 1979—read early typescripts and offered many valuable comments. I am particularly indebted to my wife Barbara, who has not only discussed my ideas with me but has also made easier my hours of work on the manuscript, sometimes under difficult conditions. I also would like to thank Darwen and Vally Hennings and John and Judy Waller for the beautiful drawings that appear throughout the book. Susan Neumann and Linda Mijangos typed the manuscript. Naturally, none of these people are in any way responsible for the final ideas that have emerged.

November 1981   G. Ledyard Stebbins

Darwin to DNA,
Molecules to Humanity

Part One

THE PROCESSES OF EVOLUTION

The Grand Canyon exposes strata that have been deposited intermittently over a period of almost a billion years. The earliest deposits, laid down between 550 million and 1.2 billion years ago, are highly contorted and visible only in the lower gorge, near the Colorado River. The upper strata, extending downward to the base of the highest cliff, were deposited almost continuously over a period of 120 million years. The high cliff consists largely of limestone—an accumulation of seashells, corals, and billions of shelled microorganisms on the floor of a quiet ocean over a period of 40 million years. (Josef Muench.)

Chapter One

THE PANORAMA
OF LIFE AND TIME

The evolutionist seeks to provide a scientific answer to questions that thoughtful people have asked since the dawn of civilization: Why is our planet filled with so many different kinds of living things? Why do the forests and fields surrounding us contain scores of different kinds of trees, flowers, grasses, and weeds—so many that an expert can recognize more than a hundred different kinds on a single afternoon's walk? Why does the air hum with the activity of thousands of different kinds of insects—ants, bees, flies, beetles, grasshoppers, butterflies, and strange-looking creatures like stick insects and praying mantises? Why is the narrow strip of shore between land and sea teeming with seaweeds, sea urchins, starfishes, snails, clams, barnacles, crabs, and small, colorful fishes? Does the gaiety of bird life that comes to our feeding trays have any meaning except as a source of pleasure to many human beings? What is the meaning of "nature red in tooth and claw," of predators such as lions and tigers that must kill and eat other animals in order to stay alive, of mice, squirrels, and other small, furry animals that spend nearly all their lives hiding from their enemies and are usually killed before they can die a natural death? Why is the world filled with ugly and harmful animals, such as scorpions, spiders, octopi, skunks, and venomous snakes?

The evolutionist's answer is that all these millions of different kinds of organisms evolved from common ancestors during the thousands of millions of years since the first appearance of life. Their evolution was opportunistic and devoid of purpose. Comparisons such as beautiful versus ugly, useful versus harmful, or good versus bad are not intrinsic to living things apart from human beings; they represent value judgments that we sometimes impose on natural phenomena, based on similar judgments that we recognize as necessary for the good of human society. The principal rules governing the evolution of life, as seen by biological evolutionists, are the following four:

1. Evolution usually proceeds gradually, at least in terms of life spans and generations. When evolutionists write about the sudden appearance of a new kind of animal, their frame of reference is a time span of millions of years. Such "sudden" changes usually require scores or hundreds of generations.

2. Given the appearance of a new *ecological niche*—that is, a new combination of environmental factors that can support life—one or more kinds of organisms will become adapted to that niche and will occupy it.

3. Animals and plants usually become adapted to new niches through a minimum number of changes in their genes; their complex, preexisting pattern of interaction with the old environment is modified only enough for them to become adapted to the new one. This basic conservatism is occasionally punctuated by more drastic changes. The diversity of animals and plants occupying similar environmental niches is due chiefly to this alternation between basic conservatism and occasional drastic and successful changes.

4. Populations of organisms continually interact with their environments. When the interaction is harmonious, this balance can be maintained indefinitely, and evolution will not take place. The balance can be upset by a change in

the environment, a drastic change in the genetic constitution of the population, or by the migration of one or
more individuals into a new habitat. A new population
might then be established if a harmonious interaction with
the new environment is acheived. When any one of these
changes in the population–environment interaction occurs,
the population either evolves or becomes extinct. As will
become clearer later in this chapter, the record of the past
illustrates that extinction has been far more common than
evolution.

WHY SHOULD WE STUDY EVOLUTION?

Nearly a century ago, humorist W. S. Gilbert reflected the excitement over Charles Darwin's new and revolutionary theory of
evolution by having one of his principal characters, Pooh-Bah in
The Mikado, pontificate: "I can trace my ancestry back to a proto-
plasmal, primordial, atomic globule"—the ultimate boast about
family pride, pedigree, and "roots." In the 1880s, the nature of the
"pedigree" of humanity was largely guesswork. Since then, a large
body of knowledge acquired by scores of scientists has enabled
evolutionists to determine the later stages of human evolution
with a high degree of confidence and to outline the general features of its earlier stages.

Evolutionists are by no means satisfied with simply learning
about the origins of humanity and the various species of animals
and plants. We also want to know *how* evolution occurs. What
processes made possible the cameralike eye, the wings of birds,
the supercomputer that is the human brain, and the beauty of
form and color in plants, butterflies, birds, and thousands of other
creatures? In recent years, scientists have been far more concerned
with research into the causes of evolutionary processes than the
determination of specific ancestries.

Is this search for knowledge about evolution merely a game that
some scientists play? Those of us who have devoted our lives to

this occupation certainly enjoy it, yet we are, of course, also convinced that basic knowledge about evolution has intrinsic value. First, such knowledge helps to satisfy human curiosity, a trait that most of us have. Why do men have deep voices and hairy faces, while women have high voices and smooth complexions? Why are male birds decked out in gorgeous color, while females of the same species are dull and inconspicuous? Why do skunks combine elegant black and white colors with a foul smell, reminiscent of the man-about-town in the last stages of inebriation? The theory of evolution helps provide answers to these kinds of questions.

A much stronger reason for learning about evolution concerns the position of humanity in the modern ecosystem. The continued evolution of our species implies more than just a simple continuation of the biological processes that preceded our emergence. Through the transmission and evolution of culture, human beings are set apart in many ways from even our closest relatives, the anthropoid apes. Those of us who wish to understand human nature and culture and the prospects for the future of our species must pay attention to both the cultural and biological aspects of human evolution.

The main theme of this book is that evolutionary change is based chiefly on the interactions between populations and their environment that become manifest by the process of *natural selection* identified by Darwin. The most effective catalysts of change are, therefore, alterations of the earth's environment. These include not only changes in physical factors, such as temperature, rainfall, mountain building, and erosion, but also the evolution of other plants and animals on which an evolving population depends. The evolution of complex animals and plants is governed to a large extent by feedback interactions with other organisms. The more drastic the physical changes of the environment and the stronger the feedback interactions with other organisms, the greater and more rapid must be the evolution of populations if they are to avoid becoming extinct.

Without knowing that they were doing so, our ancestors acquired the ability to control the evolution of other organisms. Our contemporaries, with hardly more knowledge and concern for the future of our environment, are modifying it with increasing vigor. Humanity has now reached the point of no return; we cannot go back to the days when humans adapted themselves to a natural environment rather than modifying that environment to suit themselves. Can human beings guide the evolution of species and ecosystems in directions that will replace with a new harmony of life the ecosystems that have largely been destroyed? Or will destruction continue until Spaceship Earth becomes a lifeless planet?

Since humanity cannot escape our stewardship over the entire world of animals, plants, and microorganisms, we must learn how to exercise this stewardship in the most beneficial way possible. To do this, we must acquire as much knowledge as we can about the present nature and past evolution of living organisms.

WHAT IS BIOLOGICAL EVOLUTION?

The theory of biological evolution is a way of explaining the origin of the diverse forms of life on earth. The evolution of species is not caused by the urge of individual organisms toward something better, nor is it a series of trends from the simple to the complex. The theory of evolution aims to explain the processes underlying phenomena as diverse as the grandeur of a giant redwood, the intricacy of an orchid, the complex society of a beehive or anthill, the efficiency of a tiger springing at its prey, and the capacity of human beings to create masterpieces of art, music, and literature. The theory of evolution also explains the emergence of less savory creatures—earthworms and slugs, fleas and lice, molds and microbes.

The processes of evolution encompass so many different phenomena that it is hard to define the term. Several definitions have been proposed. First, evolution is change, but by no means do all kinds of change constitute evolution. Changes in individuals, whether due to aging or other causes, are not evolutionary because individuals cannot evolve; only populations evolve. When evolutionists write about evolutionary lines, they are referring to successive populations, not individuals. Evolutionary lines are sometimes illustrated as a succession of individuals, but such an oversimplification does not tell the whole story. Every population contains within itself a large amount of genetic variability, and only those changes in appearance that have a genetic basis are part of biological evolution. For example, average Americans of Japanese parentage today are larger and mature earlier than did their parents. This change is due to improved nutrition and is not hereditary; the genes have not been and cannot be altered by such an influence. Hence, although a change has affected an entire population, this change is not an example of biological evolution.

Even some changes in the hereditary or *genotypic* makeup of entire populations are best considered to be outside the scope of evolution, although they may tell us much about the nature of evolutionary processes. For example, intensive research on populations of the European hamster in central Russia has shown that the population of black hamsters, which differs genetically from the more common brown ones, may increase during the moist, cold spring weather but may decrease again during the warm summer months. These annual, cyclic changes have occurred for one or two centuries, but they have not permanently altered the population. Similar cyclic seasonal changes have been identified with respect to the frequency of certain chromosomes in some species of the fruit fly *Drosophila*. Neither of these examples constitutes evolution.

Genuine biological evolution involves changes in several different characteristics in a species. Although each change or *mutation* by itself may be reversible, the combination of cumulative changes that constitutes evolution is so complex that, once a certain number of alterations have occurred, the population cannot return to an earlier stage in its development. If a succession of populations undergoes genetic changes in response to the challenges posed by its exposure to a new set of environmental conditions, and if it is later exposed to an environment similar to the one it originally occupied, it will adapt to the old habitat in a new way rather than regain its former characteristics. For example, when reptiles, birds, and mammals returned to the water, they became sea turtles, penguins, seals, porpoises, and whales rather than fishes.

In defining evolution, we must also ask: Are all evolutionary changes adaptive? Many evolutionists maintain that some differences between populations arise by chance and have nothing to do with adaptation. Such random purposelessness may cause changes in certain individual characteristics but not changes in entire character complexes. Continued evolution requires changes either in the kind of habitat the population occupies or in the character of the adaptations made.

I propose the following definition. Biological evolution is a series of irreversible transformations of the genetic composition of populations, based principally on altered interactions with their environment and guided by natural selection. Although individual genetic changes can be reversed, irreversibility of evolutionary changes in populations results from the low probability that many such changes can be reversed simultaneously. Evolution consists chiefly of diversification; descendants of a generalized population evolve in different directions in response to different environmental stimuli, and the genetic constitution of many populations causes them to exploit similar environments in different ways. A small number of populations evolve in one direction far enough so that they acquire characteristics that we humans perceive as novel, or qualitatively different from the ancestral condition. Evolutionary trends, therefore, fall into two categories: diversification, or branching, and continued change that ends in evolutionary novelty. The basic theme of this definition, the interaction of populations with their environment, is further developed in the next chapter. First, however, we must examine the implications of the fact that evolution is a series of historical events, and we must consider the enormously long time scale over which the evolutionary drama has been enacted.

THE TIME SCALE OF EVOLUTION

This time scale begins with the first recognizable appearance of life, about 3.5 billion years ago. Of course, the universe and solar system were evolving long before the surface of the earth became habitable for living organisms. Evolution from the first organic molecules to the first cells probably required millions of years. Since we have no records of these events, we will not consider them in this preliminary discussion of the evolution of cellular organisms.

Most physicists, chemists, and biological evolutionists agree that the evolution of organic molecules began about 4 billion years ago. The first living cell appeared about 3.5 billion years ago, and the first simple many-celled animal appeared roughly 600 million years ago. The common ancestor of apes and humans appeared some 6 million years ago, and the beginning of recorded history was about 6,000 years ago.

During the past century and a half, scientists have acquired many new facts about the age of the earth. With this new knowledge, scientists have progressively increased their estimates of the length of time that has elapsed since life began. The theologian's estimate of 5,000 to 6,000 years was first challenged by scientists during the middle of the last century. In 1868, Thomas Huxley, one of Darwin's most vigorous and articulate apostles, delivered to the working men of Norwich an address entitled "On a Piece of Chalk," now a classic of popular science. In it he reviewed with painstaking care the recently discovered evidence indicating that the chalk cliffs bordering the English Channel, the white cliffs of Dover, could have been formed only by the accumulation of millions upon millions of tiny shells fashioned by microorganisms living in the ocean. (Their modern descendants can be dredged from the bottom of the Atlantic at any point between Ireland and Newfoundland.) He realized that these remains of dead microorganisms could accumulate only very slowly, and he suggested that the formation of the deposits that form the thousand-foot-high cliffs along the south coast of England would have required at least 12,000 years to be formed, and probably much longer. If so, this suggestion would lead to calculations, which Huxley was not willing to make, that might have indicated the earth's age to be about 1 million years.

At the turn of the century, geologists and paleontologists accumulated data on all the world's strata, and their observations and calculations led them to believe that the earth is about 100 million years old. These investigations were followed by a flood of new discoveries about radioactivity. Physicists discovered that such elements as uranium and strontium, and mixtures of potassium and argon, decay at regular, measurable rates. The technique of *radiometric dating* that emerged from these discoveries is now standardized to such a degree that geologists and physicists agree almost unanimously about the results obtained. Radiometric dating tells us that the earth is about 4 billion years old, that the first living cells may have appeared more than 3 billion years ago, that organisms having cellular nuclei, chromosomes, and possibly the ability to reproduce sexually evolved between 1 billion and 1.4 billion years ago, and that primitive jellyfishes, worms, seaweeds, and other many-celled organisms first appeared between 600 and 700 million years ago. The time scale as now recognized is so long that different stages of it are hardly commensurate with each other (Figure 1-1).

DO POPULATIONS
CONSTANTLY EVOLVE?

The fossil record and calculations based on it indicate that evolution may be a succession of "quantum jumps" brought about by special environmental conditions. First, consider that many kinds of animals and plants are indistinguishable from their ancestors that lived millions of years ago (Figure 1-2). For example, the Virginia opossum has changed little during the past 50 million years. The tuatara (*Sphenodon*), a lizardlike reptile living on a few islands off the coast of New Zealand, closely resembles fossils that are 150 million years old. Modern oysters are very similar to their ancestors that lived 300 million years ago. A shelled animal known as the lamp shell (*Lingula*) is essentially the same as fossils extracted from strata that are about 500 million years old. Many similar examples are known among plants and microorganisms; the most striking of these are blue-green "*algae*," which are really bacteria. Microscopic fossils of algae from central Australia, a billion years

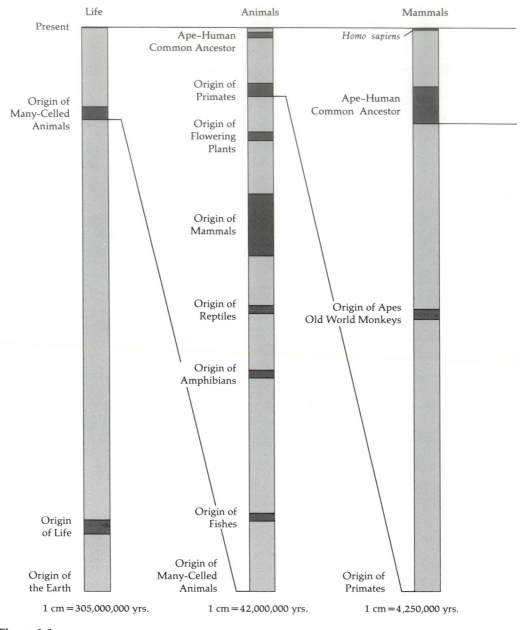

Figure 1-1.

The complete time scale of evolution can be well understood when shown in five parts, each of which is on a larger scale than the preceding one. The left column of this figure represents the entire time span from the origin of the earth to the present. During most of this time, the earth was inhabited only by primitive microscopic forms of life. The second column is an

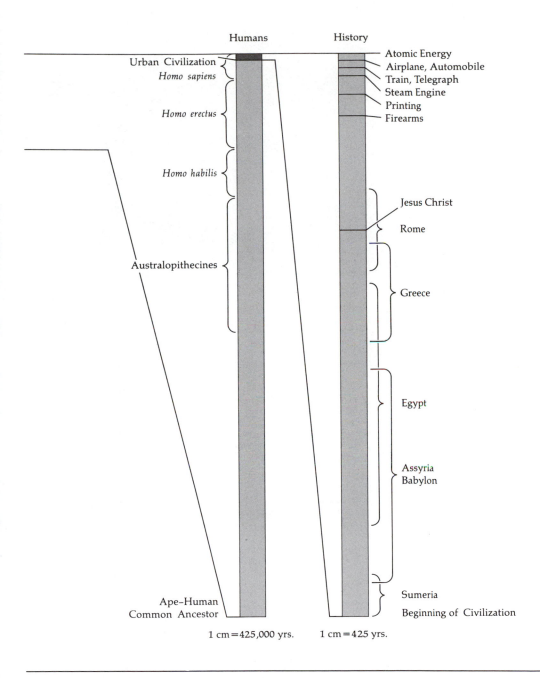

enlargement of the part of the first column that is above the diagonal line connecting the two. It shows the relative time spans during which various groups of animals and plants have existed. The remaining three columns show, in a similar way, the relative time spans of modern (placental) mammals, human beings, and recorded history.

(a)

Oscillatoriaceae
aff. *Oscillatoria*

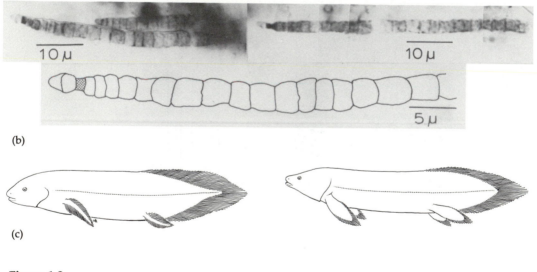

(b)

(c)

Figure 1-2.

Three examples of evolutionary stability, or very slow evolution. (a) The tuatara *(Sphenodon)*, a lizardlike animal that is now confined to a few islands off the coast of New Zealand. (Photo supplied by Nathan Cohen.) (b) Fossil remains of bacteria that are about 1 billion years old, but when seen through the microscope, they closely resemble modern blue-green bacteria. (From J. W. Schopf.) (c) Outline sketches showing the great similarity between a lungfish *(Conchopoma)* that lived 250 million years ago (Permian period) and the modern lungfish *(Epiceratodus)* of Australia.

old, are preserved so perfectly that their cellular structure can easily be seen. These cells are indistinguishable from those of modern blue-greens of the common genus *Oscillatoria*. These examples are just a few of hundreds that can be documented from the fossil record or inferred from comparative studies of related modern populations. Evidence from fossil animals and plants that lived from 1 million to 2 million years ago indicates that, during this period of time, the great majority of animals, plants, and microorganisms—with the exception of mammals—have evolved little or not at all.

We can approach this problem in another way by asking: Is the length of time that has elapsed since the first living cells appeared, or since a particular group of organisms started to evolve, too short, about right, or too long to explain the amount of evolution that has occurred, assuming that evolution in a particular direction has been continuous? If one assumes that evolution has been continuous ever since life began, one must conclude that it is an extremely slow process.

This point becomes clearer if we examine two examples of evolutionary change, the first hypothetical and the other, similar, an interpretation of an actual fossil record. First, consider a population of animals with an average weight of 40 grams, about the size of a mouse. Then assume that, generation by generation, the members of the population increase in size at a mean rate of one-tenth of one percent of the mean weight at any generation. This rate is so slow, relative to differences in weight between individuals belonging to any particular generation, that the difference between the means of two successive generations could not be calculated because of sampling errors and because slight differences in nutrition affect nonhereditary differences in weight. Nevertheless, an increase at this almost imperceptible rate over 12,000 generations would produce a population having a mean weight of 6,457,400 grams, which is about the size of a large elephant. If the mean duration of a single generation were five years, which is much longer than that of the mouse but shorter than that of the elephant, the elapsed time required for this tremendous increase in size would be 60,000 years. Compared to the human life span, this period is extremely long, as long as the time elapsed since the last Ice Age when Neanderthal men walked over the earth. Nevertheless, 60,000 years is so short relative to geological periods that it cannot be measured by geologists or paleontologists. The origin of a new kind of animal in 100,000 years or less is regarded by paleontologists as "sudden" or "instantaneous."

The rate of genetic change postulated in the preceding example, slow as it may seem, is nevertheless many times faster than any change in visible characteristics that can be inferred from the fossil record. Our second example illustrates this point. At the beginning of the Tertiary period (Paleocene and Eocene epochs), shortly after the extinction of the dinosaurs, mammals evolved at an unusually rapid rate. From a small, shrewlike animal hardly larger than a mouse, over a period of about 45 million years and in several intermediate stages, a group of elephant-sized horned beasts known as Titanotheres evolved. Except possibly for the evolution of the largest dinosaurs, this increase from mouse-sized to elephant-sized animals is the most rapid known in the fossil record. It occurred stepwise in recognizable stages represented by fossil genera that maintained a constant size for long periods of time. Unless one assumes that many populations existed and became extinct without leaving fossils, one must assume that the entire evolution occurred in a series of spurts, or quantum jumps, that appear sudden on the geological time scale but are actually slow as measured by the generations of the genetic time scale. A hypothesis of about ten such spurts is consistent with the record. This estimate requires that, between each spurt, the animals of this evolutionary line remained unchanged in size for periods lasting almost 4 million years.

We will examine the far-reaching implications of this conclusion later, following a discussion of the principal cause of evolution as recognized by Darwin and his followers. For now, we wish only to point out that, given the enormous length of the geological time scale in terms of generations of organisms, the course of evolution that gave rise to any known succession of fossils can be interpreted in either of two very different ways, each of which has radically different implications with respect to the causal mechanisms of evolution.

The two extreme possibilities are illustrated in Figure 1-3. In diagram (a), 22 species of elephants, mammoths, and mastodons are each represented by vertical lines that indicate the time span during which they lived. The lines are placed to show as accurately as possible their relationships to one another and to the two living species, the African and Indian elephants. In diagram (b), the lines representing the species of diagram (a) are connected by broken lines that represent their presumed evolutionary connections if evolution is assumed to occur continously at extremely slow rates of change per generation. In diagram (c), the same fossil species are connected by lines that show their presumed evolution if evolution is assumed to comprise a series of sudden bursts, each lasting 100,000 years at most, separated by much longer periods of stability during which evolutionary change was negligible. Note that the postulated sudden bursts might represent small changes for each generation, as was explained earlier.

Which of these two explanations is the most correct? This question has been hotly debated by evolutionists during the past decade. Traditionally, paleontologists have supported the hypothesis of gradual change, as did Darwin, the founder of modern evolutionary theory. In recent years, however, a group of younger paleontologists, particularly Niles Eldredge, Stephen J. Gould, and Steven Stanley, have compiled a large body of evidence in support of the hypothesis of *punctuated equilibria*, or sudden bursts. They represent the course of evolution as a series of sudden bursts, as shown in Figure 1-3(c). They obtained their most important evidence by forging links between the lengthy fossil record and changes in contemporary populations that have been observed and analyzed over much shorter periods of time.

These paleontologists maintain that, for many groups of organisms, the fossil record is now so well known that it is very unlikely to contain long gaps unrepresented by fossils, as is implied by the broken lines in Figure 1-3(b). The shorter gaps represented in Figure 1-3(c) are much easier to reconcile with the fossil record of elephants and other animals, particularly on the basis of concepts of change in modern populations that are discussed in Chapter Three. In addition, Eldredge and Gould have studied fossil sequences of snails in Bermuda in which gaps in a sequence are so recent that they are difficult to explain on the assumption that they represent periods during which no fossils were formed.

Steven Stanley has presented an even stronger argument against the hypothesis of extremely slow, gradual evolution. He points out that the Darwinian theory of natural selection, which is accepted by most evolutionists because of the large amount of diverse supporting evidence, will not work unless a fairly rapid rate of evolution is postulated with reference to the geological time scale.

The hypothesis of punctuated equilibria is by no means new. It is, rather, an extension and refinement of the concept of *quantum evolution* as presented by paleontologist George G. Simpson, one of the pioneers of the modern theory of evolution. The principal difference between Simpson's theory of quantum evolution and that of punctuated equilibria is that the more recent theory relates evolutionary bursts much more closely to the origin of species.

The belief in long periods of evolutionary stability has also been challenged by some evolutionists who maintain that, although

Figure 1-3.

Two interpretations of the evolution of elephants. (a) The data as recorded by paleontologists. The horizontal axis represents time in millions of years (My). The species are placed along the vertical axis in rough approximation to the amount of difference between their skeletal structure. The horizontal bars indicate the period of time during which each species lived. (b) Interpretation of the data shown in (a) according to the hypothesis of continuous evolution. The broken diagonal lines represent supposed lineages that would be represented by intermediate forms for which fossils have not been discovered. (c) Interpretation of the same data according to the hypothesis of quantum bursts of evolution and punctuated equilibria. The connections between the species are shown by nearly vertical broken lines, representing populations that were probably small and existed for only short periods of time. Note, however, that the slope of each line is enough to show that the transition from one species to the next required at least 50,000 years.

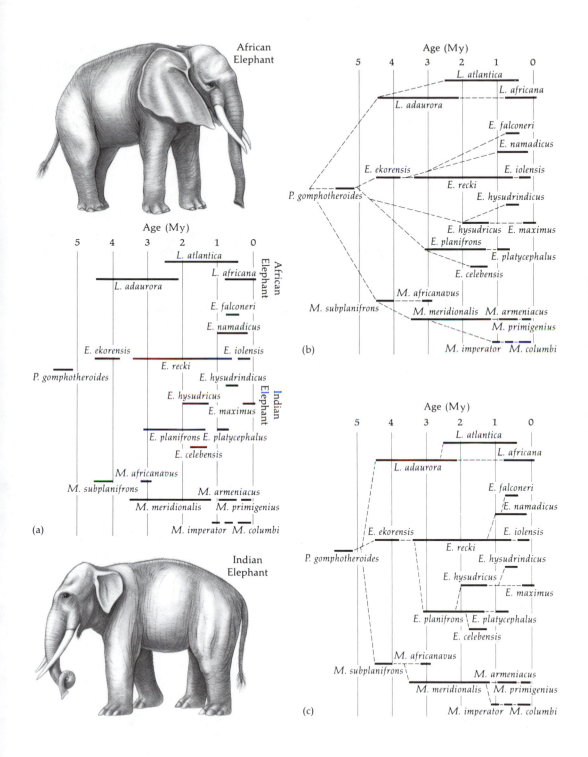

African
Elephant

Age (My)

5   4   3   2   1   0

L. atlantica
L. africana
L. adaurora
E. falconeri
E. namadicus
E. ekorensis
E. iolensis
E. recki
E. hysudrindicus
P. gomphotheroides
E. hysudricus E. maximus
E. planifrons
E. platycephalus
E. celebensis
M. africanavus
M. subplanifrons
M. meridionalis M. armeniacus
M. primigenius
M. imperator M. columbi

(b)

Age (My)

5   4   3   2   1   0

L. atlantica
L. africana   African
L. adaurora   Elephant
E. falconeri
E. namadicus
E. ekorensis   E. iolensis
E. recki
P. gomphotheroides
E. hysudrindicus
E. hysudricus   Indian
E. maximus   Elephant
E. planifrons E. platycephalus
E. celebensis
M. africanavus
M. subplanifrons
M. armeniacus
M. meridionalis M. primigenius
(a)
M. imperator M. columbi

Indian
Elephant

Age (My)

5   4   3   2   1   0

L. atlantica
L. africana
L. adaurora
E. falconeri
E. namadicus
E. ekorensis
E. iolensis
E. recki
P. gomphotheroides
E. hysudrindicus
E. hysudricus
E. maximus
E. planifrons E. platycephalus
E. celebensis
M. africanavus
M. subplanifrons
M. armeniacus
M. meridionalis M. primigenius
(c)
M. imperator M. columbi

some living animals and plants look very much like their ancient fossil ancestors, nevertheless constant changes have been taking place with respect to internal, largely biochemical, characteristics that we cannot detect. Adopting this point of view in its most extreme form, Leigh Van Valen has made an analogy with the Red Queen in *Through the Looking-Glass*, who said: "Now, here, you see, it takes all the running you can do, to keep in the same place." For some kinds of animals, the Red Queen hypothesis may be valid. Song birds, mice, and similar small, highly active creatures are continually challenged by new and different predators, and the ruling carnivores (lions, tigers, and large birds of prey) are tested to their utmost ability by the elusiveness of their respective prey. The situation is different, however, for small, secretive, or sedentary animals that survive largely because of a temendous reproductive capacity or because they are inedible—shrews, cockroaches, scorpions, oysters, many kinds of worms, and jellyfishes. These animals have already successfully met all the environmental challenges to which they have been exposed during periods of scores or hundreds of million years. Their evolutionary constancy is probably real and complete.

For plants, evolutionary constancy over millions of years has been demonstrated experimentally. An example is the plane tree, or sycamore (Figure 1-4). Plane trees in eastern North America belong to a species having relatively broad, shallowly lobed leaves; plane trees native to the eastern shores of the Mediterranean Sea have much more deeply lobed leaves. Evidence from fossils indicates that the American and the Mediterranean species have been separated from each other for at least 20 million years. Yet, when the two species were brought together during the eighteenth century in English gardens, they hybridized, and the resulting progeny are so fertile and vigorous that they are among the commonest ornamental trees in parks throughout the Northern Hemisphere. They are spontaneously establishing themselves along some rivers in California's Sierra Nevada foothills. During the past 20 million years, plane trees that were separated from each other by a distance of 4,000 miles and grew in distinctly different climates have

Figure 1-4.

Leaves of the London plane tree, a fertile hybrid between a species from North America (*Platanus occidentalis*) and one from southeastern Europe and the Middle East (*P. orientalis*). The vigor and fertility of this hybrid show that the parental species have evolved very little since their separation from each other at least 20 million years ago.

not evolved differences greater than those that distinguish breeds of cattle or races of mankind. The visible differences that distinguish them are greater than their internal, genetic differences.

EVOLUTION AND EXTINCTION

If the previous deductions are valid, if evolution is not continuous but is confined to periods of activity separated by long intervals of stability, then the next obvious question is, why? The answer, which is developed in the next chapter, is that populations do not evolve beyond the differentiation of races or closely related species unless they are faced with an environmental challenge. Populations of organisms are basically conservative; if they can survive by staying in the habitat that they have always occupied and by exploiting it in the same way their ancestors did, they will do so. They evolve new characteristics only when a changing environment forces them either to evolve or become extinct.

The vast majority of evolutionary lines, when faced with the challenge of a drastic environmental change, become extinct rather than evolve. The beginning of a new evolutionary line is the origin of a new species, which consists of individuals capable of mating and producing vigorous, fertile offspring among themselves but isolated sufficiently from other populations so that their hereditary makeup can be altered independently of and in a different way from that of other evolutionary lines. The formation of new species that are variations on a particular structural and functional theme is a commonplace fact of nature. Among doglike animals, there are wolves, coyotes, and jackals; the horse genus is represented by a few remaining wild horses, domesticated horses, donkeys, and several kinds of zebras; bird lovers in our eastern forests can recognize many species of warblers, and so on. Each of the species that exists at present is theoretically capable of evolving in its own way to form a new evolutionary line. How many of them do so, and how many become extinct after having existed for a geologically "short" period of a few hundred thousand or million years?

The following calculation leads inevitably to the conclusion that the vast majority of species become extinct without evolving any further. The sexual process, by means of which chromosome sets belonging to two parents are united in fertilization to form a new individual, probably originated about 1 billion years ago. This process is essential for members of a population to share freely their hereditary endowment and to be sufficiently isolated from other populations to evolve independently. Let us assume that a single sexually reproducing species existed at this remote time in the past and that all the populations that are descended from it have split to form two separate species at an average rate of once per 5 million years. This rate is extremely slow; in the fly genus *Drosophila*, a single immigrant population that arrived in the Hawaiian Islands between 5 million and 10 million years ago has given rise to more than 600 species. Yet, even the very slow rate of one splitting every 5 million years yields the following results. After the first 5 million years, there would have been two species; after 10 million years, four; after 15 million years, eight. If none of these species became extinct, and if each split to form two species once every 5 million years, the number of species among their descendants, after 1 billion years, would be 10^{60}, or 10 followed by 59 zeros! Since the actual number of species on earth is between 1 million and 10 million, or at the very most 10^7, the ratio of potential living species that have either become extinct or have never been formed to those that now exist might be estimated as 99.99 followed by 52 nines to 10 million.

We can only conclude that by far the commonest fate of every species is either to persist through time without evolving further or to become extinct. Based on his knowledge of fossil animals, Stephen J. Gould estimates that "more than 99.9 percent of species are not sources of great future diversity." Evolution is not a universal property of life, like self-reproduction, growth, and individual response to the environment. Its most significant changes result from unusual combinations of events. When studying the course of evolution, scientists can unravel the history of only a small proportion of the most successful species, those that have left a reasonably large number of fossils. The failures, and even species that may have been moderately successful for a while, are buried in the oblivion of bygone ages.

Nevertheless, throughout the hundreds of millions of years during which living organisms have existed on the earth, hundreds of different evolutionary lines have always been generating new species. Some of these have replaced old species that became extinct, but an even larger number have enriched the earth's biotic communities by their ability to exploit new ecological niches. The result has been a more or less constant increase in the number and diversity of living organisms ever since the first cells evolved 3.5 billion years ago.

Each new kind of animal or plant that evolves creates new opportunities for other organisms to feed on it or parasitize it. A billion years ago, the habitable regions of the earth were covered with hundreds of species of one-celled microorganisms. If new species can evolve only when new ecological niches are available for them, one might conclude that this worldwide occupation of the most diverse habitats would have spelled the end of evolution for one-celled organisms, except for minor changes and replacement of extinct forms. Actually, this was only the beginning of a long course of evolution for many of them. As multicellular animals evolved, their digestive tubes provided new habitats rich in food and well protected from enemies. Various kinds of bacteria and protozoa continued to evolve as parasites in the animals' stomachs or intestines. The evolution of bacteria and protozoa that live in the stomachs of grazing animals such as cows and sheep has continued up to the present.

Why have organisms evolved? Answering this question will give us only a partial understanding of the process of evolution. Equally important is the question: Why have many kinds of organisms survived with continuing success even though they have not evolved toward greater levels of complexity? Why did some bacteria evolve into the more complex eukaryotes while others continued as prokaryotes? Why did one kind of lungfish evolve into amphibians and move onto land while related fishes stayed in the water and changed very little? Why did a few small dinosaurs evolve into birds and populate the earth with a new form of life while all their relatives became extinct? Why did a few populations of apelike animals evolve into humans while others gave rise to chimpanzees and gorillas?

WHAT CAUSES EVOLUTION?

The theories that scientists have advanced to explain the causes of evolution have been of two very different kinds. One set of theories is based on the postulate of an internal directive force that lies within organisms themselves. Supporters of these theories maintain that the principal task of evolutionists is to discover and define such a force or "evolutionary urge." Theories of this kind are supported principally by paleontologists who study the fossils of animals long extinct, by physicists and chemists who look at biological phenomena from the point of view of their own disciplines, and by philosophers who seek to construct general laws that govern all branches of knowledge.

The second set of theories stresses the interdependence of populations and the ecosystems they inhabit. More fundamental than any of the processes that modern textbooks list as causes of evolution—such as mutation, genetic recombination, and natural selection—are the numerous kinds of interactions between populations and their environments, of which these processes are essential parts.

What do we mean by the concept of environment? Many scientists who are not naturalists or ecologists—particularly physicists, chemists, geologists and biologists—think of the environment chiefly in terms of its physical features—climate, temperature, moisture, soil, and atmosphere. For most organisms, however, by far the most significant factors of their environment are the other organisms in their ecosystem. The survival of animals depends to some degree on their adaptation to climate, but to a much greater degree on the other kinds of organisms available to them as food and on the predators from which they must protect themselves. The environmental challenges that have stimulated the great majority of animals' adaptive evolutions have been interactions with other animals and plants. The evolution of lions and tigers from primitive carnivores, of bears from doglike animals, of horses from their diminutive ancestors, of birds from small dinosaurs, and of human beings from forest-dwelling apes and monkeys has been triggered much more by coevolution along with prey, predators, and food plants than by changes in the earth's climate. Coevolution is the key to understanding most examples of long, continued evolutionary change.

These two sets of theories are by no means mutually exclusive. Many biologists believe that population–environment interactions are responsible for producing many different kinds or species of related animals or plants but that major trends of evolution are guided by an internal force. The first scientist to propose such an evolutionary dualism was Jean Baptiste de Lamarck, the author of the first complete theory of evolution. We will discuss his ideas more fully in the next chapter.

The enormous diversity of evolutionary rates that is implied in the theory of punctuated equilibria is almost impossible to explain on the assumption of a generalized force of evolution. Moreover, discoveries made by biologists of many different disciplines indicate that evolution is unitary, not dualistic. Biologists can observe experimentally demonstrated interactions between contemporary populations and their environments and from these experiments derive explanations of evolution at all levels. Evolutionists should therefore construct theories or generalizations about the causes of evolution by synthesizing data obtained from analyses of population–environment interactions of all possible kinds. This is the dominant theme of the modern synthetic theory of evolution.

WHAT ARE THE RATES OF EVOLUTION?

How fast does evolution proceed? Many evolutionists, especially those who emphasize internal genetic changes or mutation rates as the primary limiting factors, look for built-in "clocks" that govern evolutionary rates. Others, like myself, have concluded from many lines of evidence that population–environment interactions are the limiting factor and believe that the search for general rates is futile. Given the spatial and temporal complexity and diversity of population–environment interactions, one might expect many different rates, both when comparing different evolutionary lines with one another and when following a single line during several geological epochs.

Strong evidence for this latter point of view was obtained by George G. Simpson, the first paleontologist to apply careful statistical methods for interpreting the fossil record in terms of evolutionary rates. He showed both that quantum bursts of evolution have taken place and that some species, which he calls *bradytelic*, have remained constant during tens or hundreds of millions of years. Punctuated equilibrium is an extension of Simpson's ideas that, if accepted, renders the search for generalized rates of evolution meaningless.

Nevertheless, particular rates of evolution are well worth studying. Comparisons between rates of evolution of different organisms living in the same ecosystem and over comparable time spans can be most rewarding. In addition, analyses of different rates exhibited by the same evolutionary line in different geological epochs are well worthwhile. Finally, even in the same line, different characteristics have evolved at very different rates. Such differences have been found even with respect to individual protein molecules.

Let us briefly state the principal message of this chapter. Although at all times some populations are evolving somewhere in the world, evolution is not a continuous, inevitable property of all populations at all times and in all circumstances. Most evolution, particularly of strikingly new adaptive types, occurs in quantum bursts that are triggered by challenges of a changing physical and biotic environment. When such a challenge occurs, the populations exposed to it respond in one of three ways. Most populations become extinct; some adjust to the new environment with minimal change in their hereditary makeup and thus persist with little evolution over millions of years; a few populations respond by evolving entirely new adaptive mechanisms. Such newly adapted organisms may spread widely, evolve further, and evolve adaptations to still other new habitats.

What determines whether a population responds to a particular challenge by becoming extinct, continuing with little change, or evolving in a new direction? Briefly, a population's response depends on the nature of its hereditary variation, or gene pool, at the time of the challenge, and on the resulting kinds of interactions between population and environment.

Charles Darwin (1809–1882) as he appeared during the prime of his life, when he was writing *The Origin of Species*. (Courtesy of The Bettmann Archive, Inc.)

Chapter Two

DARWINISM
THEN AND NOW

How have interactions between populations and their environments caused evolution to take place? One might imagine that an environment directly alters the animals and plants that live in it and that these alterations, acquired by single individuals, are transmitted to their descendants. By this reasoning, people who live in the tropics would acquire dark skins because, generation after generation, they are exposed to the hot sun, become well tanned, and transmit to their children the darker skin. This kind of reasoning appears in folklore and myth.

As a scientific explanation, adaptation through the inheritance of acquired characteristics was first proposed early in the nineteenth century by the French naturalist Jean Baptiste Chevalier de Lamarck. For half a century, Lamarck's theory was the sole hypothesis about the evolution of adaptation. Now, however, its falsity has been fully demonstrated, and we know that genetically based evolutionary changes cannot be explained by individuals acquiring traits. To understand why Lamarck's theory is incorrect and why the alternative theory of natural selection commands credibility, we must first understand the chromosomal and molecular mechanisms of heredity.

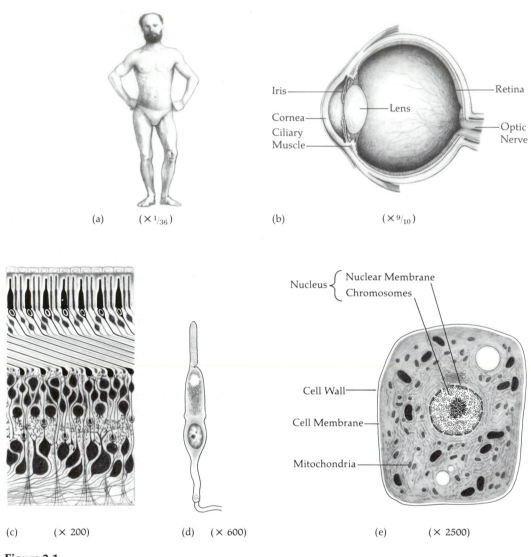

(a)   ($\times \, ^1/_{36}$)

Iris
Cornea
Ciliary
Muscle
Lens
Retina
Optic
Nerve

(b)   ($\times \, ^9/_{10}$)

(c)   ($\times$ 200)

(d)   ($\times$ 600)

Nucleus $\left\{ \begin{array}{l} \text{Nuclear Membrane} \\ \text{Chromosomes} \end{array} \right.$

Cell Wall
Cell Membrane
Mitochondria

(e)   ($\times$ 2500)

Figure 2-1.

The hierarchy of complexity of the human body. (a) A human figure ($\times$ 1/36). (b) Cross-section of an eye ($\times$ 9/10). (c) Cross-section of a small part of the retina ($\times$ 200) showing the different kinds of cells as black ellipses from which project linear processes. (d) A single cell (rod) of the retina ($\times$ 600). (e) Idealized cross-section of a single cell, as seen under a lower power of the electron microscope ($\times$ 2500). (f) Two organelles, mitochondria, and a large number of ribosomes, upon which the genetic message is translated into protein structure, as seen under a high-powered electron microscope ($\times$ 62,500). (g) A small portion of a molecule of DNA, showing its atomic structure. The black spheres represent atoms ($\times$ 150,000,000).

THE MECHANISM OF HEREDITY

The *nucleus* of a plant or animal cell contains *chromosomes*—long, coiled structures of *deoxyribonucleic acid,* or DNA (Figure 2-1). Each strand of DNA has several thousand units, the genes. Between the genes, and even within some of them, are stretches that do not contain specific information (*spacer DNA*) but are important in ways that are not yet fully understood. Each nucleus of ordinary body cells, such as those of skin, digestive organs, nerves, or blood, contains a complement of paired chromosomes. The two members of each pair are about as much alike as two differently

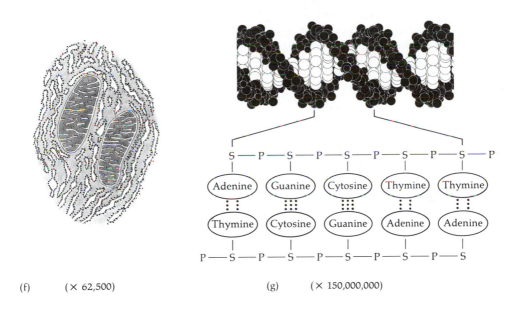

(f)   (× 62,500)   (g)   (× 150,000,000)

If the human figure (a) were enlarged in proportion to the magnification of the structures shown in parts (c) through (g), the parts of the body would be enlarged to the following sizes: (c) cells of retina: 0.245 miles; (d) single rod cell: 0.67 miles; (e) ultrastructure of cell: 2.79 miles; (f) mitochondria and ribosomes: 70 miles; (g) atomic structure of DNA molecule: 16,772 miles, more than twice the diameter of the earth! (Parts b, c, and d after Maximor and Bloom, *Textbook of Histology,* 1952. Philadelphia: W. B. Saunders. Part g from *An Introduction to Genetic Analysis* by D. T. Suzuki and A. J. F. Griffiths. W. H. Freeman and Company. Copyright © 1976.)

revised editions of the same book. In contrast, chromosomes belonging to different pairs differ from each other as much as books on different subjects. Human beings have twenty-three pairs of chromosomes; the common pomace fly, *Drosophila melanogaster*, four pairs; the peas in Mendel's experiments on the laws of heredity, seven pairs. An exception to this rule of pairs are bacterial cells, which contain only a single chromosome.

Each gene is composed of several hundred chemical molecules called *nucleotides*. The four kinds of nucleotides in DNA are designated by the letters A (adenosine), C (cytidine), G (guanidine), and T (thymidine). Genetic information is encoded in the linear order of these four nucleotides along the DNA molecule.

During the 1950s and 1960s, molecular biologists discovered three almost miraculous sequences of events that occur in the cell nucleus. First, James Watson and Francis Crick identified the double-helix structure of DNA. If the smallest human chromosome were magnified to the dimensions of commercial videotape, it would be about 118 miles long, and the largest of our chromosomes would be some 600 miles long. These extraordinarily complex molecules of DNA replicate themselves every few hours, and the copy is delivered to a new cell through a carefully regulated "dance of the chromosomes" (*mitosis*) that is as precise and well ordered as the drill of a regiment of soliders on a parade ground (Figure 2-2). This complex sequence of events is made possible by the precise action of a battery of protein molecules, the *enzymes*.

Figure 2-2.

The chromosomes during the four stages of cell division of a peony plant. (a) The chromosomes first appear as slender, twisted threads, still within the nucleus. (b) The nuclear membrane disappears, and the chromosomes line up in the center of the cell. Separation into daughter chromosomes begins at a position from which attached contractile threads of protein can pull the daughter chromosomes apart from each other. (c) Daughter chromosomes have separated and are moving to opposite ends of the cell. (d) The cell contains two sets of daughter chromosomes, each of which has the same genes as its parent, shown in (a). Subsequently, a nuclear membrane is formed around each group, and a cell wall is formed between them, converting the original cell into two daughter cells. (Photos from J. L. Walters.)

(a)

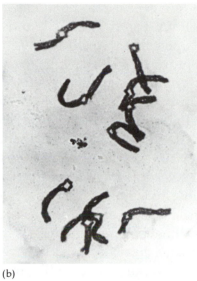

(b)

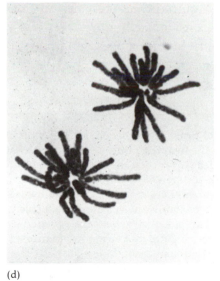

(c)

(d)

The second miraculous sequence of events begins when the chromosome opens up so that segments of the DNA that contain the code for constructing the working molecules (proteins) of the cell are exposed and *transcribed* onto molecules of *ribonucleic acid*, or RNA (Figure 2-3). Cells and their chromosomes receive signals to open that part of the chromosome that contains information appropriate to a skin cell, liver cell, nerve cell, or any other kind of cell. The signaling system that causes the cellular system to transcribe DNA onto RNA is still very imperfectly known.

The RNA moves from the nucleus to the cytoplasm of the cell, where the third miraculous sequence of events begins. Each RNA molecule attaches itself to a *ribosome*, or series of ribosomes, where it serves as a *template* for translating a *polypeptide* chain that can be converted into an enzyme. This assembly of the enzyme, the essential step in the third sequence, is the most miraculous of all. The final polypeptide chain is not just a blueprint or model. When it leaves the ribosome and moves into the fluid medium of the cell, it is a working enzyme (Figure 2-4).

We can now return to Lamarck's theory of adaptation and acquired characteristics and understand why it cannot be correct. Consider again our earlier example of skin pigmentation. Although bright sun causes certain enzyme molecules in human skin cells

Figure 2-3.

The processing of genetic information. (a) Transcription. In the nucleus, the messenger RNA is assembled on the template of the DNA molecule, and its sequence of base pairs is a mirror image of the order sequence of the DNA from which it is transcribed. (b) On the ribosome, shown as a cross-section of a ball on a doughnut, the messenger RNA and transfer RNA molecules place themselves in such a way that the latter "read" the message and translate it into the sequence of amino acids of the protein chain, which emerges from the ribosome (*upper left*). (c) A small part of a genetic message as coded by the order of base pairs in DNA (*top line*), the corresponding order of base pairs in messenger RNA (*middle line*), and the order of amino acids in the protein chain (*bottom line*). (d) The genetic code, a table showing the correspondence between triplets of bases in RNA and amino acids of protein chains. (Parts a and d from *Evolution* by T. Dobzhansky, F. J. Ayala, G. L. Stebbins, and J. W. Valentine. W. H. Freeman and Company. Copyright © 1977; Part b from *An Introduction to Genetic Analysis*, Second Ed., by D. T. Suzuki, A. J. F. Griffiths, and R. C. Lewontin. W. H. Freeman and Company. Copyright © 1981.)

(a) Transcription (Nucleus)

(b) Translation (Cytoplasm)

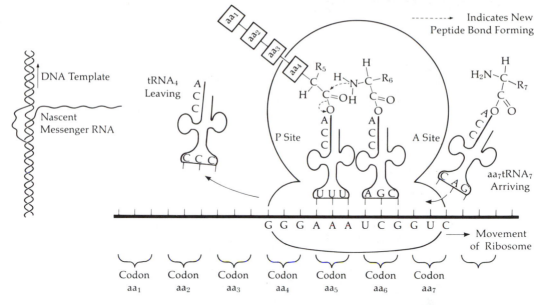

DNA Template

Nascent Messenger RNA

tRNA₄ Leaving

- - - - - → Indicates New Peptide Bond Forming

P Site

A Site

aa₇tRNA₇ Arriving

G G G A A A U C G G U C

→ Movement of Ribosome

Codon aa₁   Codon aa₂   Codon aa₃   Codon aa₄   Codon aa₅   Codon aa₆   Codon aa₇

(c) The Colinear Information

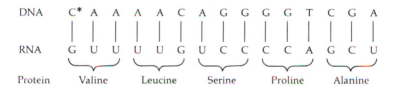

DNA	C* A A	A A C	A G G	G G T	C G A
RNA	G U U	U U G	U C C	C C A	G C U
Protein	Valine	Leucine	Serine	Proline	Alanine

(d)

Second letter

First letter		U	C	A	G	Third letter
U		UUU UUC } Phe UUA UUG } Leu	UCU UCC UCA UCG } Ser	UAU UAC } Tyr UAA Stop UAG Stop	UGU UGC } Cys UGA Stop UGG Trp	U C A G
C		CUU CUC CUA CUG } Leu	CCU CCC CCA CCG } Pro	CAU CAC } His CAA CAG } Gln	CGU CGC CGA CGG } Arg	U C A G
A		AUU AUC } Ile AUA AUG Met	ACU ACC ACA ACG } Thr	AAU AAC } Asn AAA AAG } Lys	AGU AGC } Ser AGA AGG } Arg	U C A G
G		GUU GUC GUA GUG } Val	GCU GCC GCA GCG } Ala	GAU GAC } Asp GAA GAG } Glu	GGU GGC GGA GGG } Gly	U C A G

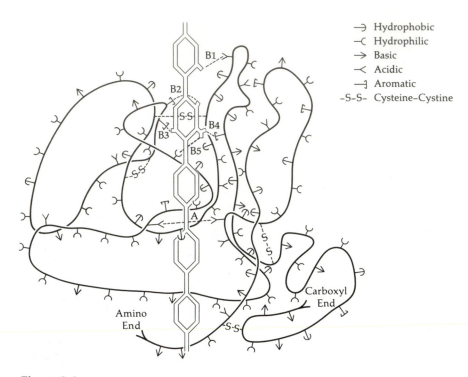

B1

B2

S-S

B3   B4

B5

-S-S-

A

S-S

Carboxyl
End

Amino
End

S-S

→ Hydrophobic
─ Hydrophilic
→ Basic
─ Acidic
─ Aromatic
-S-S- Cysteine–Cystine

Figure 2-4.

Diagram of a generalized enzyme protein molecule that contains a single chain, based on the analysis of egg white lysozyme by D. C. Phillips. When in an active state, the molecule has a complex but precise three-dimensional conformation that is determined by the chemical properties and the order sequence of the constituting amino acids. The molecule is irregularly wound around its substrate, shown as a chain of hexagons. At the central position, A, is the active site, where a chemical bond between two units of the substrate is broken, thus achieving the digestive function of the enzyme. At positions B1 to B5, certain amino acids of the protein chain having acidic, aromatic, or hydrophilic (water-attracting) properties bind to the substrate and hold it in place. The majority of the amino acids placed around the outside of the molecule are hydrophilic, thus increasing its solubility, while most of those inside are hydrophobic, or water-shedding, producing a water-free area around the active site. (From Stebbins, *The Basis of Progressive Evolution*, University of North Carolina Press.)

to work harder and produce more dark pigment, sunlight cannot alter the molecular structure of DNA. An acquired characteristic, such as tanned skin, thus is not hereditary. The cellular environment can affect only the final conformation of the enzyme molecule—its shape while at work—and not the primary structure of the molecule, the *amino acid sequence* of the protein or polypeptide chain. Only mutations that alter the DNA or RNA code can affect the enzyme's structure. Although changes in the DNA or RNA will affect the structure of an enzyme, changes in working enzymes are not transcribed onto DNA or RNA molecules and therefore do not affect the genetic information. Lamarck's hypothesis of the inheritance of acquired modifications has been discarded because no molecular mechanism exists or can be imagined that would make such inheritance possible.

NATURAL SELECTION AND MUTATIONISM

The only plausible alternative to Lamarck's theory of evolution by inheritance of acquired characteristcs is that of natural selection as proposed by Charles Darwin. Although indirect, and therefore superficially less obvious than the inheritance of acquired modifications, Darwin's explanation is almost equally simple and is based on easily observable facts. All organisms produce far more offspring than their habitats can contain. Inevitably, many individuals, particularly young offspring, die from starvation, disease, or predators' attacks. If hereditary differences exist among the individuals of a population, those that are from the beginning best equipped to cope with the environment are the most likely to survive and to produce the largest number of successful progeny. In this way, generation after generation, the hereditary characteristics most germane to survival are perpetuated. The essential validity of this explanation is now almost universally accepted.

Darwin's theory of natural selection as the principal agent of evolutionary change, brilliant as it was, had two almost fatal flaws.

First, neither Darwin nor any of his contemporaries performed experiments to show that natural selection would actually take place as a result of changed population–environment interactions. Up to the time of Darwin's death and for many years later, natural selection was an untested working hypothesis. Second, Darwin failed to identify precisely the correct mechanism of heredity. Gregor Mendel performed his famous experiments on heredity between the publication of the first edition of Darwin's *On the Origin of Species* in 1859 and its final edition in 1872, but neither Darwin nor any of his immediate followers knew anything about Mendel's findings. Ironically, when scientists later rediscovered Mendel's laws, they viewed Mendel's work as antagonistic to, rather than complementary to, natural selection.

Scientists began to doubt Darwin's theory in 1900, when Hugo De Vries of the Netherlands and Carl Correns of Germany independently rediscovered Mendel's laws of inheritance. The advent of Mendelism struck squarely at Darwin's failure to explain even the rudiments of hereditary transmission. This failure permitted the geneticists of the early twentieth century, who knew nothing about the genetic structure or the natural history of populations, to propose mutationism as an alternative to Darwin's theory. Mutationism sought to explain evolution entirely on the basis of internal changes within the chromosomes, and it almost completely disregarded population–environment interactions. To understand mutationism, we must review Mendel's findings.

Mendel showed that inheritance is based not on the transmission of a generalized substance such as blood, a common belief of scientists during the nineteenth century, but of discrete particles that we now recognize as genes or segments of DNA. He reached this conclusion from the results of two kinds of experiments (Figure 2-5). First, he chose two pea plants that differed with respect to a single, easily visible characteristic, such as tall versus dwarf stature, purple versus white flowers, or smooth versus wrinkled seeds. Crossing these plants produced a first generation (F_1) that were all like each other and resembled one of the parents. If one parent was tall and the other was a dwarf, all the offspring were tall. When these offspring were self-pollinated or mated with each other, three-fourths of the offspring of the second generation were tall and one-fourth were dwarf. Genes that govern such alternative character states and that produce simple ratios such as 3:1 in the second generation (F_2) are called *alleles* or *allelic genes*.

In the second kind of experiment, Mendel chose plants that differed with respect to two different characteristics, governed by two separate pairs of alleles; for example, he crossed tall plants having smooth seeds with dwarf plants having wrinkled seeds. All first-generation offspring were tall and had smooth seeds, a result expected on the basis of single-difference crosses. In the second generation, four kinds of plants appeared (tall and smooth, tall and wrinkled, dwarf and smooth, and dwarf and wrinkled) in the ratio 9:3:3:1. This ratio can be derived simply by multi-plying two Mendelian ratios for a single characteristic: ¾ *T* plus ¼ *t*) multiplied by (¾ *S* plus ¼ *s*), where *T* and *t* represent tallness and shortness, and *S* and *s* represent smooth and wrin-kled seeds.

We can explain the results of Mendel's experiments and similar experiments on plants and animals by referring to the hereditary transmission of genetic characteristics. This transmission involves two stages. During *meiosis*, the number of chromosomes and genes in the nucleus of a sex cell is reduced to half of the number present in body cells. During *fertilization*, two sex cells, or *gametes*, unite to form a cell that has the number of chromosomes found in the body cells of the previous generation (Figure 2-6).

The two successive cell divisions of meiosis are illustrated in Figure 2-7. The chromosomes of the gametes divide in such a way that a sex cell, egg or sperm (pollen in plants), receives only one member of each pair. During fertilization, each sex cell contributes a member to each pair of chromosomes. In this way, the offspring that develops from the fertilized egg receives one member of each chromosome pair from each parent.

Genes that determine a given single characteristic are located at comparable positions, called *loci*, on the two members of a chro-mosome pair. Consequently, when one crosses a tall pea plant and a dwarf pea plant, the offspring each receive a gene for tallness from one parent and a gene for shortness from the other. Off-

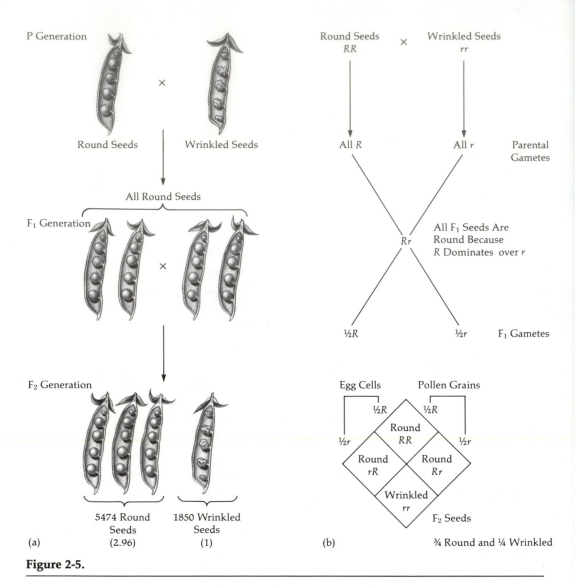

Figure 2-5.

Diagrams showing the results of Mendel's experiments on peas, and their interpretations. (a) When Mendel crossed true-breeding lines, one having round seeds and the other having wrinkled seeds, the offspring of the first generation (F_1) all had round seeds. The second generation (F_2) offspring included 3/4 with round seeds and 1/4 with wrinkled seeds. (b) Mendel's interpretation of the F_1 was that, in the development of each plant, the factor for round was dominant over that for wrinkled and thus was entirely responsible for the appearance (phenotype) of the seeds. Half of the gametes (eggs or pollen) of each F_1 plant contained a factor for round, and half contained a factor for wrinkled. As a result of random union between these gametes, three kinds of F_2 plants were produced. One-fourth were true-breeding round; one-half were round in appearance but segregating, and one-fourth were wrinkled. He obtained similar results when he crossed a plant with yellow seeds with a plant with green seeds. (c) When Mendel crossed a plant with round, yellow seeds with a plant with wrinkled, green seeds, all the F_1 plants, as expected, had round, yellow seeds. In the F_2 generation, he obtained 9/16 round, yellow; 3/16 wrinkled, yellow; 3/16 round, green; and 1/16 wrinkled, green. He explained these results by assuming that the same kind of segregation was taking place for both kinds of factors (genes) and that the two pairs of factors were

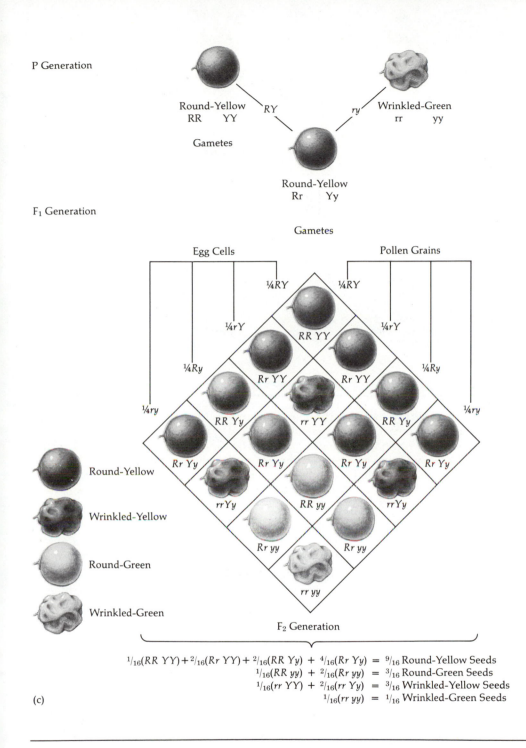

P Generation

Round-Yellow $\quad$ RY $\qquad$ ry $\quad$ Wrinkled-Green
RR $\quad$ YY $\qquad$\qquad$\qquad$ rr $\qquad$ yy

Gametes

Round-Yellow
Rr $\quad$ Yy

F₁ Generation

Gametes

Egg Cells $\qquad\qquad\qquad$ Pollen Grains

¼RY $\qquad$ ¼RY

¼rY $\qquad$ ¼rY

RR YY

¼Ry $\qquad$ ¼Ry

Rr YY $\qquad$ Rr YY

¼ry $\qquad$ ¼ry

RR Yy $\qquad$ rr YY $\qquad$ RR Yy

Rr Yy $\quad$ Rr Yy $\quad$ Rr Yy $\quad$ Rr Yy

rrYy $\quad$ RR yy $\quad$ rrYy

Rr yy $\qquad$ Rr yy

rr yy

Round-Yellow

Wrinkled-Yellow

Round-Green

Wrinkled-Green

F₂ Generation

$$\frac{1}{16}(RR\ YY) + \frac{2}{16}(Rr\ YY) + \frac{2}{16}(RR\ Yy) + \frac{4}{16}(Rr\ Yy) = \frac{9}{16}\ \text{Round-Yellow Seeds}$$
$$\frac{1}{16}(RR\ yy) + \frac{2}{16}(Rr\ yy) = \frac{3}{16}\ \text{Round-Green Seeds}$$
$$\frac{1}{16}(rr\ YY) + \frac{2}{16}(rr\ Yy) = \frac{3}{16}\ \text{Wrinkled-Yellow Seeds}$$
$$\frac{1}{16}(rr\ yy) = \frac{1}{16}\ \text{Wrinkled-Green Seeds}$$

(c)

segregating independently of each other. The four kinds of gametes that result from this independent assortment can combine in 16 different ways, as shown at the bottom of part c. (Parts a and b reprinted by permission from Francisco Ayala and John Kiger, *Modern Genetics.* Menlo Park, Calif.: The Benjamin/Cummings Publishing Company, 1980, pp. 18–19. Part c after Stent and Calendar, *Molecular Genetics,* Second Ed., W. H. Freeman and Company. Copyright © 1978.)

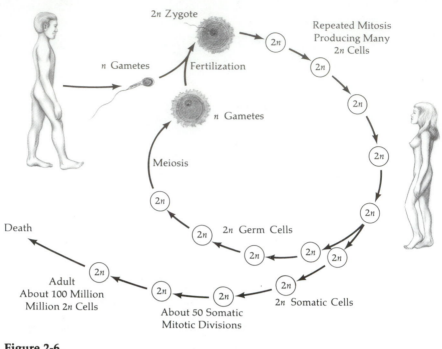

Figure 2-6.

The cycle of alternating chromosome numbers in an animal or human. The body or somatic cells all contain the diploid or unreduced chromosome number, which in humans is $2n = 46$. The gametes—eggs and sperm—contain 23 chromosomes as a result of the two meiotic divisions (Figure 2-7). Union of gametes (fertilization) restores the original diploid number. (From *Biology*, Third Ed., by G. Hardin and C. Bajema. W. H. Freeman and Company. Copyright © 1978.)

Figure 2-7.

The two cell divisions, collectively known as meiosis, by means of which the chromosome number is reduced in the formation of gametes. Photographs are from sperm-producing cells of a male grasshopper that has 17 chromosomes, forming 8 pairs and a single unpaired chromosome. The female of this species has 9 pairs of chromosomes. (a) The chromosomes appear first as a dense mass of very slender, unpaired threads within the nucleus. (b) Still within the nucleus, the threads pair to form 8 bands and a contracted, unpaired chromosome. (c) The pairs contract while still within the nucleus, and their paired nature becomes more apparent. (d) The chromosome pairs line up on a mitotic spindle within the cell and (e) separate to opposite poles, where they condense (f). At the second division of meiosis (g), two mitotic spindles are formed, side by side, upon which are aligned the reduced number (9) of double chromosomes. The chromosome halves then separate (h) to form four groups of single chromosomes (i), around each of which forms a nuclear membrane. Each of the resulting four structures will become a sperm cell. (Photos from J. L. Walters, University of California, Santa Barbara.)

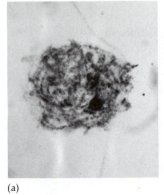
(a)

(b)

(c)

(d)

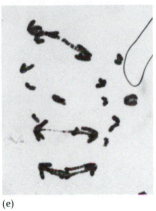
(e)

(f)

(g)

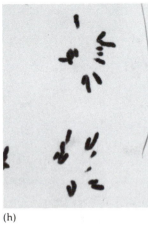
(h)

(i)

spring that contain two different genes that code for a character-istic are *heterozygous* for that pair of genes; offspring that have two similar genes are *homozygous*. Heterozygous cells develop into plants that appear tall because, during development (and only then), the gene for tallness dominates that for shortness and determines the final stature. Heterozygous offspring produce two kinds of sex cells; one-half carry a gene for tallness, and the other half carry a gene for shortness. When their sex cells unite with each other at random, statistical probability asserts that one-fourth of the fertilized eggs, or *zygotes*, that result from their union will have two genes for tallness, like their tall grandparent; one-half will have both kinds of genes, like their parents; and one-fourth will have two genes for shortness, like their short grandparent.

Mendel's results from crossing two different characters are valid only if the genes for these characters are located on different pairs of chromosomes and these two chromosome pairs behave inde-pendently of each other during meiosis and during fertilization. The almost infinite variety that results from hereditary transmis-sion in a cross-breeding population is due to the fact that many different chromosome and gene pairs *segregate* and *recombine* simultaneously and independently of one another. Furthermore, each individual in these populations is heterozygous for hundreds or thousands of different gene loci.

Mendel's experiments on peas are ideal examples of precise, carefully controlled scientific experimentation. Individual plants in his garden cultures did not show subtle differences among each other as do most wild flowers in nature; plants belonging to one of his parental strains were as alike as "peas in a pod." Moreover, Mendel carefully selected the genetic characteristics he worked with. If a factor did not show clear-cut differences among the sec-ond-generation progeny of a cross, he rejected it. His method was the essential first step for discovering the laws of heredity. Never-theless, slavish devotion to the Mendelian laws in their narrowest form, rather than the principle of particulate inheritance that is derived from them, blinded many of his followers to the true relationships between heredity and evolution. They assumed, without question, that observations of natural populations should yield the conspicuous differences and precise ratios that Mendel found in his carefully selected garden varieties.

In 1901, Hugo De Vries, one of the rediscoverers of Mendel, challenged Darwin's contention that evolutionary change is based on almost imperceptible hereditary differences between individuals. De Vries' studies of the wild evening primrose led him to propose that evolution occurs as a series of one-step mutations that produce an individual conspicuously different from both of its parents. We now know that De Vries' choice of the wild evening primrose was unfortunate in that the plant exhibits strikingly atypical chromosome behavior.

De Vries' mutation theory received strong support from the experiments of the Danish geneticist Wilhelm Johannsen. Like Mendel, Johannsen worked with cultivated plants that were self-fertilizing and had been bred for constancy. From experiments with garden beans, Johannsen concluded that, within a *pure line* (a variety selected for its genetic constancy), visible differences between individuals are due to the effects of the environment and are not inherited. For several generations, he planted only the largest beans in a harvest from plants of a pure line. This kind of *aritificial selection* failed to increase the average size of the beans in successive progenies. Following the precepts of De Vries, Johannsen expressed the belief that most natural species are a collection of *biotypes*, genetically constant pure lines that cannot be affected by artificial or natural selection. William Bateson, the eminent British geneticist largely responsible for making Mendel's work known and recognized among English-speaking scientists, incorporated a similar opinion into his textbook on genetics, the first published in the English language, and many later textbook writers followed suit. At the beginning of the twentieth century, most geneticists accepted the mistaken notion that, contrary to Darwin's opinion, natural selection is ineffective in most natural populations, except for eliminating obviously defective individuals. They believed the basis of change to be continuously recurring mutations, or *mutation pressure.*

Additional support for the mutation theory was offered by Thomas Hunt Morgan, the most eminent geneticist of the early twentieth century. His research on the pomace fly, *Drosophila melanogaster*, showed that mutations can easily be found and consist of alterations of genes that occupy particular positions on the fly's chromosomes. Many mutations produce such drastic effects as changes in eye color, body color, wing shape, and deformities of various parts of the body. Morgan believed that mutations could occasionally produce in one step a completely new and successful type of organism and thus account for the appearance of new species.

REACCEPTANCE OF DARWIN'S THEORY

During the late 1920s, the influence of Mendelians and mutationists on laboratory genetics continued to increase, and by the 1930s it dominated all such research. I first became acquainted with Darwin's theory in 1926, when natural selection was at its lowest ebb of disapproval. At the time, biology students were often asked to read what was considered to be the most authoritative history of biology, by Erik Nordenskiöld of Sweden. He wrote: "To raise the theory of natural selection, as has often been done, to the rank of a natural law, comparable to the law of gravity established by Newton is, of course, completely irrational, as time has already shown; Darwin's theory of the origin of species was long ago abandoned. Other facts established by Darwin are all of second-rate value."

When I was an undergraduate at Harvard in 1926, both of the professors whose lectures in introductory biology I attended, Oakes Ames and George H. Parker, told the same story. They said, in essence, that natural selection was a completely inadequate explanation of evolution. The true cause of evolution, in their opinion, lay partly in the occasional occurrence of *sports*, or mutations—that is, in individuals who differ drastically from both of their parents in respect to one or more characteristics and who transmitted these new characteristics to their progeny. Natural selection, they believed, had merely the negative function of eliminating maladapted sports. They admitted that other causes of evolution might exist but maintained that these were entirely unknown.

During the 1930s and 1940s, however, such evolutionists as Ronald Fisher, Sewall Wright, Julian Huxley, Theodosius Dobzhansky, and Ernst Mayr realized that populations of laboratory flies and cultivated garden pea plants are very different from populations of animals and plants in nature. When they applied the techniques and principles developed by geneticists in the laboratory and the garden to natural populations, they quickly found that the theories and concepts developed by De Vries, Johannsen, Bateson, and Morgan were inadequate. The study of natural populations required that new principles be added to the principles of Mendelian heredity. A new integrated and comprehensive theory was needed.

The first extension of Mendelism was the principle of blending, or quantitative inheritance. When dark-skinned and light-skinned people marry and produce children, their offspring do not segregate into sharply defined dark and light categories, as did progeny from Mendel's crosses between purple-flowered and white-flowered pea plants. Among people who are descended from frequent intermarriages between dark- and light-skinned ancestors, every shade of intermediacy between dark and light skin can be found, and proportions that resemble Mendelian ratios cannot be detected. The Swedish geneticist Herman Nilsson-Ehle studied a similar phenomenon when he crossed a wheat variety having white kernels with another having red kernels. In the second-generation progeny from these crosses, he detected five different shades of color—red, yellow-white, and three shades of pink. The number of individuals in each color class agreed with expectations based on Mendel's second law governing the simultaneous transmission of characteristics controlled by two pairs of genes located on two different chromosomes. In an extension of this work, Harvard geneticist Edward M. East analyzed progeny of hybrids of tobacco plants having very long flowers and others having very short flowers (Figure 2-8). In spite of the apparently blending inheritance, his results were similar enough to those of Nilsson-Ehle that he interpreted them as the results of Mendelian segregation of particulate genes. He assumed that his tobacco flowers differed with respect to a large number of different gene pairs that control flower length, each pair on a different chromosome or at a different locus on the same chromosome.

To explain individual variation within a population, East produced a large family of tobacco plants, all of which had exactly the
same genes as their common parent. These plants were just as
much alike as the bean plants in Johannsen's studies. East raised
their progeny for several generations, without selecting for either
constancy or change. Finally, he obtained a population in which
each plant differed slightly from its neighbors, even though all
had been raised in as constant an environment as he could provide. He concluded that, during several generations, a few of the
genes in each plant had been slightly altered so as to give rise to
a somewhat different adult plant. He ascribed the changes to
mutations, a hypothesis similar to those postulated by De Vries
and Morgan. But, whereas the mutations detected by Morgan
were changes in genes that produced a great alteration of outward
appearance, the mutations in East's tobacco plants produced only
slight effects. When he published these results in 1935, East
remarked: "The deviations forming the fundamental material of
evolution are the small variations of Darwin. We return to the

Figure 2-8.

The inheritance of a quantitative character, flower size, in descendants of a
cross between two races of a tobacco species, *Nicotiana longiflora*, one having
very long flowers and the other having much shorter ones. Since the plants
can be self-fertilized, the parental lines approach the condition of pure lines,
similar to those of Mendel's peas. Numbers across the top of the diagram
indicate the length of the flowers in millimeters; those along the side indicate
the number of individuals in each size class. *Top row*: The two parental lines.
Second row: The number of individuals in each size class found in the first
generation (F_1) hybrids. The variation is only slightly greater than that within
the parental lines and is due mostly to environmental effects rather than
genetic differences between the F_1 plants. *Third row*: Values recorded for the
sum of two second generation (F_2) progenies. The greater spread of variation
in this generation as compared to the F_1 is due to simultaneous segregation for
several different gene pairs, each of which has a small effect on flower length.
Fourth, fifth, and sixth rows: Values recorded in families of later generations (F_3,
F_4, F_5) that were produced by self-fertilizing selected plants of the previous
generation. A reduced amount of segregation lowers the range of variation of
successive families and may bring about a reversion toward one or other of the
parental types. (Data from E. M. East.)

Darwinian idea, modified by the demonstration of alternative inheritance."

East was by no means the only geneticist who recognized that modifications of Mendelian theory to accommodate quantitative, apparently blending inheritance were necessary to explain evolution. Even T. H. Morgan, a staunch supporter of the importance of mutation theory, remarked in an address in 1923: "Today we agree with Darwin that such extreme variations as those he called

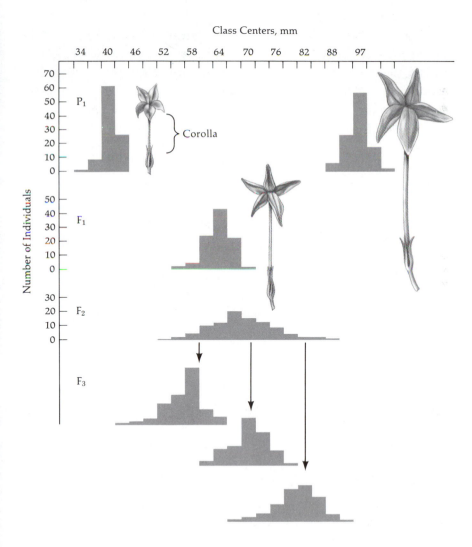

sports would rarely, if ever, have contributed to the formation of new types in nature. But we also know that minute differences also arise as mutants, and that these are inherited in the same way as are the large mutant changes. It is also now clear that these smaller mutant variations must be the small heritable variations of Darwin."

Strangely enough, when Morgan, nine years later, published a book that expressed his final opinion on evolution, he completely reversed his point of view: "A mutationist might well insist that the essential part of Darwin's theory of natural selection is not survival, but Darwin's postulate that individual variations, everywhere present, furnish the raw materials of evolution. This the mutationist would deny." We do not know why he changed his mind. Perhaps he denied the importance of those mutations that have small effects because their existence had not been experimentally demonstrated, as had the mutations that have large effects, such as those he found in *Drosophila*. Perhaps he was swayed by the importance that he attached to his own research. Because of Morgan's prominence, his book was widely read, and his dismissal of natural selection had a much greater influence on the course of science in the 1930s than did his earlier opinion or the contrary opinions of other geneticists.

Concepts of evolution based on Mendelian theory had to be modified even more when geneticists discovered that genetically uniform pure lines—the model for Mendel's crosses that Johannsen and Bateson believed to be frequent in nature—were found to exist rarely, if at all, in natural populations of most species. Contrary to the predictions of these early geneticists, numerous recent experiments show that populations of outcrossing plants, such as corn or maize as well as *Drosophila*, chickens, and mice, respond to artificial selection by continuous changes in their hereditary composition over twenty, fifty, or even a hundred generations. In *Drosophila*, successful artificial selection has been carried out for at least 51 different traits, including body and wing size, numbers of bristles on various parts of the body, resistance to DDT, egg production, ability to fly toward or away from a light, and ability to crawl upward or downward. In a recent summary of this experimental work, geneticist Richard Lewontin remarked: "There appears to be no character—morphogenetic, behavioral, physiological, or cytological—that cannot be selected in *Drosophila*."

Johannsen's experiments are perfect models of laboratory genetics, and his results are applicable to artificially pure varieties of garden plants. The mistake that he and his followers made was to assume that natural species resemble pure lines with respect to their gene pool, their degree of hereditary variation. Like many biologists of their time, they assumed that different kinds of organisms resemble each other closely enough so that experiments carried out on one population provide generalizations that apply equally well to all kinds of animals and plants. Experiences of modern geneticists and evolutionists have taught us that any biological theory of broad significance must be tested on a variety of animals, plants, and microorganisms before it can be accepted as valid.

CONTEMPORARY THEORY

Other geneticists were reevaluating the Darwinian concept for entirely different reasons. Almost simultaneously, Ronald Fisher and J. B. S. Haldane in England, Sewall Wright in the United States, and S. S. Chetverikov in the USSR realized that the genetics of individual families, as studied by Mendel and pursued by Morgan, could not explain evolution in large natural populations. All four geneticists based their conclusions on a sound and thorough knowledge of mathematics and statistics, disciplines of which geneticists and evolutionists before their time were woefully ignorant. The theory developed by these statistically minded geneticists includes six main assumptions, which we now must consider.

Hereditary Variability.   All populations of sexually reproducing, cross-breeding organisms contain a large store, or gene pool, of heritable variability. Some of this variability expresses itself in the form of visible differences among individuals, such as blue eyes or brown eyes, or curly or straight hair. Other genetic differences among individuals belonging to the same population can be recognized only by means of biochemical tests, such as those that distinguish between blood groups and genetically determined resistance to various diseases. Moreover, all individuals that belong to cross-breeding populations may be heterozygous for two corresponding genes or alleles, as are the first-generation off-spring of a Mendelian cross. The variability of the gene pool in a population is measured in two ways, by the proportion of gene loci at which two or more alleles exist among its individuals and by the average proportion of loci at which individuals are heter-ozygous. The latter proportion is important because it determines the number of different kinds of sex cells (eggs or sperm) that individuals can produce. For instance, the first-generation off-spring of Mendel's crosses involving a single pair of allelic genes were heterozygous at one locus and could produce two kinds of sex cells, one carrying the allele for tallness and the other the allele for short stature. The first-generation offspring of crosses involv-ing two pairs of alleles can produce 2^2 or 4 kinds of sex cells (tall and smooth, tall and wrinkled, dwarf and smooth, dwarf and wrinkled). When larger numbers of loci and pairs of alleles are involved, the potential number of different kinds of sex cells increases exponentially, that is, 2^3, 2^4, 2^5, 2^6, and so on. The alternative alleles at any gene locus may produce widely different visible effects, effects so small that they can be detected only by special methods, or a variety of intermediate effects.

The most common kind of variability now recognized consists of small modifications of the structure of enzymes. These are revealed by a technique known as *electrophoresis*. Proteins extracted from cells are placed on an electrostatic field and are stained with dyes that are specific for a particular kind of enzyme. When the current is on, the protein mixture moves across the field. Enzymes having similar staining properties are found to travel different dis-tances, and they appear as parallel bands (Figure 2-9). This dif-ferential mobility across the electrostatic field reflects different electric charges on the protein molecule, charges based on differ-ent arrangements of their amino acid components. These differ-ences, in turn, reflect different sequences of nucleotides in the

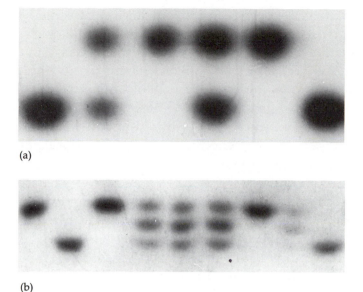

(a)

(b)

Figure 2-9.

Stained bands that record the movement of particular
enzyme proteins across an electrophoretic field, due to the
charge present in the molecule. The more heavily charged
proteins, which move the farthest, form the upper line in
each of the figures. (a) Gels containing two bands, formed
by proteins that consist of a single chain,
phosphoglucomutase in *Drosophila pseudoobscura*. Flies
homozygous for a slow-moving protein *(columns 1 and 6)*
have bands only at the lower position; those homozygous
for a fast-moving protein *(columns 3 and 5)* have bands
only at the upper position; hybrids between them, which
are heterozygous for the two alleles, have bands at both
positions *(columns 2 and 4)*. (b) Gels formed by an enzyme,
malate dehydrogenase in *Drosophila equinoxialis*, that
consists of two or more protein chains. *Columns 2 and 9:*
homozygous slow; *columns 1, 3, and 7:* homozygous fast;
columns 4, 5, 6, and 8: heterozygotes containing three
bands, the middle one of which is made up of slow chains
combined with fast chains. (Photos courtesy of Francisco
J. Ayala.)

DNA of the genes coding for the enzymes. Electrophoretic studies, combined with genetic experiments that follow traditional Mendelian methods, show that from 40 to 60 percent of the gene pairs present in many populations possess two or more distinguishable variants. In those populations, any individual is heterozygous for two different mutated states with respect to 10 or even 15 percent of the gene loci present in that individual. If the total number of gene loci present in an individual is 10,000, an average number for species of mammals, the mean number of heterozygous loci in an individual may be as high as 600, so that the individual might produce 2^{600} genetically different kinds of sex cells—a number many times larger than the total number of atoms in the universe! The generalization that genetic variation in populations is essentially infinite in quantity is supported by solid data. Evolutionary response to changing environments is limited not by the quantity of available genetic variation but by its quality.

Genetic Recombination.   The amount of variability present in populations is greatly increased by genetic recombination. Each time an animal forms an egg or sperm, meiosis causes the chromosomes and the genes contained in them to be dealt out at random (Figure 2-7). New combinations are delivered to each fertilized egg because many of the gene loci in the unfertilized egg contain different alleles from those contributed by the sperm. The potential number of genetically different kinds of offspring that a particular mating can produce is the product of the number of different kinds of eggs and different kinds of sperm produced by the parents.

The importance of genetic recombination for continued evolution becomes evident when one compares groups of populations that have nonsexual reproduction with those that have sexual reproduction. A good example among plants is a comparison of blackberries, raspberries, thimbleberries, and their relatives (Figure 2-10). Among the blackberries or brambles of Europe and the eastern United States are hundreds of recognizable varieties. They form large amounts of seed, but seedlings from a single parent show no sign of genetic segregation because, during seed development, meiosis does not occur. Each egg cell has exactly the same genes as the cells of the parent plant. In contrast, raspberries, as well as thimbleberries and the blackberries that grow along the Pacific Coast of North America, reproduce sexually. A single raspberry bush, containing seeds that have been fertilized by genetically different kinds of pollen grains, produces seedlings of which no two are exactly alike. They possess individuality just as do brothers and sisters among animals.

European and eastern American blackberries are among the most successful of plants. They often invade disturbed ground and form dense, impenetrable thickets. Evidence from both geographical distribution and fossil records indicates that they are at least 10 million years old. Yet, during all this time, they have given rise only to endless variations on the blackberry theme. Meanwhile, their sexually reproducing cousins have branched out to produce raspberries, thimbleberries, cloudberries, dewberries, and many other berries. Horticulturists have added to their diversity by breeding loganberries, boysenberries, and several other cultivated varieties.

In many other groups of plants, such as dandelions, hawkweeds, cinquefoils, and hawthorns, sexually reproducing and asexual populations exist side by side. The asexual populations have a high degree of ecological success but a limited capacity for evolving new kinds of adaptations. Their sexual relatives may often be more restricted in their present geographical and ecological distribution, but they have evolved a greater diversity of adaptive complexes and strategies.

Other kinds of plants reproduce sexually, but their seeds are produced by self-fertilization, involving union of pollen and egg cells of the same flower. If self-fertilization is relieved by occasional outcrossing, and if the resulting hybrids are adaptively superior, a large amount of genetic variability can be maintained in the population. If, however, reproduction is exclusively by self-fertilization, the gene pool becomes exhausted, and the ability for the population to evolve is strongly curtailed. Adherents of the modern synthetic theory of evolution agree with Johannsen, Morgan, and other classical geneticists in believing that genetically pure lines, such as garden varieties of beans or peas, are incapable of

(a)

(b)

(c)

(d)

Figure 2-10.

Relative amounts of genetic diversity achieved by sexually reproducing and asexually reproducing species belonging to the blackberry–raspberry genus (*Rubus*). Genotypes shown in parts a to d are sexually reproducing; those in parts e to h produce seeds that give strictly maternal genotypes, since in them meiosis and fertilization do not take place. (a) Western thimbleberry

(e)

(f)

(g)

(h)

(*R. parviflorus*). (b) Western raspberry (*R. leucodermis*). (c) Dewberry
(*R. trivialis*). (d) Cloudberry (*R. chamaemorus*). (e–h) Four asexually
reproducing clones of blackberries (*Rubus*, subgenus *Eubatus*). (Drawn from
specimens in the Herbarium, University of California, Berkeley.)

evolution by natural selection. Modern population genetics, however, shows that nearly all natural populations are, in the words
of Morgan, "genetically impure," so that they are susceptible to
natural selection for new adaptive combinations of genes.

Populations Versus Individuals.   Populations, not individuals, evolve
by responding through natural selection to the challenge of changing environments. Nevertheless, every new gene originates via
mutation in a single individual. These new genes can affect evolution only if, in successive later generations, their frequency
increases so that they become spread throughout the population.
Consequently, in order to affect evolution, the smallest genetic
changes must undergo a two-step process—origin via mutation
and spread via natural selection or, less often, the effects of chance
sampling.

The unimportance of single mutations that do not greatly
increase their frequency in populations can be recognized from
observations of many populations in nature. Among a population
of plants having pink, purple, or yellow flowers, white-flowered
individuals are occasionally seen. Populations of other species
consist almost entirely of plants having five petals, but individuals
having four, six, or seven petals occasionally appear. Genetic tests
of progeny from such plants show that the aberrant condition is
inherited and so represents a change in one or more genes. Such
aberrant individuals, or sports, have been observed in many populations of animals both wild and domestic, but only by means of
artificial selection will these sports give rise to distinctive varieties.

Constancy of Gene Frequency.   Frequencies of genes in populations
remain constant unless altered by mutation, selection, or the
effects of chance sampling. The basis of this assumption is a fundamental theorem of population genetics, developed independently by the British mathematician G. H. Hardy and the German
physician W. Weinberg. The Hardy–Weinberg law states that frequencies of genes remain constant in a randomly mating population unless altered by mutation, selection, or the effects of chance
sampling, and it also predicts the relative frequencies of homo-

zygous and heterozygous individuals for each pair of genes. The validity of this law has been tested by direct observation and by progeny tests for a large number of populations belonging to various species of plants and animals, including human beings. Actual data consistently agree with predictions made on the basis of the Hardy–Weinberg law.

Changes in Gene Frequency.   Mutation rates are too low to have a significant effect on gene frequencies, except over very long periods of time. The effects of chance sampling are important only under special circumstances. Consequently, in most populations of animals and plants, changes in gene frequencies are brought about chiefly by natural selection.

Numerous examples illustrate how natural selection acts to alter populations in nature. One of the most striking examples results from the use of such insecticides as DDT to control the spread of insect pests, particularly flies and mosquitoes. Soon after this practice began, many observers noticed that doses which at first killed almost all the pests were later ineffective in controlling them. Laboratory tests showed that new populations of flies had arisen that inherited a high resistance to the pesticide. When resistant strains were crossed with susceptible ones, the presence of genes for resistance was detected. Each individual gene, however, was responsible for only a part of the total resistance; highly resistant flies were produced only when genes from different strains were combined by artificial hybridization followed by selection. As a result of selection in the classical Darwinian sense, previously rare genes, which provide high resistance, increased in frequency. Selection for insecticide resistance is possible only if a population is relatively large. Attempts to induce DDT resistance by applying the insecticide to relatively small populations of flies consistently fail. Apparently, mutations causing high resistance are so rare that small, semiartificial populations lack them, although they are present in much larger, natural populations.

Adaptive Complex.   Most adaptive characteristics are based on traits coded on many different genes, and they can be altered only by the occurrence and establishment of changes in many genes. Mutations of single genes that produce large effects are usually detrimental. Consequently, during the initial stages of natural selection, a population's response to a changing environment relies on genetic variability already present. Mutations are important chiefly because natural selection reduces the frequency of unfavorable alleles, often eliminating such alleles and thereby reducing the size of the gene pool available to selection in later generations. New alleles produced by mutations can restore the quantity of variation present in the gene pool while at the same time altering its composition. Without mutation, genetic variation would, after many generations, become so much reduced that the population could not respond any further to natural selection.

The assumption that most adaptations reflect changes in many different genes is supported by analyses of the evolution of insecticide resistance and by experimental evidence that reveals, almost directly, that the presence of single gene differences have very small effects on the appearance or functioning of an organism.

A highly effective method for detecting such mutations was used by geneticist Charles Yanofsky while working with the bacterium *Escherichia coli* (Figure 2-11). He spent several years analyzing the structure and action of tryptophan synthetase, an enzyme responsible for the synthesis of the amino acid tryptophan. By chemical analyses, he determined the exact sequence of amino acid units in the molecule. He then compared the normal sequence with the sequences present in various mutants that he had produced artificially. Two of the mutations, involving the replacement of glycine by either arginine or glutamic acid at a particular position along the chain, produced mutated enzymes that were inactive. Yanofsky then asked whether enzymatic function could be restored by means of artificially induced mutations, and, if so, would restoration of function always involve the restoration of glycine to the position it originally occupied? He recovered many functional derivatives, or *revertants*, as mutants from the nonfunctional strain, but only some of them had glycine in its original position. In others, function was restored by mutations at new

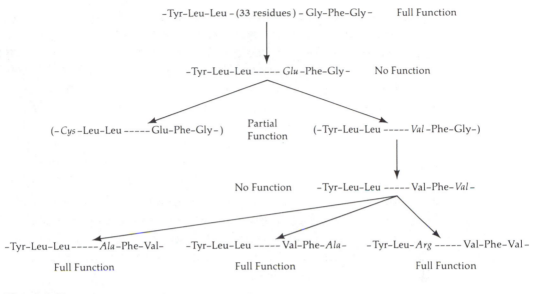

Figure 2-11.

Diagrams showing the complexity of the relationships between protein structure and function within a single enzyme and demonstrating by an indirect method the possibility of occurrence and the nature of mutations having small effects on enzyme activity. The enzyme is tryptophan synthetase in the bacterial species *Escherichia coli*. *Top row*: Two sequences of three residues each, separated from each other by 33 residues that were not studied, because no mutational changes in them had effects that could be detected. This normal sequence has full function. *Second row*: Sequence in a mutation that was produced by X-radiation and caused the enzyme to lose its function. *Third row*: Two revertant lines isolated from mutation experiments performed on the nonfunctional strain shown in the second row. The nullifying effect of the first mutation, which substitutes glutamic acid for glycine in the fourth of the six positions illustrated, is partly counteracted by a second mutation, which in one instance (*at right*) substitutes valine for glutamic acid at the fourth position, and in another instance (*at left*) substitutes cysteine for tyrosine at the first recorded position. *Fourth row*: The sequence of one of the revertants becomes inactive via a mutation that substitutes valine for glycine at the sixth recorded position. *Fifth row*: The functionless sequence of the fourth row can be transformed into a functional sequence by either one of three different mutations. *Right*: Alanine for valine at the fourth position; *middle*: alanine for valine at the sixth position; *left*: arginine for leucine at the third position. By comparing the sequence of the fifth row (*right*) with the original one, the conclusion is reached that a mutation that substitutes alanine for glycine at the fourth position would have no detectable effect on enzyme activity, whereas one substituting valine for glycine at that position would have only a small effect. Such mutations could never be detected by direct methods. (Data from Yanofsky, Horn, and Thorp. From Stebbins and Lewontin in *Proceedings of the Sixth Berkeley Symposium of Mathematics, Statistics, and Probability*, Vol. 5, 1972, University of California Press.)

positions on the tryptophan synthetase gene. More interesting, however, were mutations that restored function by placing amino acids other than the original glycine at the position of the initial change, for instance, alanine. Yanofsky's experiments show that mutations having small effects can be recognized by the reverse mutation method, provided that the amino acid sequence of the protein is completely known.

Have mutations with small effects often been established in natural populations? An answer to this question can be derived by three methods. One method is to hybridize individuals belonging to different races within the same species and to determine whether the second-generation progeny show clear-cut Mendelian segregation, such as a 3:1 or 9:3:3:1 ratio for a particular character by which the parents differ, or whether variation with respect to this character shows a continuous spectrum. Analyses of such laboratory hybridizations almost always demonstrate that a given characteristic results from the activity of many different genes, most or all of which have small effects. One can infer, therefore, that differences with respect to these characteristics originated by the accumulation of many mutations, each of which altered only slightly the appearance and functioning of the organism. This conclusion supports the concept of genetic variability held by Darwin, East, and Fisher rather than that advocated by De Vries, Morgan, and their followers.

A second way to discover past occurrences of mutations having small effects is by molecular comparison. Biochemists select for study protein molecules that, on the basis of similarities in their amino acid sequences, most probably evolved from the same ancestral molecule. The comparison consists of noting differences between animal or plant species with respect to particular amino acids located at comparable positions along the molecule. For example, *hemoglobin* is the oxygen-carrying molecule found in all vertebrates. When extracted and analyzed in test tubes, hemoglobins derived from vertebrates as different as horses, apes, and birds are found to have very similar properties. Nevertheless, horse hemoglobin differs from human hemoglobin with respect to 43 amino acids, and other hemoglobins are even more different.

During the evolution of vertebrates, thousands of mutations have occurred in hemoglobin molecules and have become established in various populations. Most of these mutations have a relatively small effect, if any, on the active molecule. Mutations having large effects, such as the sickle-cell gene, have doubtless occurred many times, but most such mutations are absent from or infrequent in present populations due to adverse natural selection.

The severity of the functional change in an organism effected by a mutation is not directly related to the actual change in the DNA molecule. Indeed, some changes in the molecular structure of DNA have no effect at all on structure of the protein for which it codes. For example, Figure 2-12 shows three mutations of a DNA triplet of nucleotides. Although the three mutations appear to be very similar—each involving a change in one nucleotide—nevertheless, they have vastly different effects on the organism.

Thus, molecular biology shows that individual mutations by themselves are rarely the agents of recognizable evolutionary change. Mutations that produce large effects on an organism's function usually produce so much disturbance that they lower the reproductive fitness of the individual, so that the individual or its descendants produce fewer viable progeny, and thus the mutation is eliminated or reduced to a low frequency by natural selection.

Original triplet	Original amino acid	Mutated triplet	New amino acid	Effect of the mutation
CAT	valine	CTT	glutamic acid	In beta chain of hemoglobin at sixth position, produces sickling; semilethal
CAT	valine	CGT	alanine	Always slight, since valine and alanine have very similar chemical properties
CAT	valine	CAG	valine	None, since both triplets code for valine

Figure 2-12.

Chart showing how mutations that are almost identical, except for the position of the base pair that becomes altered, can produce either negligible, slight, or drastic effects, depending on the amino acid for which the DNA triplet codes and the position of the amino acid on the protein (polypeptide) chain.

Occasionally, when a population is faced with a great environmental challenge, one or more mutations having large effects may produce a new organism that is better adapted to the new environment. Such events, although rare, are of profound importance for the evolution of new kinds of animals, plants, and microbes.

A GENERALIZED SYNTHETIC THEORY
OF EVOLUTION

The six assumptions we have reviewed form the basis of the modern synthetic theory of evolution. This theory is synthetic in that it is the product of a large number of scientific discoveries by geneticists, biochemists, paleontologists, biogeographers, cellular biologists, and other specialists. Theodosius Dobzhansky, in *Genetics and the Orgin of Species*, first published in 1937 and revised in 1941, 1951, and 1970, provided the first model for a successful synthesis of facts from genetics, selection theory, quantitative population biology, and direct studies of natural populations.

The modern synthetic theory can be summarized as follows. All populations that are capable of continued evolution by virtue of sexual reproduction and continued or occasional outcrossing between genetically different individuals possess a gene pool of genetic variability that is essentially infinite in quantity, although it may be highly restricted in quality. To quote J. B. S. Haldane, one of the founders of the theory: "The human species is capable of evolving in many directions, but it will never produce a race of angels; genes for wings and for moral character are not present in human populations."

Evolution takes place through successive interactions between populations and their environment, mediated by natural selection. If the environment, over long periods of time, remains constant relative to the adaptive capacity of the population—that is, if the population–environment interaction remains unaltered—then natural selection will remove the least adaptive individuals, generation by generation, and so increase the perfection of the adaptation until no more improvement is possible. Evolution then ceases, and the population remains constant until a significant change in the environment occurs.

If the environment is changing, some populations will draw on the reserves of genetic variability in their gene pools, consisting chiefly of gene differences having small effects, and will evolve in a direction that better suits the new environment. The particular new adaptation that is evolved will depend on the nature of the gene combinations that appear the soonest, and the adaptation may be triggered by particular mutations that occur at a critical time during the adjustment process. Consequently, populations having different gene pools may evolve in different directions in response to the same kind of environmental change.

This part of the theory can be illustrated by the following example. Imagine an animal community in which live a running mouse, a hole-digging pregopher, a somewhat bristly preporcupine, and a somewhat smelly preskunk. A new, highly efficient predator migrates into the community from another region. Which animals will survive this drastic, potentially disastrous change in the environment? The mice that have the greatest speed and agility, the pregophers that have the strongest digging paws, the preporcupines that have the longest, sharpest bristles or spines, and the preskunks that smell the worst have the best chances for survival. In each population, natural selection will cause the fittest individuals to survive, and thus the frequency of their genes will increase in the population's succeeding generations.

During the course of this change, the gene pool becomes depleted as individuals who have genes that were adaptive in the old environment but not in the new one are killed by the predator before they can reproduce. Variability is restored by the occurrence and establishment of mutations that contribute to the new adaptation, particularly those having small effects.

If two or more populations that were originally part of the same interbreeding system become isolated from each other during a period of environmental change, either through extinction of connecting populations or migration to new habitats, they are almost certain to evolve in different directions. After many generations of such divergent evolution, their gene pools and chromosomal structure may become so different that individuals belonging to these populations either are unable to mate with each other or their mating yields inviable or sterile offspring. In this way, new species come into being, as we will discuss more fully in the next chapter.

SELECTION, STRUGGLE, AND PURPOSE

In presenting his theory, Darwin made a serious mistake in characterizing natural selection as a "struggle for life" or "struggle for existence." Perhaps he felt that such strong words were necessary to convey to his readers the general principle. In popularizing Darwin's work, however, some nineteenth-century authors resorted to even more lurid language, such as "nature red in tooth and claw." Herbert Spencer, an economist and philosopher, and others proposed a theory of "Social Darwinism" to justify the economic and social inequality in postindustrial societies. They distorted Darwin's principle to derive "laws" of economic fitness and survival.

Almost all contemporary evolutionists have discarded such phrases as "the struggle for existence." To what extent does natural selection depend on the outcome of violent struggles or lethal combat? The answer is, very little. One cannot deny that life in nature is hard and brutal. The most common fate of a young field mouse, squirrel, or fledgling bird is to become dinner for an owl, a hawk, or a bobcat. Only the luckiest of them ever become adults and parents of a new generation. Even lions, the "king of beasts," have to suffer through many days of hunger when game is scarce. These hardships and untimely deaths are, however, interactions between different species, between a species and its biotic environment. Natural selection acts as different individuals of the same population display different rates of survival and reproductive capacity. Between such individuals, struggles to the death are rare. Even when two male deer or mountain sheep are competing for the favor of females, they do not, as a rule, engage in lethal struggle. They may often fight with each other for a while, but when one of the fighters is evidently losing, he usually submits and is allowed to leave unharmed. In most species of fishes, insects, and plants, which constitute the majority of known organisms, active struggle between individuals belonging to the same population is completely absent.

Contemporary evolutionists often describe evolution as a series of games—rather than a series of purposeful competitions—that are played because they are inevitable. Different individuals or populations adopt different strategies in response to their environment. Some of these strategies may be highly successful for a short while, but later lead to extinction or stagnation. Others may lead the populations through temporary reductions in size, followed by expansion based on a novel way of playing the evolutionary game. At any particular point in time, to predict which strategy will lead to the greatest eventual success is difficult or impossible.

Heads of four species of Hawaiian honeycreepers. These birds differ so greatly from each other that if they were found on the mainland and were not connected by intermediate species, ornithologists would regard them as members of different families. Their evolutionary connections are shown in Figure 4-8. (Drawing by Vally Hennings after photographs by S. Carlquist.)

Chapter Three

CHANCE, DIRECTION, AND NECESSITY

In 1971, Nobel Laureate Jacques Monod published a book entitled *Chance and Necessity*. The work has two main themes. First, the direction of mutations is entirely random with reference to their functional or adaptive value; there is no internal force or guiding principle that has directed the course of evolution from bacteria through amoeba, worms, fishes, reptiles, and mammals to human beings. This is the aspect of chance. Second, all mutations, even before they can be incorporated into the genetic information of a single individual, must run the gauntlet of the internal or cellular environment, which can be likened to a room full of complex machinery whose parts are related to each other in a pattern of enormously detailed, exquisite harmony. The activity of the cell itself provides a finely graded sieve through which most mutations cannot pass, and thus they are rejected soon after they occur. Every change in the DNA blueprint must add to the efficiency of operation of the cellular machinery or at least not detract from it. This is the aspect of necessity. According to Monod, the evolution of all the marvelous plant and animal structures that we are familiar with has resulted from the combined action of these two factors.

The publication of Monod's book engendered a violent storm in French intellectual circles, particularly among the followers of the Jesuit Pierre Teilhard de Chardin, who had written some twenty years earlier a very different account of evolution, *The Phenomenon of Man*. (The book was not published until after his death, in 1955.) Teilhard saw evolution as a steady progress upward from the cosmos through the solar system, the planet earth and all the myriad plants and animals on it, to mankind as the pinnacle of evolutionary creation. Like T. H. Morgan and other early twentieth-century biologists, he believed that natural selection could do nothing except get rid of undesirable mutations. Teilhard gave no precise definition of the creative force in evolution, but he pointed to some kind of internal guiding principle; in this he differed from Darwin and Monod.

The followers of Teilhard were highly vocal in their opposition to Monod's ideas. In 1973, Pierre Grassé acknowledged the truth of Monod's principles but stated that chance and necessity in Monod's sense were responsible only for the excrescences or complications of evolution—the proliferation of hundreds of species of bacteria, of the fly genus *Drosophila*, of streamlined fishes, snakes, and so on. Recognizing that the origin of all major groups of animals is buried in the past and can be made known directly only through the fragmentary evidence in the fossil record, Grassé concluded that the processes responsible for these origins can never be known with certainty. His view was endorsed enthusiastically by scientists brought up in the classical tradition, like the zoologist Jean Rostand, a member of the Académie Française, who wrote: "When I see a cricket running or the flight of a dragonfly, this is enough to make me feel closer to Pierre P. Grassé than to Jacques Monod." But these accolades failed to impress either the molecular biologists or the experimental, population-based evolutionists, and the controversy rages on.

The evolutionists' debate resembles the old story of the blind men who tried to describe an elephant, each using his knowledge of a different part. As modern evolutionary science progresses, more and more aspects of the evolutionary "elephant" are revealed, with increased knowledge of the nature of variation in populations and of interactions between populations and their environments. A theoretical framework is provided by the guiding principles of epigenesis, autocatalysis, and coevolution, which will be explained and discussed in the next chapter. They form a bridge between understanding evolution at the level of populations, races, and species and understanding it at the level of major trends of evolution taking place over hundreds of millions of years.

CHANCE VARIATION AT THE LEVEL OF POPULATIONS

We learn to recognize people by their outward features—tall or short, fat or thin, long nosed or short nosed, with curly hair, and so on. Although many features can be modified by environment or personal habit, the majority are governed by the individual's genetic heritage. Differences with respect to individual characteristics are the most obvious expressions of the genetic variability within human populations that the evolutionist calls the *gene pool*. With respect to superficial characteristics, populations of nearly all animals are as variable as those of humans. Anyone who regularly feeds sparrows or other birds learns to recognize some individuals as though they were friends or neighbors. If we walk through a forest, picking leaves as we go, and then compare the leaves in our collection, we can easily recognize the individuality of each tree. Hidden biochemical differences in the structure of enzymes or other proteins are just as numerous as visible differences in nature.

How much of this variation is adaptive, and how much is due to chance accumulation of mutations and other genetic differences that are neutral from the standpoint of adaptation and natural selection? This problem has plagued evolutionists for half a century, ever since population geneticist Sewall Wright developed his theory of the relative effects on populations of mutations, selection, and random alterations of gene frequencies. The term *drift*, which he used for these random changes, is now firmly established in the evolutionist's vocabulary.

During the 1960s, much less emphasis was placed on drift than in the 1930s, when Wright's theory was new and was being hotly debated. Then came the use of electrophoretic techniques to reveal protein differences, and with it there arose a whole new school of evolutionists who maintained that most, if not all, biochemical variation in gene pools is neutral with respect to adaptation and natural selection, or else consists of harmful genes, genetically *recessive*, hiding behind their normal counterparts. The most extreme advocates of this new school called their discipline "non-Darwinian evolution." Like T. H. Morgan, William Bateson, and other biologists of the early twentieth century, they believed that they had a substitute that would reduce Darwin's theory to insignificance. The controversy between the non-Darwinians and those who believe that most of the biochemical variation in the gene pool—both morphological and biochemical—is to some extent adaptive has been raging for more than a decade.

How valid is this controversy? The majority of the outward differences by which we recognize our neighbors, the students in a class, or the different trees of a forest are trivial and largely neutral with respect to adaptation. The same is true of much biochemical variation, for example, among the proteins in the blood of humans and other mammals. At the social level, we recognize people by means of facial and other anatomical features, but police officers and detectives do this more accurately on the basis of trivial, neutral differences in fingerprint patterns, while medical scientists and other human biologists achieve the greatest possible accuracy on the basis of immunological blood groups. More than 20 genes belonging to nine different groups are now known to govern such differences in human populations. The best known of them—the A, B, O group—has been investigated carefully in attempts to find the adaptive properties of differences. Although some strong correlations have been found between certain blood groups and the incidence of particular diseases, these are by no means universal. Most human population geneticists think that a large proportion of individual differences with respect to blood groups are neutral, at least under present conditions. There is reason to believe that the situation is materially the same in other species of mammals, and it may well be true of many of the recognized differences between the electrophoretic properties of enzymes as well.

Nevertheless, not all variation in the gene pool with respect to either visible characteristics or biochemical ones is neutral. Differences that make some people better athletes, artists, musicians, or mathematicians are determined at least to some degree by genes that affect size and strength, muscular coordination, keenness of sight, perception of sound, or those parts of the brain associated with manipulation of ideas and figures. Comparisons of identical (monozygotic) and nonidentical (dizygotic) twins, although they encourage skepticism about the genetic basis of many differences in intelligence and emotional traits, nevertheless force us to conclude that many other characteristics are largely controlled by genetic differences.

Several experiments have shown that some enzymatic differences within the same or related populations are probably adaptive. Key enzymes controlling cellular activity in some species of fishes and butterflies as well as in the cattail plant have been extracted and their kinetic properties examined. Populations living in colder climates have been found to possess enzymes that are most active at relatively low temperatures, whereas related populations belonging to the same species but living in warmer climates have enzymes that perform better at higher temperatures. On the other hand, many of the species tested in this way have not shown such clear-cut differences.

In a much larger number of examples, the distribution of enzyme variants is well correlated with the climatic distribution of the species, but the reasons for these correlations are not obvious. All such examples suffer from a difficulty associated with the way in which genes are packaged on chromosomes. Often, two "families" of genes (or alleles) that control very different functions are placed so close to each other on the same chromosome that two particular genes, one at each locus, are almost always passed to the same egg or sperm as a single unit during the process of meiosis. In genetic terms, they are said to be "closely linked." One of these genes may have a strong effect on adaptation and survival, and the other may be selectively neutral. Given this situation, the neutral gene will "hitchhike" on the chromosome in the same direction as its adaptive neighbor. Consequently, when a significantly adaptive pattern is recognized in the geographic distribution of a particular enzyme variant, one can never be sure without further information whether the gene that controls this variant is itself responsible for the pattern or whether it is riding on the coattails of an unknown adaptive gene.

Eventually, research that integrates the disciplines of genetics, physiology, biochemistry, and ecology may provide more evidence on this matter. Meanwhile, we need only remember that the amount of genetic variation stored in the gene pool is so great that even if only 10 percent of it is adaptive (or neutral), the possible number of adaptively (or nonadaptively) different genetic combinations is far greater than the number of individuals in the population. This generalization holds for both morphological and biochemical variation.

REASONS FOR THE EXISTENCE OF STORED GENETIC VARIATION

A reader who is aware of the tremendously long time during which evolution has taken place might wonder at this point whether most populations have maintained rich stores of genetic variability throughout these millions of years. Indirect evidence suggests that they have done so. Rich gene pools have been found in animals belonging to most of the major groups or phyla (they include species such as *Limulus*, the horseshoe crab, which has evolved little during the past 200 million years). The populations of some modern species contain little genetic variability, but this condition may well be recently derived, and such populations probably have a limited evolutionary future. Modern theory supports the notion that the more generalized species of the past could tolerate a greater amount of genetic variability than the more complex modern forms, and many evolutionists believe that mutation rates were far greater in early, primitive organisms than in existing species.

This reasoning supports the belief that large gene pools of variation have existed throughout the evolution of organisms. There never has been a time when many new mutations were needed to enable populations to respond by natural selection to new environmental challenges. Those populations that had the genetic reserves to meet a challenge evolved; the others became extinct. New mutations appeared and became established during the course of changing gene frequencies. Their principal role was to replenish the store of variation depleted by natural selection.

What factors have gone into the maintenance of variability in spite of natural selection? Part of the variation consists of genes that have deleterious effects if present in an individual in a double dose (homozygous condition), but do not affect normality if accompanied by a corresponding gene that codes for a normal function (heterozygous condition). To explain: Mendel's first experiment showed that, if a tall pea plant is crossed with a short one, the offspring of the first generation are just as tall as the taller parent; the gene for shortness has no apparent effect and is said to be recessive. In humans, many genes that in the homozygous condition are responsible for serious diseases are likewise recessive. If the individual has only one gene that codes for the deleterious effect, and if it is accompanied by a corresponding gene that codes for normal function, he or she is said to be a heterozygous carrier of the deleterious gene.

This condition is easily explained. Both the corresponding (or allelic) genes code for proteins with almost identical molecular structures. These genes often differ from each other by only a single nucleotide pair, and the proteins for which they code differ by a single amino acid unit out of the scores or hundreds of units present on the protein chain. The substitution of a single amino acid at a sensitive position on the chain can inactivate the protein. If a cell in a heterozygote contains one normal and one defective kind of protein, the normal molecules, by doubling their activity, can perform the required enzymatic cellular function well enough to produce a normal individual in spite of their defective counterparts.

Heterozygous carriers of defective genes play an important role in the genetic structure of populations because in them harmful recessive genes are protected from adverse natural selection. Such genes can remain in the population for many generations at a fairly high frequency. If the normal pattern of mating is outcrossing (mating between unrelated parents), a deleterious gene present in the heterozygous condition will sometimes be closely linked with an adaptive gene and so will be more or less permanently shielded from selection. These genes constitute much of the concealed genetic load that all outcrossing populations possess.

As a result of many experiments, geneticists have concluded that, among populations of the fly *Drosophila*, few if any individuals are free of chromosomes that produce more or less serious defects in the homozygous condition. Animal breeders have revealed their presence in cattle, chickens, and other kinds of domestic animals as well as in the laboratory rat by enforced mating between brothers and sisters over many generations. Indirect evidence indicates that all humans, like animal populations, carry a load of potentially harmful genes whose damaging effects may appear only after mating between closely related individuals such as brothers and sisters, parents and offspring, or, less commonly, first cousins. These harmful genes are of little evolutionary importance, but they are the price all organisms must pay because the effects of mutations are random with respect to viability. Far more important is the fact that deleterious, neutral, and beneficial mutations are not three separate, discontinuous groups. An entire spectrum exists, from genes that are always harmful in the homozygous condition, through those that are only slightly deleterious, to genes that, when present either doubly or singly, code for a normal function. Recently, population biologist T. Ohta has suggested that a large number of biochemically detectable genes, including many that had been believed to be "neutral," that is, equivalent in function to corresponding genes having normal effects, actually have effects that are adaptively inferior but are *almost* equivalent to their normal counterparts. If her hypothesis is correct, the superiority of one gene over another can be estimated only in a particular, constant environment. As soon as the environment changes, the relative adaptive values of many genes are certain to change also. For instance, genes of insects that code for resistance to insecticides will increase in frequency as long as the chemical is being applied, but their frequency is correspondingly reduced in the absence of the insecticide. Many of these genes are inferior in the absence of insecticides because they reduce the growth rate or fecundity of the insect. In the insecticide environment, they become superior because they permit the population to survive.

Another reason why different corresponding genes or alleles can be retained in the gene pool is that they sometimes interact with each other in such a way that the heterozygous condition is superior to the homozygous for either gene. An example in humans is the gene for sickling (or sickle-cell anemia) and its normal allele. Individuals who are entirely normal (homozygous for the normal allele) are superior to carriers of sickling except in regions where malaria is a common disease. In such an environment, the greater resistance of the carriers to malaria increases their adaptiveness in spite of their physical inferiority. In malaria-free regions, the gene for sickling, which is completely lethal when doubled and deleterious if present singly, quickly disappears from the population. In regions were malaria is common, it persists indefinitely because of the superior disease resistance of its heterozygous carriers.

In *Drosophila*, several examples are known of heterozygote superiority for genes that are not in any way associated with diseases or serious defects but that as homozygotes produce fewer offspring than do heterozygous combinations of different genes. In the majority of examples, several different genes are involved that occupy different positions on corresponding chromosomes. The superior flies are, in these examples, heterozygous not only for different individual genes but for different orders of the genes on chromosomes that permit clusters of genes to be inherited as if they were single Mendelian units. Indirect evidence indicates that similar conditions exist in many different species of animals and plants.

In many species, the most common basis for the richness of

gene pools may be the unevenness of most habitats. Even a cultivated field, which the farmer tries to make as regular as possible, has some relatively wet depressions and dry hillocks. Plant breeders have planted mixtures of closely related varieties of the same crop species year after year in the same field. A common result is that markedly inferior varieties quickly disappear from the mixture. Others that are less inferior dwindle in amount but never disappear completely. The explanation, which the evolutionist calls "frequency-dependent selection," is that the latter varieties are inferior to their most successful relatives when present in high frequencies but are favored by natural selection when present in low frequencies. In most instances, this is due to their adaptive superiority in certain uncommon microhabitats.

Stored genetic variation is also maintained within populations when the most adaptive condition with respect to one or more characteristics represents a balance between opposing selection pressures. This situation can best be understood if one realizes that *natural selection does not necessarily produce evolution*. Both experiments with artificial populations and observations in nature have shown that, if the population–environment interaction remains constant, evolution will not take place. Natural selection is still acting, and under some conditions selective pressures may be strong; but if those individuals that deviate the most from the mean or average of the population for some characteristic such as body size are eliminated, and regardless of whether they are larger or smaller, the average size of the population will remain the same.

A balanced condition of variation exists if, for a characteristic such as size that is controlled by many gene loci, an individual possesses approximately equal numbers of genes that code for increased size as compared to those coding for smaller size. In any individual, these may be distributed as opposite alleles at the same locus (heterozygous condition) or as two alleles for small size at one locus, balanced by two for large size at another locus (balanced polymorphism).

One can easily see how a changed environment—one that might favor a larger or a smaller individual—could sort out new combinations of genes from a population of individuals having such a balanced condition and thus enable the individual to respond to natural selection even though no new mutations had occurred. Several experiments in which artificial selection for quantitative characteristics has been applied to mice, *Drosophila*, corn, and other organisms have succeeded in the manner predicted by this model.

Finally, genes can be retained in populations because of their adaptive neutrality, that is, because they are neither favored nor rejected by natural selection. We will discuss this more fully as we go on.

SOME MISCONCEPTIONS ABOUT MUTATIONS

The opinion expressed throughout this book, that mutations are most important as a way of replenishing the gene pool as it becomes depleted by natural selection, is now supported by a wealth of experimental and observational evidence. Nevertheless, the old idea that mutations direct evolutionary change is still held by many biologists and is expressed most often in popular accounts of evolution. Other misconceptions about mutations are also widespread.

One misconception, which follows from the mistaken idea that mutations can establish the rate and direction of evolution, is that bursts of rapid evolutionary change are produced by increases in the rate of mutations. Some population geneticists have sought out "hot spots" on the earth's surface where radiation intensities are unusually high, hoping to find evidence of rapid evolutionary rates caused by the effect of radiation increasing the mutation rate. One particular spot for these investigations has been certain radioactive sands along the south coast of India. The results of these investigations have been negative; increased rates of evolution have not been observed. These results are to be expected on the basis of the interaction–selection hypothesis. Rapid evolution results from a strong challenge generated by a rapidly changing environment and the presence of organisms with gene pools

capable of meeting the challenge. Since the most rapid evolutionary changes documented by the fossil record are very slow according to genetic standards—that is, in terms of the numbers of generations involved—mutations occurring at the usual rates can keep
up with the depletion of the gene pool brought about by natural
selection.

Another misconception is that mutations require extra energy
for their production. According to current molecular theory, no
more energy is needed for spontaneous mutations than for the
normal replication of a DNA molecule. They are mistakes in copying (Figure 3-1) analogous to the mistake that results when one
strikes the wrong key on a typewriter—an act that takes no more
extra energy than striking the right one. The notion that energy
is needed to produce mutations probably comes from a misinterpretation of the classic experiments of geneticist H. J. Muller, who
showed that high doses of X rays greatly increase mutation rates.
When these experiments first became known, the way in which
genes replicate was not understood. Some geneticists suggested
that the effect of the radiation was to break up the gene structure
so that it could be reorganized into a new pattern. Once geneticists
realized that genes are transmitted by a copying process, like the
printing of a newspaper, this idea was seen to be invalid. Much
later, genes were discovered that greatly increase the spontaneous
mutation rate in bacteria growing under normal conditions, without using extra energy. Geneticist E. C. Cox has provided much

Figure 3-1.

Diagram showing how mistakes in copying or replication of DNA can bring
about substitution of one nucleotide base pair for another. Due to a "wobble
effect," in which an atom of hydrogen is for a brief instant displaced from its
usual position, a molecule of adenine will very rarely mate with a molecule of
cytosine (C) instead of its usual mate, thymine. This mismatch introduces
cytosine into a position on the DNA molecule where thymine would be
expected. At the next replication, the out-of-place cytosine joins with its
normal mate, guanine, thus substituting a guanine–cytosine nucleotide pair for
an adenine–thymine pair that existed in the parent molecule. (After F. Stahl,
from G. Ledyard Stebbins, *Processes of Organic Evolution,* © 1966, pp. 28–29.
Reprinted by permission of Prentice-Hall, Inc., Englewood Cliffs, N.J.)

(a) Changes in States of Bases

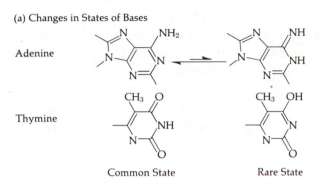

Adenine

Thymine

Common State   Rare State

(b) How Changes in State Produce Mismating of Bases

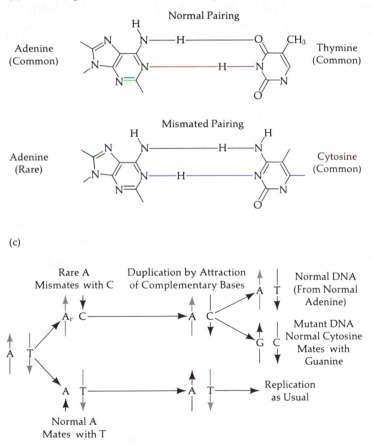

evidence that these "mutator" genes are associated with the enzyme system that regulates the replication of DNA and in fact may owe their power to defects in this system. Normal replication includes a certain amount of "proofreading" that eliminates deleterious mutations. In primitive cells that existed before modern replication and "proofreading" systems had become perfected, mutations may well have been thousands of times more frequent than in modern organisms. In fact, present evidence suggests strongly that mutations in today's organisms are a small residue of a much greater genetic instability that existed in the earliest forms of life. As the amount of order and degree of integration of living bodies has increased over long periods of evolutionary time, natural selection has favored genetic systems that have increasingly lower mutation rates.

GENE POOLS, ORDER, AND ENTROPY

The fact that orderliness of living systems has increased during evolutionary time has caused some critics of evolutionary theory to declare that evolution cannot be based entirely on natural causes. They maintain that progressive increase in order can be accomplished only with the aid of a supernatural guiding force. Their argument is based on a well-known generalization or law of physics—the second law of thermodynamics. They interpret this law to mean that under natural conditions the entropy of a system always increases. Since *entropy* is the antithesis of order, increase in entropy means decrease in orderliness.

This interpretation is false and is based on a misinterpretation of the second law. For anyone not familiar with physical chemistry, including most biologists, the concept of entropy is difficult to understand and interpret. In fact, it has often been misinterpreted, particularly by some who fear that the earth's resources are being consumed more rapidly than they are being replaced and that the point of no return is near.

The second law of thermodynamics and the concept of entropy can be applied only to isolated systems that form a "universe" within themselves and do not interact with other systems. If such isolated systems are sufficiently large and complex, different parts of them may interact with each other. Some of these parts may receive energy from other parts and, by diverting some of the energy received, may increase temporarily the amount of order present in that interacting part or individual.

The solar system is, to a large degree, an isolated system, but the earth definitely is not. Still less isolated are the species of organisms that cover the earth. Green plants can exist only because they receive solar energy and convert a small part of this energy into complex form. Animals depend for food on the organized structures of plants or other animals that they eat. The greater complexity that some evolutionary lines have evolved consists merely of the diversion of a somewhat larger proportion of the energy obtained into temporary structures—their bodies—that are more complex in form and activity. Eventually, all these complex bodies increase their entropy or state of disorder to a point where they die and decay.

INTERACTIONS BETWEEN CHANCE AND ADAPTIVE DIVERSIFICATION

Both mutations and gene shuffling, or recombination, generate variation that is random with respect to adaptation. Nevertheless, most differences between populations are either obviously adaptive or can be associated indirectly with adaptation by appropriate observation or experiment. This paradox is more apparent than real. We are dealing with two different levels in the hierarchy of variation—the level of variation *within* populations is different from that of differences *between* populations.

There is a precedent in physics for randomness at one level and direction at another: When the air about us is still, the molecules of its component gases are constantly moving in random pathways, but we are unaware of this movement. Wind, which is generated by an inequality of barometric pressure in the environment, produces directed movement that we can detect but does not alter the random molecular movements that are continually taking place. Similarly, random changes in the genetic composition of individuals are taking place in all populations, whether they are evolving or not. If these changes produce individuals that differ widely in some characteristics from the mean or average of the population, such individuals tend to produce fewer offspring than their more normal relatives. The frequencies of their genes are reduced in favor of genes that produce more normal individuals. This is the process of *normalizing natural selection* (Figure 3-2). It tends to keep populations genetically constant within the limits of chance variation, as long as their environment is not changing. Changes of the environment generate environmental challenges, to which populations may respond by directed changes in gene frequencies that constitute evolution; these responses, however, do not suppress the random fluctuations that are continually occurring. *Direction is superimposed on randomness.*

Figure 3-2.

Effects on the amount and kind of variation in gene pools of populations brought about by genetic segregation, mutations having small or moderately large effects, and natural selection. Each small circle represents an individual, and the line attached to it represents genetic deviation from the position of its parent, which can be brought about by gene recombination, mutation, or both factors acting together. Arrows represent selection pressures that restrict the scope of variation found in the population. (a) The situation in populations that are in harmonious balance with their environment, in which normalizing selection prevails. Depending on fluctuations between more severe and more tolerant environmental conditions, selection can either reduce or increase the amount of variation present in the gene pool. Under these conditions of balance, increased competition reduces variation but does not change its mode, and evolutionary change is minimal or absent. (b) The situation in populations that are being subjected to different kinds of environmental challenges, brought about by changing population–environment interactions. In this diagram, two successive generations are shown, the first represented by a solid line and the second by a broken line. Strong selective pressure in one direction and weak or no pressure in the opposite direction eliminate genotypes that survived in the previous generation (*circles*) while permitting survival and reproduction of individuals (*triangles*) that in the previous environment were inadaptive. Large arrows beside the figures indicate directions of divergent evolutionary change.

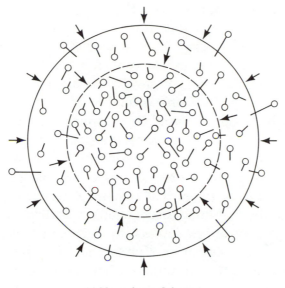

(a) Normalizing Selection

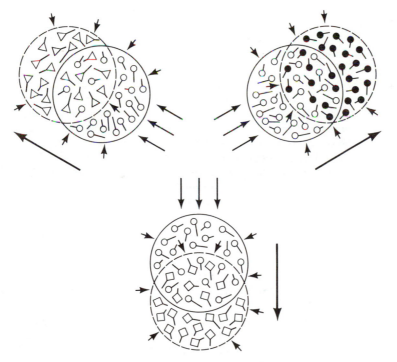

(b) Directional and Divergent Selection

A different situation appears when populations are reduced to a very small size, such as fifty individuals or fewer. This can happen in two ways. An environment that is occupied by a widespread species may change in a direction that is generally unfavorable, so that the species can survive in only one or a few corners of its previous range. If, for instance, the climate is becoming drier, forests may be largely replaced by savannas or by other kinds of open country. Forest-loving species, unless they carry in their gene pools combinations of genes permitting their adaptation to savanna conditions, will be reduced to a few small colonies in the restricted patches of forest that remain. Alternatively, a species containing individuals that can occasionally be dispersed over long distances may send out explorers into regions that are far from the original habitat but similar enough so that the explorer can colonize the new area, thus founding a new population. This is the way in which oceanic islands like the Galápagos and Hawaiian Islands were colonized by plants and animals. Moreover, patterns

Figure 3-3.

Diagram showing how the founder principle, based on changes that accompany the founding of new populations by one or a few colonizers, can fix neutral or nonadaptive differences in association with adaptive traits. Each pair of letters represents an individual. A_1 through A_5 represent characteristics that adapt the organism to successive habitats 1 through 5. B, C, D, E, F, G, and H represent different neutral characters that are associated with the adaptive ones. (a) The original population sends out colonizers to similar habitats, each of which happens to contain a different sample of neutral characters from that present in the parental population. (b) As the populations established by the founders increase in size, they acquire different adaptations to a changing environment. The neutral characters, however, perpetuate themselves without change, and so become widespread in the new, larger populations. (c) Further environmental changes establish wider adaptive differences between the populations that are descended from the founders, while neutral mutations alter the store of nonadaptive variability. Meanwhile, the changed environment has caused the original population to become much smaller, and finally extinct. In this way, through a succession of stages, a single population having only one adaptive mode can give rise to two or more populations that differ from each other with respect to both adaptive and adaptively neutral traits.

of geographical distribution of related species suggest that inter-
continental migrations of this nature have occasionally occurred.

The results of such colonizations were first recognized by Sewall
Wright, and they were applied to actual situations by Ernst Mayr.
They are highly complex (Figure 3-3). First, the small size of the
population enforces inbreeding between related individuals and
so exposes in homozygous condition recessive genes that in the
heterozygous condition were shielded by their corresponding
dominants and had no effect on the population. Most of these
recessive genes are deleterious and will retard or even prevent the
success of the colonizers or the survivors of change. Nevertheless,
given the long span of evolutionary time and the possibilities for

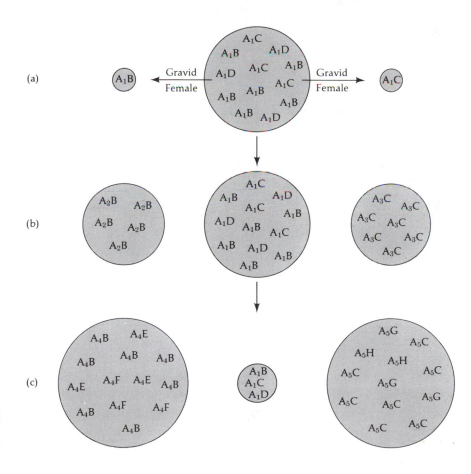

repeated colonization, some of the small populations may become homozygous for genes that were slightly deleterious in the old environment but that equip their bearers to become adapted more quickly to new and different environments. When this happens, the reviving population may evolve a new adaptive strategy.

Here chance enters the picture, because the most successful and fertile founders of the new populations may possess, in addition to the adaptive complex, distinctive genes that are neutral in effect. Such genes gain a free ride to widespread success because they happen to be present in well-adapted, fecund individuals. An illuminating analogy is the frequency of certain surnames in large human communities as compared to small ones. A few common surnames appear frequently in the telephone directories of large cities, whereas relatively unusual surnames appear frequently in directories of small towns. The frequency of surnames in the telephone directories of large cities reflects chiefly the history of migrations. For example, the five most frequent surnames in Kansas City are, in order of frequency, Smith, Johnson, Jones, Brown, and Williams. In Manhattan, the five most frequent names are Smith, Brown, Williams, Johnson, and Cohen. The last of these surnames reflects the influx of European immigrants during the past century. In San Francisco, the five most frequent surnames are Lee, Wong, Smith, Johnson, and Williams, reflecting a history of immigration from Asia. The small communities of New England, which were founded by a few families and into which there has been little immigration, feature relatively unusual surnames. In the village in Maine where I spent much of my childhood, the surnames Clement and Jordan were by far the commonest among the permanent residents. In another community a few miles away, Bracy and Hancock were the most common surnames, and in still another, the surname Fernald predominated.

The evolutionary significance of the surname analogy is simply that surnames by themselves have no adaptive value, and differences between surnames that belong to the same language family do not reflect any general historical trends. If unusual surnames are commonest in small, isolated communities, this reflects accidents or chance events in their past history. Nevertheless, the differences between the surnames that dominate individual small villages are not based entirely on chance; they reflect the material success and procreative activity of certain males among the original founders of a community. These qualities, in turn, reflect genetic and cultural advantages that the founders possessed. The chief element of chance is the *association* between these advantages and a particular surname. Similarly, when populations of animals and plants pass through fluctuations often called *bottlenecks*, which consist of reduction followed by a later increase in numbers, chance associations are established between adaptive and neutral characteristics. If the latter are particularly conspicuous, they *appear* to form the basis for the differentiation of populations.

Examples of this kind abound in populations of animals and plants. Consider, for instance, the color patterns in the plumage of male birds. The possession of such patterns is highly adaptive. It serves to advertise the presence of the male, to attract females, and particularly to establish territory and to ward off potential intruders. The existence of different patterns in separate but related species is also valuable. Females can distinguish males that belong to their own species from foreign males. This helps them to avoid mismatings that often give rise to sterile offspring. Nevertheless, the evolution of the particular color pattern possessed by males of a given species is due as much to chance as to natural selection.

Chance and adaptiveness are blended to a remarkable degree in the "flags" that most species of ducks carry on their wings (Figure 3-4). Most ducks fly in flocks that consist entirely of birds belonging to the same species. A young bird must learn quickly to recognize members of his or her own species and to join them. As an aid to this recognition, each species possesses a distinctive, brightly marked patch on the base of the wing. If this patch is obliterated or altered, the ability of birds to find and join the right flock is greatly hindered. The particular design of the flag corresponds to the surname in humans. Its only prerequisite is that all members of a flock carry the same design. If mallards had evolved a flag like those now found in the black ducks while black ducks had evolved a different pattern, both species would be as well

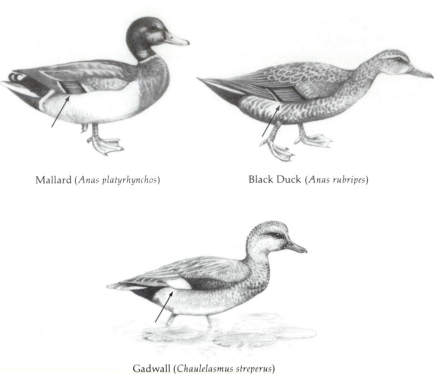

Mallard (*Anas platyrhynchos*)   Black Duck (*Anas rubripes*)

Gadwall (*Chaulelasmus streperus*)

Figure 3-4.

Males of three species of ducks, showing the distinctive "flags" on their wings that help to keep the flocks together. (From E. H. Forbush, *A History of the Game Birds, Wild Fowl and Shore Birds of Massachusetts and Adjacent States.* Massachusetts State Board of Agriculture, 1912.)

adapted as they are now. The only adaptive feature is the *difference* between the species.

HOW NEW EVOLUTIONARY LINES ORIGINATE

Many new evolutionary lines originate from more generalized ancestors and then evolve in particular new directions for varying lengths of time, usually millions or tens of millions of years. Their development and maturation involves increasing specialization. The end of the line is often reached when adaptive specialization for a particular habitat or way of life has become so great that the population cannot respond to new environmental challenges and so becomes extinct. This kind of evolutionary sequence is usually followed by several simultaneously radiating and diverging lines, most of which become extinct after undergoing a small amount of change. The apparent "straight-line evolution" (orthogenesis) that was described in many textbooks written forty to a hundred years ago is the result of the unusual success of a few adaptive gene combinations in successive geological epochs. As is shown in Figure 1-3 (and is supported by additional data in the next chapter), the "straight lines" that these textbooks present are produced by selecting a very few out of many different kinds of organisms that have evolved and then drawing arbitrary lines between them.

Populations of animals evolve in new directions through distinctive adaptive strategies. They can do this only if they are well isolated from other populations that are adopting different strategies and thus are protected from contamination by hybridization, which introduces alien genes and tends to break down the adaptive system. Related populations that live in the same region yet remain distinct through lack of hybridization are usually recognized as distinct species. Familiar examples are wolves and coyotes, black bears and grizzly bears, horses and donkeys, and song sparrows and Lincoln sparrows.

The first step in the origin of a new evolutionary line is the origin of a distinct species. This can happen in many ways. The following examples will be instructive.

The hybrid between horse and donkey is the mule, produced by inseminating a mare with sperm from a male donkey. The hybrid offspring is healthy, strong, and useful, but it is almost completely incapable of producing offspring, so that exchange of genes between populations of horses and donkeys is virtually impossible. Although horses and donkeys possess very similar genes, which can work together in harmony to produce vigorous progeny, the pattern of nucleotide sequences, particularly with reference to spacer DNA between genes, is quite different in each kind of animal. When the critical stage of meiosis is reached in the testes or ovaries of a mule, the chromosomes that came from the horse parent find no match among those derived from the donkey. Chromosomes derived from opposite parents can pair only if the nucleotide sequences of their DNAs are essentially similar in overall pattern, and precise pairing is a necessary prelude to the delivery of a complete, viable set of genes to eggs or sperm. Its absence in the mule causes the reproductive cells to have highly abnormal gene combinations and so to be nonfunctional.

A different basis of genetic isolation exists between two species of frogs that were once thought to belong to the same species and thus share the common name of leopard frog. If a leopard frog native to Vermont is crossed with one native to Florida, tadpoles are produced, but these soon die before maturing into adult frogs. The reason for their death is suggested by the different conditions under which the two species normally reproduce as well as by the abnormalities of the hybrid tadpoles, which are different depending on whether the female parent comes from Vermont or from Florida. When the two species are raised in controlled environments, the tadpoles derived from Vermont eggs fertilized by Vermont sperm are found to grow best at cool temperatures and to develop rapidly, because they are adjusted to the cool spring climate and rapid progression from spring to summer that takes place in their native home. Tadpoles of Florida origin on both sides develop best at warmer temperatures and mature more slowly. Hybrids obtained by fertilizing Vermont eggs with Florida sperm develop rapidly but have greatly enlarged heads and very small tails. If Florida eggs are fertilized by Vermont sperm, the tadpoles develop more slowly and have very small heads combined with large tails. Apparently, adaptations of each species to the temperature regime and seasonal cycle that is characteristic of its native home has sorted out gene combinations that differ so strongly as to produce disharmonious development when combined into a hybrid. The reciprocal differences probably result because the earliest stages of development in a frog's egg are controlled almost entirely by proteins that are coded by the mother's genes and are present in the egg before fertilization. The new proteins, if they are coded by genes belonging to a differently adapted male, clash with those already present in the developing egg. The conflict between these disparate elements is lethal.

As a third example, the mallard duck (*Anas platyrhynchos*) and the pintail (*A. acuta*) are common in most parts of our country, particularly in the central states. The two species are easily distinguished by a number of distinctive characteristics. Hybrids between them are occasionally found in nature but are rare. Nevertheless, tame mallard drakes, when kept in an enclosure with pintail ducks, will mate and produce vigorous, fertile offspring. They are apparently kept apart under natural conditions by their different behavior patterns. Both fly in flocks consisting entirely of drakes and ducks belonging to the same species, so that the chance of a lone mallard drake finding an equally lonely pintail duck are small. Moreover, mallards habitually search for their mates and pair with them during the autumn months; by mid December, most pairs have established bonds of affection that keep them from looking elsewhere. Pintails, on the other hand, do not seek mates before the middle or latter part of December; thus, even if an individual should become separated from a flock of his or her own species, the chance of finding an unmated mallard rather than another pintail would be very slight. In their usual habitats, these ducks are prevented from exchange of genes between species as effectively as are horses and donkeys, although the incompatibility barrier between them is behavioral rather than genetic.

A final example involves flies and gnats of the genus *Drosophila*, which contains more than a thousand different species. They have been intensively studied by many geneticists, so that more is known about how species of this genus originate and are kept apart than about any others. Dobzhansky devoted particular attention to a pair of species that inhabit forests of the western United States, *Drosophila pseudoobscura* and *D. persimilis*. Adult male flies of these two species look so much alike that they cannot be identified without careful dissection of their genitalia, while adult females cannot be told apart at all, unless one studies the details of their chromosome structure and the chemical properties of certain enzyme systems. Both of these characteristics provide infallible criteria for separating the two species, regardless of sex.

Drosophila pseudoobscura and *D. persimilis* are separated from each other by a combination of barriers to gene exchange. Like the mallard and pintail ducks, males and females belonging to different species of *Drosophila* will mate under artificial conditions, though they are more likely to mate with members of their own species. Natural hybrids are virtually unknown. Hybrids from artificially induced interspecific matings become vigorous adults. Males are completely sterile, but their hybrid sisters will produce a few offspring when mated to fertile males belonging to either one of the parental species. These backcross offspring are, for the most part, weak, sterile, or both. Even in the laboratory, exchange of genes between the two species is difficult or impossible.

The weakness and sterility of the *Drosophila* hybrids have a complex basis. Sterile hybrid males have very small testes; apparently, the development of their reproductive organs suffers from disharmony between programs of development that are coded by different gene complexes, just as does the development of the tadpoles derived from hybrid frogs' eggs. In addition, comparisons between the detailed structures of giant salivary chromosomes shows that *D. pseudoobscura* and *D. persimilis* differ greatly from each other with respect to the order of genes along certain regions of their chromosomes. These differences are comparable to the differences between chromosomes in horses and donkeys.

The behavioral differences between the two species, which reduce the frequency of hybrids in the laboratory and virtually prevent their occurrence in nature, are as effective as those that separate mallard ducks from pintails, and they are more complex. First, *Drosophila persimilis* lives mainly in cooler climates than *D. pseudoobscura*. The two species occur in the same habitat in only a few localities, such as pine forests on the western slope of California's Sierra Nevada Mountains. There they are partly separated by the adaptation to different temperatures. Flies belonging to the species *D. persimilis* feed and copulate chiefly at dawn, when the air is cool. Both species hide under bark and in other crannies during the warm, bright daylight hours. In the twilight, when temperatures are still fairly warm, one can catch single flies and mating pairs belonging to the species *D. pseudoobscura*.

For this species pair and other pairs of related *Drosophila* species, elaborate courtship patterns are a barrier to mating with the wrong species. Herman Spieth put female *Drosophila pseudoobscura* flies into a bottle with two kinds of males—some of the same species and some belonging to *D. persimilis*. He found that all the males eagerly sought after all the females, regardless of species. The females, however, were more discriminating and responded to males of their own species by performing with them a complex but stereotyped ritual of touching antennae, flicking wings, dancing about, and finally copulating, whereas they rebuffed the advances of males from the other species. During the past thirty years, these and similar observations have been verified many times for all species of *Drosophila* that have been investigated.

Some species of *Drosophila*, particularly those that live in the Hawaiian Islands, have even more elaborate patterns of courtship behavior. The so-called picture-wing flies, about the size of a house fly, have wings mottled with black in a definite pattern, which is distinctive for each of the scores of species. The male displays his finery before his prospective mate in a manner similar to the feathered display of a male peacock or bird of paradise. Moreover, the males of some species lay claim to a particular part of a tree branch for courtship, driving competing males away from this territory and inviting females to enter their domain by their display activity.

Zoologist Robert Colwell has shown recently that even tinier creatures display courtship behavior. Mites, diminutive relatives of ticks, are so small that they are barely visible to the naked eye. Several of them together can cling to a single leg of a *Drosophila* fly. In studying mites that are carried from one flower to another on the beaks of hummingbirds, Colwell has found that male mites display "agonistic behavior"—a term that refers to sparring or fighting between males for the possession of females. They also caress the bodies of the females before mating with them. Apparently, specific patterns of competition between males and courtship of females has such a high adaptive value that it has arisen independently hundreds of times, in animals as different from each other as mites, flies, birds, elephants, whales, and humans.

The origin of species appears to depend on a complex and diverse series of processes. It can never be accomplished in a single event by one all-encompassing mutation, as some geneticists formerly thought possible. The logical question, then, is: How can population–environment interactions, acting through the medium of natural selection, build up effective barriers between species? A second question, equally relevant, is: Can such barriers be built up by chance establishment of genetic or behavioral differences? Like so many other questions in biology, neither of these questions has an easy answer, but review of the examples just described may provide helpful clues.

Why are the chromosomal differences between horse and donkey responsible for the sterility of the mule? It is hard to see how the order of genes on chromosomes can affect the differential adaptation of these two species. In spite of the mixed order sequence that is present, the genes of horse and donkey can collaborate to direct the entire sequence of developmental processes needed to produce a vigorous adult mule. But trouble comes at meiosis, when chromosomes from different parents would normally pair with each other and deliver a viable packet of genes to eggs and sperm cells. The formation of a new arrangement of genes and spacer DNA, through breakage of chromosomes and reunion of broken parts in a new order, affects the development of an individual little, if at all, but can reduce its fertility. Nevertheless, geneticists learned through many experiments in which orders of genes were changed through breakage and reunion of chromosomes that heterozygosity for a single structural change may have such a slight effect on fertility that it would not interfere with the evolutionary success of a vigorous, well-adapted individual.

The question then becomes: How could such chromosomal changes increase in frequency until they become predominant in a population? Two answers are possible. First, genes placed near each other on a chromosome are much more likely to be inherited as single Mendelian units than genes that reside on completely different (nonhomologous) chromosomes. Several examples are known of gene clusters that direct the development of different parts of an adaptive complex and are inherited as a single unit because they are close together on the same chromosome. If the challenge of a new environment renders adaptive a new combination of genes, rearrangements of chromosomal parts that put these newly adaptive genes close to each other may possess a high adaptive value and so become fixed in the population. Alternatively, if chromosomal changes accompanied the reduction of a population to small size, new gene order might become established entirely on the basis of chance or by chance association with adaptive traits. In addition, randomly produced changes in spacer DNA, particularly addition or removal of large segments, could greatly reduce the fertility of interspecific hybrids.

There is no way to know which possibility best explains the horse and donkey example. An interesting and perhaps relevant fact is that the wild donkeys from which the domestic animal was derived once inhabited mountain ranges that rise from the deserts of the Middle East, Arabia, and southern Iran. Migrations across the deserts that separate the mountains would have been rare, and most donkey populations would have been isolated from each other and self-contained. However, after a migration, a new population could become established on a previously uninhabited mountain. During such establishment, the *founder principle* would operate. As defined by zoologist Ernst Mayr, this principle states that neutral characteristics, such as new chromosomal arrangements, can become established in combination with characteristics adaptive to the new habitat. In this way, divergent order sequences of gene arrangement could have arisen, leading to the reproductive isolation between modern horses and donkeys.

In our example of leopard frogs, the species must have separated from each other as a result of adaptation to different climates, accompanied by drastic modifications of their development patterns. Their present reproductive isolation is a by-product of divergent natural selection in the past.

The related species of *Drosophila*—*D. pseudoobscura* and *D. per-similis*—probably have existed for millions of years. The isolating barriers that now separate them have probably been acquired during a relatively long period of separation. More recent studies of closely related *Drosophila* species found in our southwestern deserts and especially in Hawaii suggest that the first kind of barrier to arise may be a repatterning of the chromosomes, as in the horse–donkey example, but more often is a divergence in courtship patterns. The initial separation began with a combination of differential adaptation and chance events, or in some instances was entirely the result of chance.

Once species of animals have become isolated from each other so that their hybrids are partly or wholly sterile, natural selection will reinforce the reproductive isolation by strengthening the original barriers and adding new ones. In several groups of animals, such as tree frogs and crickets, males attract females by a call that is distinctive in pitch and quality of tone for each species. If two species have overlapping ranges, races of the species that live side by side in the region of overlap usually have more distinctive calls than do races of the same species that live elsewhere.

Once a species has begun to evolve in a new direction and its independence from other species is assured by reproductive isolation, it may give rise to descendant species of an entirely different kind. By far the great majority of newly evolved species, however, become extinct without giving rise to any descendant species. What conditions or processes are needed to bring about successions of progressively more complex species, as in the sequence jawless fish → jawed fish → amphibian → reptile → mammal → human? The answer to this question is the subject of the next two chapters.

Rattlesnakes and their prey, ground squirrels, are an excellent example of coevolution in geologically recent (Miocene, Pliocene) epochs. Rattlesnakes prey almost entirely upon small mammals, whose presence they detect by special heat receptors below their eyes. Squirrels have evolved speed and agility, which enable them to be almost caught by the striking snake and still survive, even when lightly hit by the snake's fangs. (Drawing by Darwen Hennings after a painting by Richard G. Coss.)

Chapter Four

SMALL-SCALE AND LARGE-SCALE EVOLUTION

The processes discussed in the last two chapters can take place rapidly enough so that evolutionists can document them by historical records. Moreover, some of them can be greatly speeded up under controlled experimental conditions. The selection process can be greatly accelerated by substituting artificial selection for natural selection. For instance, several experiments with species of *Drosophila* have shown that selection of the most discriminating females can greatly increase the amount of behavioral reproductive isolation that exists between two closely related species. The hypothesis that passage through bottlenecks of greatly reduced population size favors the origin of species can be tested by setting up controlled population cages in which the number of individuals in a cage is greatly reduced at regular intervals, and the few survivors are then allowed to build up a new large population. The hypothesis that changes in the structure and number of chromosomes play an important role in the origin of new species can be tested by subjecting cells to radiation or chemical reagents that are known to produce chromosomal changes, and then raising populations from the altered individuals. By using such experimental techniques, investigators can cause artificial populations to acquire altered gene pools hundreds or thousands of times more rapidly than their natural counterparts can.

Consequently, evolution at the level of populations and species can be analyzed by experimental and quantitative methods that are quite comparable to those used by research workers in other fields of biology for analyzing contemporary organisms. Even the most confirmed fundamentalists believe in evolutionary change at this level.

When we come to the larger aspects of evolution, such as the origin of life, of many-celled animals, of mammals, or of humans, the picture becomes quite different. Changes at this level have required millions of years, and each individual change has occurred only once. Hence the chance of repeating the sequence under controlled conditions is almost nil. Evolutionists working at this level have tasks similar to those of scholars in the discipline of ancient history. By accumulating circumstantial evidence of various kinds, followed by extrapolation and deduction, they must piece together the best possible theory about the sequence of events that actually happened. This theory must then be tested by making predictions based on it and then looking for additional evidence that will either support the theory or cause it to be revised. Sometimes drastic revisions become necessary, as happened when geophysicists and geologists had gathered evidence to show conclusively that continents have not always occupied their present positions but have slowly moved over the earth's surface following the principle of plate tectonics.

THE HIERARCHY OF BIOLOGICAL CLASSIFICATION

To understand large-scale evolution, the reader must become acquainted with the hierarchy of classification that has been constructed to show relationships between major groups of organisms. The system was proposed in its modern form by the great classifier of the eighteenth century, the Swedish naturalist Carl von Linné, whose name is often written in its Latin form, Carolus Linnaeus. The hierarchy resembles the pyramid of authority in an army, a formal church, or a bureaucracy. The basic or lowest unit is the *species*. Individuals belonging to the same species have similar though by no means identical genes. Those belonging to different species differ from each other with respect to a large number of genes; the exact number varies according to the closeness of relationship between the species concerned. Moreover, species of animals—and to a lesser extent those of plants and microorganisms—are separated from each other by barriers to gene exchange that either prevent hybridization in nature or render the products of such hybridization inviable or sterile.

Related species are grouped together into a *genus* (plural, *genera*), genera are grouped into *families*, families into *orders*, orders into *classes*, classes into *phyla* (singular, *phylum*), and phyla into *kingdoms*. Ideally, each of these categories should contain only organisms that have been derived from a single ancestral species that was different from the ancestor of other categories having the same rank in the hierarchy. Actually, biologists have great difficulty in determining whether or not this is so for any particular genus, family, or other category, particularly when the fossil record is imperfect or lacking. In many groups, therefore, the classification that is adopted and used is a compromise between biological knowledge, speculation, convenience, and tradition. Sometimes, classifiers find it convenient to add even more ranks to the hierarchy, and they speak of subgenera or tribes as intermediate between species and genera, of subfamilies or superfamilies, and so on.

As an example of the way the system works, we give the classification of the chimpanzee (*Pan troglodytes*), our nearest relative, according to the traditional system.

Pan troglodytes

Category	Chimpanzee	Other categories of same rank
Species	troglodytes	*Pan paniscus*
Genus	Pan	Gorilla
Family	Pongidae	Hominidae (humans)
Order	Primates	Rodentia, Carnivora
Class	Mammalia	Reptilia, Aves (birds)
Phylum	Chordata	Mollusca (shellfish) Arthropoda (joint-legs)
Kingdom	Animalia	Plantae

THE FOSSIL RECORD

The course of evolution at higher levels, which gave rise to genera, families, and higher categories of the hierarchy of classification, is interpreted on the basis of the fossil record. The age of fossils is determined partly by their position in strata that have been laid down one on top of the other so that fossils found in lower layers are older than those found in upper layers of a series. Originally, the length of time during which a particular stratum was being laid down was estimated by considering its thickness and composition. Coarse strata of sandstone, which are usually formed by overflow of rapidly running rivers, are laid down hundreds of times more rapidly than limestone or chalk, which are formed from extremely slow disintegration and accumulation of the remains of organisms that have limey shells or other hard parts. Recently these estimates have been revised by use of radiometric dating. This technique verified the succession of strata as previously deduced but showed that estimates of time lapse were much too short. During the past 100 years, these estimates have been codified into a sequence of named time intervals that form another hierarchy. The shortest intervals generally recognized are called *epochs*; the intermediate ones, *periods*; and the longest, *eras*. The geological heirarchy of strata, now almost universally accepted, is shown in Figure 4-1.

Critics of evolutionary theory have often objected to this interpretation of the fossil record, saying that it involves a circular argument. The epochs are determined, they say, on the basis of the fossils found in them, while the age of a fossil is estimated by identifying the epoch to which the stratum containing it belongs. This objection is invalid, because the relative age of the great majority of fossils is determined by their position in a long

Figure 4-1.

Chart showing the names and approximate duration of the geological epochs, periods, and eras. (From Don L. Eicher, *Geologic Time*, 2nd ed., © 1976, p. 152. Reproduced by permission of Prentice-Hall, Inc., Englewood Cliffs, New Jersey.)

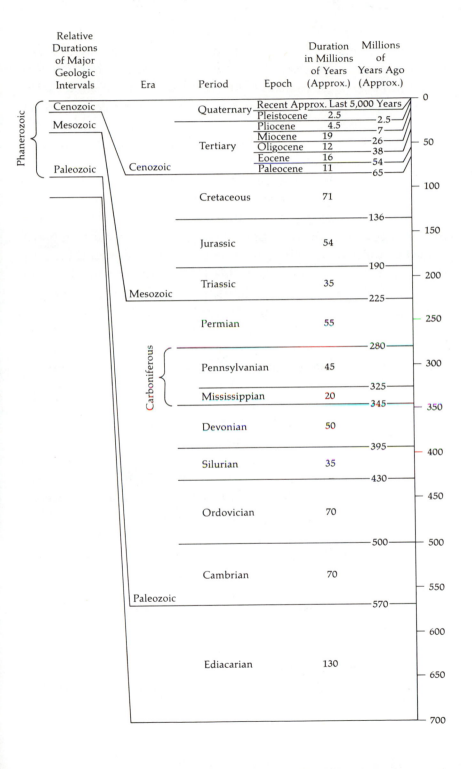

Relative Durations of Major Geologic Intervals	Era	Period	Epoch	Duration in Millions of Years (Approx.)	Millions of Years Ago (Approx.)
					0
Cenozoic		Quaternary	Recent Approx. Last 5,000 Years		
			Pleistocene	2.5	2.5
Mesozoic			Pliocene	4.5	7
			Miocene	19	26
		Tertiary	Oligocene	12	38 — 50
			Eocene	16	54
Paleozoic	Cenozoic		Paleocene	11	65
					100
		Cretaceous		71	
					136 — 150
		Jurassic		54	
					190 — 200
		Triassic		35	
	Mesozoic				225
					250
		Permian		55	
					280 — 300
	Carboniferous	Pennsylvanian		45	
					325
		Mississippian		20	345 — 350
		Devonian		50	
					395 — 400
		Silurian		35	
					430 — 450
		Ordovician		70	
					500 — 500
		Cambrian		70	550
	Paleozoic				570
					600
		Ediacarian		130	650
					700

Phanerozoic

sequence of strata that lie conformably one on top of the other. The few exceptions to this rule are found in mountainous regions, where the strata are visibly folded back upon each other or where mountain building by earthquake movements has pushed part of a lower layer into a position above one that formerly was on top of it.

Further dating of strata on the basis of fossils contained in them is necessary when comparing sequences found in widely separated regions of the earth. Nowhere is the entire sequence, from Cambrian to recent, displayed conformably in a single locality. Nevertheless, many regions contain conformable sequences that include up to one-fourth or one-third of the total record. In North America, for instance, the entire sequence can be pieced together from three large chunks, the oldest being in the northeastern states and the eastern Great Plains, the middle one in the Rocky Mountains and adjacent regions, and the youngest on the coastal plain, from Texas to Florida and northward to New Jersey. The upper strata of the northeastern segment contain fossils that are comparable to those in the lowest strata of the Rocky Mountain segment, the upper strata of which contain fossils comparable to those found in the lowest strata of the coastal plain segment. When sequences of strata found throughout the world are compared with each other, a similar consistency is found.

Unfortunately, the fossil record is woefully incomplete. After considering all known facts, paleontologists David Raup and Steven Stanley concluded: "Regardless of which estimates we accept, we must agree that the fossil record yields but a tiny fraction of the species that have lived, and that even within phyla whose members are relatively easy to preserve, fossilization and ultimate discovery by paleontologists are rare events."

Even more important, the record is strongly biased, because some species have a much greater chance of being preserved than others. To be preserved, an animal needs to possess some kind of hard parts, such as a shell or skeleton; it must be living at or near a place where it will be protected from decay; and because of the low probability of preservation it must be a relatively common, widespread species. Animals having large, hard skeletons, such as dinosaurs and the larger mammals, are much more likely to be preserved than small, fragile ones, such as insects. Small mammals, such as the shrews that were near the ancestral line of modern mammals, as well as most rodents, are represented in the fossil record almost entirely by teeth and other fragments of easily preserved skulls. Plants are never preserved in their entirety, since they break up as soon as they die. The fossil record of land plants must be pieced together by comparing leaves, pollen grains, fragments of wood, fruits, seeds, and flowers or flower clusters.

Some of the facts presented above can be verified by a perceptive reader on the basis of personal observations. Wandering through a forest or over a grassy plain, one occasionally runs across the remains of a cow or deer that has died and is decaying. Nevertheless, nearly all these remains are only a few years old; in a few years they will have broken down completely. Species that have become extinct during historical times are rarely represented by remains that are becoming fossils. Has anyone ever seen the fossilizing remains of a passenger pigeon in the eastern forests or of a grizzly bear in the California foothills? Yet, only 150 years ago these animals were common. If common species have become extinct so recently without leaving fossils, can we expect to find many fossils of rarer species that lived thousands of years ago?

Furthermore, interpretation of the existing fossil record suggests that the most critical species—those that formed links between the major categories of modern animals and plants—were small, relatively uncommon, and often lacked hard parts. Consequently the chances of their preservation are even less than for common, more highly evolved species that were typical representatives of a family, order, or class. The transitions between major phyla of animals occurred more than 500 million years ago, during the Cambrian or earlier periods. Moreover, extrapolation from existing fossils suggests strongly that they were soft-bodied and lacked hard parts. Their virtual absence from the fossil record is to be expected.

Given these severe limitations, gaps in the fossil record—particularly those parts of it that are most needed for interpreting the course of evolution—are not surprising. What is more surprising and encouraging, however, is the presence of some series of fossils forming elegant, almost complete transitions between major categories. The transition from reptiles to mammals lasted for more than 100 million years and included scores of known species and their near relatives. Some of these transitional forms, particularly the mammallike reptiles of Africa, are represented by hundreds of individuals, and thousands more lie buried in the strata, waiting to be uncovered. Thus no reason exists for assuming that, because a form transitional between two major categories has not yet been found, it did not exist and never will be found. Fossil hunting and interpretation is an ongoing research discipline that is certain to provide even more clues about the course of evolution at the level of higher categories.

BRIDGES BETWEEN SMALL-SCALE AND LARGE-SCALE EVOLUTION

One of the most important discoveries of evolutionists during the past 50 years is that, at the level of populations and species, both the processes and the course of evolution can to some extent be duplicated by experiments conducted under strictly controlled conditions. But can extrapolation from these studies of contemporary populations lead to an explanation of evolution on a larger scale, which has given rise to entirely new kinds of organisms? To explain the course of evolution from amoeba to man, must biologists discover principles or processes that cannot be recognized at the level of populations and species?

Some recent opponents of evolution, as well as classical zoologists like Pierre Grassé, have maintained that the investigations of contemporary evolutionists are insufficient to explain long-continued trends or the evolution of major groups of organisms. They accept the Darwinian explanation for the origin of races and species but reject it for the origin of different "kinds" of animals and plants. They are, however, vague about the distinction between species belonging to the same kind and those that belong to different kinds. Most evolutionists argue that this vagueness is inevitable and that in fact no reason exists for separating evolution into the two categories microevolution and macroevolution, supposedly governed by different processes. In this chapter and the next, we will review evidence supporting both the unitary concept of evolution and the validity of extrapolation from known facts about the origin of species to interpret evolutionary changes on a larger scale that are documented by the fossil record. Nevertheless, evolutionists do recognize four processes that are not obvious at the level of populations and species. These are *differential survival of species and higher categories, epigenesis, autocatalysis,* and *coevolution.* As these four processes become better understood, the extrapolation from lower to higher levels of the evolutionary hierarchy will become increasingly clear.

DIFFERENTIAL SURVIVAL OF SPECIES AND HIGHER CATEGORIES

There are two ways by which an evolutionary line can evolve a succession of species that are progressively more complex and advanced—species transformation and species replacement by differential survival. In *species transformation*, populations evolve by means of natural selection of new gene combinations that replace old ones in response to new environmental challenges. Sometimes these changes progress sufficiently to transform an old species into a new one. In *species replacement by differential survival*, a new species "buds off" from a population of an existing species, forming an isolated population; the bud then undergoes a series of gene substitutions guided by natural selection in response to a new environmental challenge, and during this process it acquires new chromosomal rearrangements or other characteristics that isolate it reproductively from the ancestral species. Ancestor and descendant then exist contemporaneously for a varying time span, after which the descendant is able to proliferate additional species that occupy still different ecological niches, while the ancestral species either becomes extinct or evolves very little.

Some evolutionists have used the term *species selection* for the sequence of events here called *differential survival*. This term may be misleading. When natural selection brings about changes in the gene pools of populations, one set of alleles replaces another. The process is direct, and it is brought about by a single direction of environmental change. Differential survival, however, is an indirect sequence of processes. The final events that bring about the extinction or stagnation of one species and the adaptive radiation plus progressive differentiation of the derived species may involve quite different selection pressures.

This difference is well illustrated by a new interpretation of the final stage in the evolution of modern horses. The evolution of the horse line during 40 million years, from the diminutive Eohippus (*Hyracotherium*) to modern horses, asses, and zebras, has been interpreted successively in three different ways. Around the turn of the century, paleontologist Henry Fairfield Osborn saw the evolution of this line as a continuous, progressive, and deterministic change guided from within, in which environmental conditions simply ensured the success of an already determined series of changes. During the middle of the present century, another paleontologist, G. G. Simpson, conceived of the evolution of the horse in terms of a series of adaptive radiations, chiefly involving transformation of species rather than the budding off of new species and subsequent differential survival. A third interpretation, consistent with the concept of punctuated equilibria, follows Simpson in accepting adaptive radiation and rejecting an autonomous internal drive, but places emphasis on speciation by budding followed by differential survival. Which of these explanations is most consistent with the facts?

A recent analysis by paleontologist W. Shotwell of the remains of horses that existed in the Great Basin region of the western United States during the Pliocene epoch (about 2 million to 8 million years ago) shows that intermediate horses with three toes and modern horses with one large hoof existed in that region for about 5 million years, but they did not occupy the same habitat. Three-toed horses lived in parklike country dotted with many trees and shrubs, while one-hoofed horses lived on open grassy plains. On the open plains, horses can best escape from predators by galloping rapidly in a straight line, for which a single large hoof is best adapted. In parklands, escape depends on successful dodging among trees and bushes—here the extra toes of the three-toed horse could have been helpful in pivoting for sharp turns, thus allowing the horse to survive in a particular adaptive niche. One-toed types were first able to become established and spread as open plains appeared and became more abundant. Three-toed horses became extinct not because of unfavorable competition with one-toed horses but because the habitat to which they were adapted was shrinking and was occupied by entirely different animals that could exploit it better. Deer, whose hoofs act as vacuum cups in holding the animals to the ground and whose bone–muscle structure is ideally suited to dodging and leaping, must have been far more successful in escaping predators than were three-toed horses; moreover, deer can easily subsist by browsing on shrubs when grasses are scarce, whereas horses are less versatile feeders. If this scenario is correct, the extinction of three-toed horses and the spread of one-toed types could be characterized as differential survival rather than species selection.

Is differential survival following on speciation the usual means by which evolutionary lines progress? This question is difficult or impossible to answer solely on the basis of fossil evidence because there are so few examples in which the ecological relationships of the species are clear cut. Nevertheless, analyses of the ecological relationships of existing species that seem to be on the verge of extinction, particularly those with more common and successful relatives, may provide clues.

A good example is the native cypress of California. Ten species of cypress trees exist in the state, of which eight are so rare that they are on the verge of extinction. In the vicinity of Monterey, one can find two species. Monterey cypress inhabits granitic headlands on the coast and has as its nearest neighbor Monterey pine; Gowen cypress is adapted to hardpan clay soils two to three miles from the coast, where it is associated with a different species of pine, the Bishop. The two species of cypress do not compete with each other in any way. One can speculate that, if the climate should become rainier, Monterey cypress would become extinct because of unfavorable competition with Monterey pine, while Gowen cypress would succumb to the superior vigor of Bishop pine. The same kind of climatic change might enable one of the more common species, Sargent cypress, to spread into areas that are now too arid for it. In short, a paleobotanist living ten million years hence could easily reach the conclusion that Sargent cypress spread at the expense of the other species.

The kind of example just presented is much more common than contemporary examples that can be interpreted as favoring direct replacement of a less complex or efficient species by a related species that is more complex or efficient. Due to the complexity of ecosystems, species replacement is usually an indirect process. The appearance of direct continuity may be due to the strong action of epigenetic constraints, which are discussed in the next section. They greatly restrict the directions in which evolution can progress.

If this interpretation is correct, evolutionists who wish to explain change at the macroevolutionary level must compare different but related evolutionary sequences and seek to explain why they progressed in different directions and at different rates.

Why did some species of fish evolve jaws and so give rise to a host of modern fishes, while others retained the jawless mouth that the common ancestor of fishes possessed and thus evolved into only two anomalous, parasitic forms—the lamprey and the hagfish? Why did one group of lungfish evolve feet that propelled them onto the land, thus giving rise to all land vertebrates, whereas other related lungfish retained slender fins and evolved very little between ancient and modern times? Why did a few small dinosaurs become the precursors of birds, whereas much more numerous, conspicuous, and aggressive dinosaurs became extinct? Why did a few mouse-sized populations of mammallike reptiles evolve into mammals, while their larger cousins became extinct, presumably in unfavorable competition with dinosaurs? Why did the common ancestor of great apes and humans give rise in one direction to our aggressive, dominant species and in another direction give rise to gorillas and chimpanzees, now on the verge of extinction? Before attempting to answer some of these questions, we will examine the three key concepts of epigenesis, autocatalysis, and coevolution.

EPIGENESIS, AUTOCATALYSIS, AND COEVOLUTION

The word *epigenesis* (derived from the Greek prefix *epi*, meaning "above" or "on top of," and *genesis*, meaning "origins") is used by biologists to mean a succession of events, each of which depends on the completion of the previous stage in a sequence. A familiar example is the development of an animal from a fertilized egg through the various stages of embryonic life to a fetus and finally to an independent offspring. By analogy, the later development of an intelligent small child into an adult mathematician or engineer could be called epigenetic in that it involves two interacting threads—the maturation of the brain and the succession of teaching and learning environments to which the child is exposed. Between the ages of one and two years, the child's brain has developed enough so that it can learn to talk. At ages three to four, it can understand numbers and counting; at six to eight, arithmetic can be learned with relative ease; and at twelve to fourteen, the more abstract concepts of algebra can be understood, followed later by trigonometry and calculus. Nevertheless, the maturation of the brain provides no more than the opportunity for the child to learn the successive steps that lead to a full understanding of mathematics. The child must be taught—in an epigenetic succession—numbers and counting, simple arithmetic, algebra, trigonometry, and calculus. The later steps can be learned and understood only by the student who has mastered the earlier ones.

Analogies between epigenesis in evolution and epigenetic development of an individual break down with respect to direction. Individual development and human learning are avenues toward predetermined goals. The capacity for developing into a normal adult exists in the fertilized egg and needs only a succession of expected environments to bring it out. The embryonic mathematician or engineer is working toward a goal that has been set by himself and others, such as parents or teachers. With regard to evolution in organisms other than humans, however, epigenesis appears to be an automatic series of processes, lacking foresight and affecting natural selection generation by generation.

The theory of epigenesis does provide a definite answer to the question of how random mutations, interacting with random shifts in the environment, can produce the directional trends in evolution that are documented by the fossil record. The selection process eliminates genes, not only if they reduce adaptation to the external environment but also if they interact unfavorably with genes that are already present and so reduce the adaptiveness of established gene combinations. This means that organisms become adapted to new environments—even ones quite different from the old ones—as much as possible by modifying preexisting adaptive strategies rather than by evolving new strategies. The more complex the organism, the more likely it is to eliminate mutations that might reorient its evolution in an entirely new direction, and the more likely it is to build new adaptive patterns on the foundations of already existing gene combinations.

If epigenesis is such a strong conservative force, restricting evolutionary adaptation to a few directions, what force can act to break through its conservatism and bring about drastic new adaptations, such as the evolution of amphibians from fishes, birds from reptiles, and humans from apelike ancestors? In comparison with evolutionary changes along epigenetic lines, such breakthroughs are rare events, probably the result of highly unusual combinations of circumstances. One of these is the challenge of a radically new environment to a population capable of responding adaptively. Another factor of crucial importance may be the capacity of the population itself to alter the environment so as to bring about new conditions that provide a drastic challenge. This is known as *autocatalysis,* a word with Greek roots that means "self-stimulation." Evolutionists borrow the term from chemistry, where it means the stimulation of a new chemical reaction by the product of a preceding reaction.

One of the best examples of such stimulation is the evolution of microorganisms that require oxygen for their cellular activities. We know that the atmosphere of the primeval earth was devoid of oxygen. In sharp contrast to most modern bacteria, the earliest cellular organisms derived energy not from oxidation or "burning" their food but from much slower, more inefficient ways of breaking down organic compounds. Among these organisms, however, were bacteria that could *photosynthesize*—that is, capture light energy, store it, and release oxygen into the atmosphere. This process, now carried out in the modern world by plants, was responsible for increasing the concentration of oxygen in the earth's atmosphere to its present level of about 20 percent.

To an organism that lives in an atmosphere not containing it, oxygen is a deadly poison. Hence the earliest photosynthesizers, by raising the level of the earth's oxygen, generated an environmental challenge that could be met only by revolutionary changes in cellular activities. The microorganisms that evolved successful responses to this challenge became far more efficient than their ancestors. By using oxygen to break down sugars and other large molecules, they gained energy for synthesizing their own compounds and for reproduction. Modern bacteria can pass through a cell generation in 20 minutes and produce thousands of millions of offspring in a few hours only because they are able to take advantage of the potent chemical activity of oxygen. Reactions based on oxygen are likewise essential for the motility of organisms, whether it be the thrashing of a tiny flagellated cell through the water, the darting of a trout through a mountain pool, or running by a human being.

From the viewpoint of life as a whole and of interactions between organisms, most of evolution appears to be based on autocatalysis. Whenever a new kind of dominant animal or plant evolves, its appearance stimulates the evolution of other kinds of organisms in new directions. *Coevolution*, which means the evolution of two unrelated kinds of organisms through specific interactions with each other, is nearly universal among higher animals and plants. Interactions between predator and prey, between plants and grazing or browsing animals, between flowers and their pollinators, or between parasites and hosts provide most of the directional trends that are documented by the fossil record. Every important step from fishes to humans is an example of coevolution. Why did the earlier fishes acquire jaws that replaced jawless sucking mouths? Probably because animals such as shrimps and similar crustaceans had become large enough to serve as food for an animal that could capture and crush them. Why did lungfishes or early amphibians first venture onto the land? They may have been forced to do so because drought dried up the shallow lakes in which they lived; with equal probability, however, we may postulate that they were already used to a diet of animals such as shrimps, and so found the ancestral insects and spiders that were establishing themselves on the land succulent morsels worth following into a new habitat. What was the advantage of the amniotic type of egg, which enabled the fetus to develop within its own watery environment, far from the watery habitat that its ancestors required? Perhaps this type of egg allowed the early reptiles to lay their eggs in places that were safe from aquatic predators. Why did the ancestors of mammals profit from the evolution of warm blood, well-differentiated teeth, and sensitive hearing? Probably because this enabled small reptiles to remain active and feed at night, when they could more easily escape from the dominant dinosaurs. Why did the early ancestors of humans become dependent on the use of tools? It may have been that in the open savanna, the most easily available foods were plant roots or bulbs that could be dug up with tools, and animals that could be killed or driven off by a well-aimed stone.

Examples of coevolution were well known to Charles Darwin, and they figure prominently in his work on natural selection. Rabbits or hares evolved swifter descendants in response to the evolution of swifter predators, such as wolves; the increased swiftness of the prey in turn challenged wolf populations in such a way that fleeter individuals, or entire wolf packs, were the most successful in obtaining food and producing offspring. The long nectar-containing spurs that adorn orchids and other flowers evolved more or less concurrently with the long probosces of butterflies and moths that crosspollinate these blooms in the effort to obtain nectar. Interaction between horses and the grasses upon which they fed gave rise by a series of reciprocal challenges to the high-crowned teeth of modern horses and to the hard, firm texture of the leaves of grasses that grow on open plains or steppes. Adjustments of parasites to their hosts came about through similar successions of reciprocal challenges. Most, if not all, examples of continued evolution in a particular direction are based on such coevolutionary, reciprocal challenges between unrelated kinds of animals or plants that inhabit the same ecosystem.

EVOLUTIONARY CHANGE
IN THE EARTH'S BIOTA

There are three main kinds of evolutionary change—transformation of species in separate lines of evolution, enrichment of diversity, and displacement and replacement.

Over millions of years, some or all populations of a species
become so altered by the combined action of genetic variation and
natural selection that they are transformed first into new species
and later into new genera. Until recently, many evolutionists, particularly those most familiar with the fossil record, regarded such
transformation as the principal kind of evolutionary change, but
this is now a matter for debate, particularly among younger evolutionists.

Enrichment of diversity occurs when new species arise by budding from isolated marginal populations of an ancestral species,
after which progenitor and descendant species live contemporaneously but occupy different ecological niches. Enrichment is
most evident when a group undergoes adaptive radiation in
response to opportunities offered by new habitats. Modern examples of enrichment are found on oceanic islands or in large lakes.
In the fossil record we have trilobites in the early Paleozoic Era,
reptiles at the end of the Paleozoic, ammonitic cephalopods during
the Mesozoic, flowering plants during the Cretaceous period, and
mammals at the beginning of the Tertiary period. In some major
groups of organisms, alternating phases of enrichment, stabilization of diversity, and decline can be recognized. Nevertheless,
a general trend toward enrichment of the earth's biota appears to
have taken place from the dawn of life until about 80 million to
100 million years ago (Cretaceous period), after which the total
number of living species apparently leveled off.

The third kind of change—displacement and replacement—
takes place when whole groups of organisms become extinct and
are replaced by descendants of a comparatively small number of
species belonging to an entirely different group. At the level of
individual evolutionary lines, replacement of one species by an
apparently more progressive one has given the appearance of
steady transformation in a particular direction over several million

years. The most massive displacement and replacement of the earth's biota is contemporary—that is, it began in human prehistory, when humans began to cultivate fields and domesticate animals, and is continuing today at an accelerating pace. Species of wild animals and native plants are becoming rare or extinct and are being replaced by domestic races of animals and cultivated varieties of plants, as well as by numerous and varied species of pests and weeds. Earlier examples of wholesale replacement are well known. About 80 million years ago, dinosaurs became extinct and were replaced eventually by mammals. The Mesozoic ammonoids were replaced by the advanced and newly diversifying bony fishes. During the latter part of the Paleozoic Era, trilobites and other archaic crustaceans were replaced by modern forms such as shrimps, lobsters, and crabs.

The paramount importance of enrichment, displacement, and replacement extends to the level of populations and species. Natural selection and differential survival are not identical processes, but they resemble each other in many ways. The immediate causes of enrichment, displacement, and replacement are to be found by probing the nature of biotic communities and ecosystems through a multidisciplinary approach that emphasizes both evolution and ecology.

DIRECTION, NECESSITY, AND EXTINCTION

The common phrase the "evolutionary tree" of a familiar animal or plant conveys the impression that evolutionary ancestry resembles the growth of a tree, in which the trunk, already recognizable at an early age, leads directly to the modern form of the animal or plant. This impression is unrealistic. The evolutionary "tree" of most common animals and plants begins in the form of a "shrub," of which several branches are about equally prominent. Only after the evolutionary line has existed for some time do one or a few "leaders" reveal one or two distinctive directions along which evolution will proceed. The best known example of an evolutionary line that followed this course of direction is the horse family and its ancestors, of which the dog-sized eohippus (*Hyracotherium*) is the generally accepted ancestor. This animal, however, was itself a radiant from a more generalized ancestor and shared many characteristics with less famous contemporaries (*Homogalax, Palaeomoropus*) that were contemporary radiants from the same ancestral source. This "bush" of early radiants gave rise to several "leaders" that became the trunks of widely divergent evolutionary lines, including one that led to the rhinoceros, another to the tapir, and still others to creatures that flourished for millions of years only to become extinct (for example, *Chalicotheres*). In each of these lines the earliest side branches led to animals with diverse skeletal architectures, such as horses with four or three toes, and to smaller as well as larger body size. With the passage of time, each of the main lines became increasingly stereotyped, and the branches became shorter and less divergent. The horse line entered the modern world in the form of three very similar animals—horses, donkeys, and zebras—that are distinguished chiefly by different body size, different proportions of organs, and adaptation to different terrain. The same sort of change can be seen in most other families of mammals and birds that have managed to adapt and thus survive over millennia.

Extinction because of too narrow specialization overtook several lines of animals known only from the fossil record. One such example is the sabre-toothed "tiger," which lived on the plains of temperate regions from 35 million to less than a million years ago. This predator had canine teeth that extended into long, swordlike blades, making it appear more ferocious than any other living relative. The sabrelike teeth were associated with a widely opening jaw, forming an ideal stabbing mechanism for penetrating the armorlike hides of mastodons, giant sloths, and the other thick-skinned, slow-moving creatures that formed its exclusive diet. But during the Ice Age, the huge herds of elephantlike animals and giant sloths largely succumbed to drastic climatic changes, surviving only in tropical regions inaccessible to the sabre-tooths. Because it was heavier and slower than others of the same family, and so narrowly specialized that it could not adjust to attacking alternate prey, the sabre-tooth became extinct, while the lion, for example, survived.

ESCAPE FROM SPECIALIZATION

The evolutionary success through specialization and final extinction of sabre-tooths because of too narrow specialization exempli-
fies a common pattern. Nevertheless, as is so often true in biology,
other patterns can be found that form remarkable exceptions to
this common one. One of the most striking of these exceptions is
the evolution of snakes. These reptiles are closely related to lizards, which they resemble anatomically in many respects. Snakes,
however, first appeared in the middle Cretaceous period, about
100 million years ago, while lizards had existed in essentially modern forms for at least 120 million years previous to that time, and
primitive mammals had been on the scene for 90 million to 100
million years. Moreover, the bodies of the earliest lizards deviated
little from those of the earliest reptiles, which appeared 300 million
years ago. In other words, by the time snakes started their evolution, all other families of reptiles, including modern lizard families, were in existence and had diverged from each other. The
ancestors of snakes were related to a family of lizards (Varanidae),
whose modern representatives include the gigantic "dragons"
found on Komodo Island off the coast of Southeast Asia.

The history of snakes, particularly in earlier forms, is fragmentary because these animals are rarely preserved as fossils. Nevertheless, paleontologists agree that during the Cretaceous period
certain varanoid lizards took to burrowing in the ground or under
dense shrubbery, so that longer bodies and shorter limbs, or even
the limbless condition, began to have increasing adaptive value.
Some of these forms still persist as specialized lizards, and others
persist as subterranean snakes that in some anatomical characteristics are less specialized than their surface living relatives but in
other respects are more so. Apparently, the earliest snakes literally
dug themselves into a hole from which there appeared to be no
evolutionary escape. They were even more specialized than modern lizards. They no longer walked or crawled; instead, they and
their snake descendants "swam" over the ground, much as a fish
swims through the water. The eventual escape arose through further specializations. One was loosening the articulation of the jaws

and making them double jointed, so that snakes could swallow small animals whole. Another was the evolution of teeth that point backward, which helps the snake to swallow its prey (Figure 4-2).

A third specialization is a thin, extensible body wall. When a snake has engulfed a whole animal, the carcass can be digested at leisure and serve as a meal for days or weeks to come. Once these specializations had been achieved, selection pressure for longer, thicker bodies was greatly increased, since it permitted the snake to prey on increasingly larger animals. The oldest known snakes were large—similar to modern pythons—and may well have been able to kill fairly large prey by coiling around them and constricting them to death.

Interestingly enough, the major diversification of snakes took place during the Tertiary period, contemporaneously with the adaptive radiation and diversification of the modern mammals on

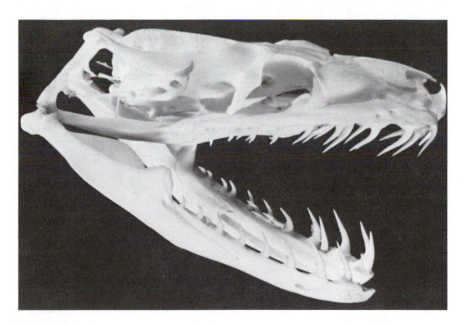

Figure 4-2.

Head of a nonvenomous snake, showing the teeth that point backward, enabling the animal to impale its prey. (Photo by Nathan Cohen.)

which they preyed. Although this period is regarded by most zoologists as the Age of Mammals, paleontologist Steven Stanley has pointed out that it could with equal justification be called the Age of Snakes, because snakes diversified at a rate exceeding that of most families of mammals, surpassed only by mice, rats, and other rodents. (Incidentally, the simultaneous rapid evolution of snakes and mammals is an excellent example of the stimulation of evolutionary rates by coevolution.) At the beginning of the Tertiary period, pythonlike forms gave rise to lithe, swift-moving descendants resembling racers, blacksnakes, and king snakes. Shortly thereafter appeared the ancestors of cobras, coral snakes, and the numerous poisonous species for which explorers in the Australian bush must keep a constant lookout. Vipers and pit vipers evolved even later. The rattlesnake family (Crotalidae) is the most recently developed of all, having appeared only about 20 million years ago (Miocene epoch), after most mammals and birds. In comparatively recent epochs, rattlesnakes have evolved two new and highly distinctive organs—the familiar rattle on the end of the tail, made up of the horny residues of shed skins, and an extremely sensitive warmth detector consisting of two pits on the front of the head, between the nostrils and the eyes (Figure 4-3). Using this detector, a rattlesnake can recognize the bodily warmth of an unseen prey and strike without warning.

The second kind of "escape hatch" from overspecialization is retardation of development with respect to many or most bodily characteristics, a phenomenon known as *neoteny*. This complex topic has been extensively reviewed by S. J. Gould. The example that interests us most is the comparison of human development to that of chimpanzees and gorillas. The human gestation period, from pregnancy to birth, is a few days longer that that of chimpanzees, but other processes of development are delayed much more. A chimpanzee baby acquires its first teeth when it is less than three months old; a human baby acquires teeth at about six months. Gorillas are sexually mature at six to seven years, humans at thirteen to fifteen years. The development of the human head, including the skull, brain, and face, is even more retarded. When we are born, our head and brains are still growing at rates comparable to those of a fetus in apes and monkeys; the closing of the

Figure 4-3.

Head of a rattlesnake, showing the temperature-sensitive pits on the front of the head. (Photo by Nathan Cohen.)

sutures between the skull bones, which occurs in apes before sexual maturity, is delayed in humans until 25 to 30 years of age. The elongation of the lower part of the face that gives gorillas and chimpanzees their characteristic protrusion of jaws and lips is absent in humans. The shape of the head in adult chimpanzees and gorillas is very different from the head shape of their young; by contrast, head shape in humans changes relatively little from infancy to adulthood (Figure 4-4). The head profile of a baby chimpanzee resembles the head of a human baby or adult much more than it resembles the head of an adult belonging to its own species.

This retardation of development, particularly of the skull and face, was the most important anatomical change that led to the human way of life, because it permitted evolution to proceed in an entirely new direction. The more rounded human head fits better upon an upright body than does the elongated head of apes. More important, the imperfect development of the skull and brain at birth allows learning to take place while the brain is still devel-

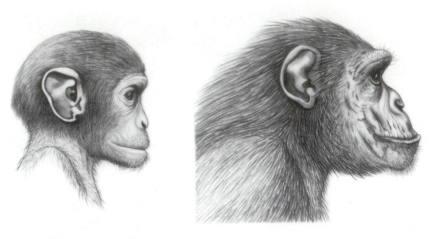

Figure 4-4.

Profiles of a young and an adult chimpanzee. The young has a head shape similar to that of a human baby; the adult is very different from an adult human. (From Naef in S. J. Gould, *Ontogeny and Phylogeny*, Harvard University Press. Reprinted by permission.)

oping, and so makes the process much more susceptible to outside stimuli than is possible in apes and other animals. In addition, the general retardation of the rest of the human body, making the baby almost helpless for weeks or months after birth, strengthens ties between human parents and offspring, thus contributing greatly to the preeminent importance of cultural transmission.

EVOLUTIONARY DIRECTION BASED ON GENES AND PROTEINS

During the past twenty years, the direction of evolution has been explored at the molecular level. The most valuable technique for this purpose is determining the exact sequence of amino acids in protein molecules—the order of "words" in the "sentence" of information that they carry. Since this order reflects to a large extent the order of nucleotides in the DNA of genes coding for the proteins under study, the biochemist can analyze the direction in which particular genes have been evolving in terms of the actual mutations that have occurred. These analyses have been used to analyze evolutionary rates and directions, and their theoretical implications have been explored thoroughly, particularly by Emile Zuckerkandl and Linus Pauling, pioneers in this field. The following account owes much to their research.

One effect of amino acid sequencing has been to provide a new dimension for an old question—that of homology, or similarity due to common origin. Even before Darwin, biologists relied on details of form and developmental pattern to decide whether organs found in different kinds of animals or plants were derived from a comparable organ in some common ancestor or had evolved independently. When anatomists compare skeletons of forelimbs of mammals, birds, reptiles, and amphibians, they can recognize each bone by its position and association with other bones, regardless of whether the limb is a foreleg, an arm, a wing, or a paddle like that of the seal. They conclude logically that all forelimbs of land vertebrates are derived by descent with modification (to use Charles Darwin's term) from that of the earliest amphibian, and they call the limbs *homologous*. On the other hand, the legs and wings of insects, with totally different structures and organizations, have undoubtedly been derived independently from those of completely different ancestors and are regarded as *analogous*.

The determination of homologous relationships runs into difficulty in many particular examples. In a broad sense, the wings of birds and bats are homologous, in that both have been derived through descent with modification from the forelimb of a primitive reptile. The wing function has, however, been *acquired independently* in the ancestors of the two groups; this is evident from ways in which leglike forelimbs have been modified for the function of flying (Figure 4-5). Although this example is obvious, others are not. Anatomists and paleontologists have long held divergent opinions with respect to the porcupines of North America and those of Eurasia and Africa, some believing that all the animals derive from the same porcupinelike ancestral stock, and others insisting that the resemblances between modern porcupines are most easily explained by convergent evolution. The New World and Old World forms were derived from different kinds of ancestral rodents that evolved toward each other because of the adaptive advantages in certain environments of a porcupinelike body and way of life.

Homology, or the lack of it, can be determined more accurately and decisively for proteins than for anatomical structures (Figure 4-6). The principal reason is that protein chains have a much larger number of recognizable units than do anatomical structures, and each unit can be changed in a limited, definite number of ways. This permits biochemical evolutionists to determine exactly the degree of probability that two molecules having similar sequences have acquired this similarity by convergence from ancestors having dissimilar proteins rather than by divergence from a common ancestor. Since every position along the chain could contain any one of 20 different kinds of amino acids, the chance that two proteins would independently acquire the same amino acid at a corresponding position is 1 in 20. If two adjacent positions are considered, the number of possible sequences for them is 20×20, or 400, so that only once in 400 times would independent evolution be expected to produce exact similarity for these positions in unrelated proteins. Most protein chains prove to contain between 100 and 160 amino acids, whereas organs that are compared with respect to homology rarely contain more than 20 parts. Consequently, the information on which biochemists base their decisions is more extensive and precise than that available to anatomists.

Figure 4-5.

Homologies and analogies of wings. The wings of birds and bats are homologous in that both of them are descended with modification from forelimbs. However, the way the bones of the forelimbs have been modified is totally different in the two examples, so zoologists conclude that birds and bats acquired wings independently of each other. These structures are homologous as forelimbs but analogous as wings. By contrast, the wing of an insect does not resemble a limb in any respect. It is completely analogous to the wings of all other kinds of animals. (From *Evolution* by T. Dobzhansky, F. J. Ayala, G. L. Stebbins, and J. W. Valentine. W. H. Freeman and Company. Copyright © 1977.)

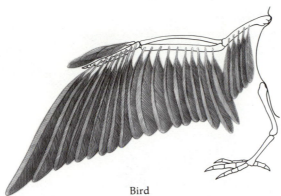

Bird

Bat

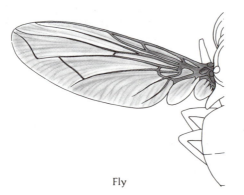

Fly

(a) Hemoglobin

Alpha	V	–	L	S	P	A	D	K	T	N	V	K	A	A	W	G	K	V	G	A	H	A	G	E	Y	G	A	E	A	L	E	R	M
Beta	V	H	L	T	P	E	E	K	S	A	V	T	A	L	W	G	K	V	–	–	–	N	V	D	E	V	G	G	E	A	L	G	R	L
Gamma	(V.	H.	L.	T.	P.	E.	E)	K	T	A	V	N	A	L	W	G	K.	V	–	–	–	N	V	D	A	V	G	G	E	A	L	G	R	(L.
Delta	G	H	F	T	E	E	D	K	A	T	I	T	S	L	W	G	K	V	–	–	–	N	V	E	D	A	G	G	E	T	L	G	R	L

1   2   5   10   20   30

(b) Cytochrome c

Pseudomonas c551	E	–	D	P	E	–	–	–	–	–	V	L	F	K	N	K	G
Rhodospirillum c2	E	G	D	A	A	A	G	E	K	V	–	–	–	–	S	K	K
Horse c		G	D	V	E	K	G	K	K	–	I	F	V	Q	K	–

C	V	A	C	H	A	I	D	–	–	–	–	T	K	M	V	G	P	–	–	–	–	–
C	L	A	C	H	T	F	D	Q	G	G	A	N	K	V	G	P	N	L	F	G
C	A	Q	C	H	T	V	E	K	G	G	K	H	K	T	G	P	N	L	H	G

(c)

	1		5			10			15		
Mouse Nerve Growth Factor	S	–	S	T	H	P	–	V	F	H	M	G	E	–	–	F	S	V	C	D	S
Horse, Proinsulin	F	V	N	Q	H	L	C	G	S	H	L	V	E	A	L	Y	L	V	C	G	E

	20			25		30			35		
	V	S	V	W	V	G	D	K	T	T	A	T	N	I	K	G	K	E	V	T	V
	R	G	F	F	Y	T	P	K	T	R	R	E	A	E	D	L	Q	V	G	Q	V

Figure 4-6.

Homologies between various protein chains. (a) The first 32 amino acids in the protein chains of human hemoglobin, showing similarities and differences between the alpha, beta, gamma, and delta chains. Identical amino acids are enclosed in a box with dark shading; amino acids shared by three out of four chains are boxed in light shading; those that are shared by only two or are different in all four chains are not shaded. Note that amino acids 18 and 19 of the alpha chain are not found in the others; they apparently represent an addition to that chain after divergence from the common ancestor. Each letter stands for a different amino acid. (b) A similar comparison between the first 37 amino acids in the chains of the enzyme cytochrome c from two species of bacteria (*Pseudomonas c* and *Rhodospirillum c*) and the horse. Note that *Rhodospirillum* resembles the horse more than *Pseudomonas* does. (c) A comparison between two protein chains that belong to molecules found in different organisms (mouse, horse) and having different functions (nerve growth factor, proinsulin). Although the similarity between these molecules is much less than that between the others that have been compared, nevertheless the probability that this similarity could have arisen by chance alone is infinitesimally small (4 in 10 million). (From C. Petit and E. Zuckerkandl, *Evolution,* Hermann, Paris.)

Exact comparisons between proteins provide an independent check on outward form and internal anatomy as guides to evolutionary relationships. This approach usually supports previously accepted systems and reinforces accepted opinions; nevertheless, important exceptions are found. Most botanists have regarded fungi—molds, mushrooms, rusts, and other disease-producing parasites—as plants, albeit somewhat peculiar ones. By comparing the amino acid sequences of a particular protein (cytochrome c) extracted from a great variety of animals, plants, and microorganisms, biochemists E. Margoliash and W. M. Fitch have concluded that fungi are not closely related to either plants or animals and may be biochemically more similar to animals than to plants. With respect also to cytochrome c, turtles and birds are more like each other than either is to the rattlesnake—although zoologists have long placed turtles and snakes in the class of reptiles and have placed birds in a different class.

Evidence from biochemistry strongly supports the Darwinian conception of evolution as a continuous succession of forms. When either proteins or DNA are compared, a complete spectrum is found, from species that are exactly alike with respect to some of their proteins but show slight differences with respect to others, through species that are progressively more different, to species that are separated by so many biochemical differences that no relationship could be recognized between them were it not for the existence of the intermediate forms. Moreover, with respect to such an important molecule as the DNA of genes, species that most zoologists and laymen alike would place close together as variants of the same general "kind" (or genus) can differ more widely from each other than animals that are generally regarded as being very different from each other. Figure 4-7 shows two species of *Drosophila* flies that look much alike but are recognized by specialists as very different with respect to form, anatomy, and chromosomal structure. These fly species have about 25 percent of their DNA sequences in common. By contrast, cows and sheep resemble each other with respect to 89 percent of their DNA sequences; humans and New World monkeys resemble each other with respect to 84 percent; and humans and chimpanzees, 97.5

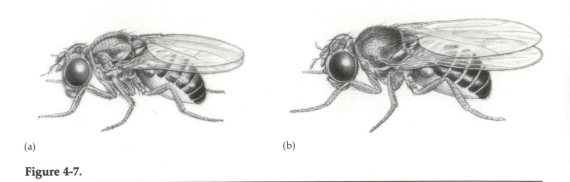

(a)   (b)

Figure 4-7.

Two species of the fly genus *Drosophila* that belong to different species groups. (a) *D. melano-gaster;* (b) *D. funebris*. Experts can easily tell them apart. With respect to DNA, they have only 25 percent of their DNA sequences in common; 75 percent of the sequences are different. Since humans and chimpanzees differ from each other by only 2.5 percent of their sequences, the difference between the genes of these two flies is 30 times greater than that between ourselves and our cousins, the apes. (From J. T. Patterson in University of Texas Publication 4313, 1943.)

percent. Differences between the genes of these two fly species are 30 times greater than differences between humans and chimpanzees.

One might argue that, although the differences between humans and apes are few, they are more important than those between species of flies because they could include the appearance of entirely new kinds of genes. Careful biochemical investigation fails to support this notion. The same kinds of proteins can be isolated from the cells of chimpanzees as from human cells. Some of these proteins are identical; others are different with respect to immunological affinities, electrical charges, or amino acid sequences. After much searching, no new enzymes have been found in humans that do not already exist in apes.

Evidence from comparative sequences of particular proteins supports the theory that most genes are far older than the modern species that contain them, just as are bones, stomachs, lungs, and many other anatomical features. New genera, families, and other major groupings of animals, plants, or microorganisms arise through Darwinian descent with modification of their individual proteins and other macromolecules. The genes that code for these macromolecules have been gradually altered by the continuous occurrence and establishment in populations of successive mutations. New genes have appeared but rarely. The only known process by which new genes come into being is through the transformation of old genes in such a way that they acquire new functions.

Although biochemistry has provided much new and valuable evidence about evolution, only a relatively small number of animal proteins have been isolated and analyzed. Information is still needed about the proteins of many cells, tissues, and organs, the proteins of cellular membranes that regulate the entrance of substances into the body, and those associated with muscular contraction and the transmission of nerve impulses. Moreover, the processes of bodily growth and regulation are still so poorly understood that they provide little evolutionary evidence. One of the most important biological mysteries is how genes regulate the development of higher animals and plants.

SELECTION, COEVOLUTION, AND EVOLUTIONARY RATES

Interactions between chance and natural selection place a lower limit on the rate at which populations can change in a particular direction. According to Steven Stanley, "We can invoke selection as the primary source of change only if its operation is highly eipsodic and therefore strong enough, over short intervals, to be realistically envisioned." Selective pressures so weak as to be barely perceptible would be overbalanced by random fluctuations in gene frequencies caused by mutation and genetic recombination, which are constantly occurring in all populations. Because natural selection is the only way in which adaptive change occurs, such evolution proceeds at rates detectable at least over periods of several generations. Data obtained by paleontologist R. Lande on mammals during the Tertiary period, when these animals were evolving rapidly in terms of the geological time scale, show that evolution was actually so slow in terms of change per generation that if one accepts the hypothesis of continuous change rather than quantum bursts or punctuated equilibria, the differences could have resulted from only about one selective death per million individuals per generation. Change caused by differential deaths at this extremely slow rate could just as well be the result of chance as of natural selection. The mammals analyzed included horses, which are often cited as an example of directional adaptiveness. Stanley concludes, therefore, that the evolution of mammals during the Tertiary period must have followed the pattern of punctuated equilibria, a conclusion with which I agree.

Examples of coevolution as a way of stimulating evolutionary lines to evolve more rapidly have already been mentioned briefly. Nevertheless, this phenomenon is best understood if we follow the history of different kinds of organisms in one particular region during several million years. A good location is the plains area of the central United States, where changes in animals and plants are well documented by fossil evidence.

About 15 million years ago, during the Miocene epoch, three-toed horses by the thousands were grazing on prairies dominated by grasses with hard-shelled seeds and tough leaves, resembling modern spear and rice grasses (*Stipa, Piptochaetium, Oryzopis*). The streams were bordered by trees, among which cottonwoods (*Populus*) and hackberries (*Celtis*) were prominent. Eight million to ten million years later, these horses had been replaced by species that walked and ran on a single toe or hoof (as do modern horses) but that still differed from modern species with respect to other anatomical characteristics. The spear and rice grass genera had also evolved new species, which differed from those of the previous epoch in the size and shape of their fruits. The cottonwoods and hackberries of the stream margin forests had lived on, essentially unchanged. Several million years later—from 1 million to 2 million years ago—a massive ice cap covered the northern plains. Parts of the central plains were converted into arctic tundra, which was inhabited by buffalo. The horses and the grasses of the spear grass–rice grass group migrated southward, as did the cottonwoods and hackberries. When the climate became warm again, horses similar to the modern species returned, but in fewer numbers. They were competing with large numbers of buffalo and antelope, and they soon became extinct, perhaps because of hunting by early humans. Grasses also returned to cover the plains, but these were of different kinds, known as grama grass, buffalo grass, and bluestems—types that had evolved in Mexico and on the southeastern Gulf Coast and Atlantic coastal plain. The grassy plains had been reestablished, but their animal and plant life had become totally different from what it was before the Ice Age. The stream margin forests also returned, with cottonwoods, hackberries, and other species that even now are much like those that were there 5 million to 15 million years earlier.

The coevolution of horses and grasses before the Ice Age sprang from the constant interaction between these organisms. Horses that had the hardest, longest teeth were best equipped to eat and digest the tough grasses; thus they obtained more food and bore more offspring than their relatives with weaker teeth. Clumps of grass with the toughest leaves could best resist grazing and so produced the largest amount of seed. Hereditary changes or mutations among the horses that equipped them with better teeth spread through the horse populations, as did mutations for tougher leaves through the populations of grasses. Through interactions of this kind, the coevolution of horses and grasses was continually stimulated.

The changes that occurred during the Ice Age were so drastic that they put both the horses and their associated grasses at a comparative disadvantage to the more recently evolved buffalo and the grasses belonging to three actively evolving groups—grama grasses, buffalo grass, and bluestems. A complete turnover of the grazing fauna and grass flora was inevitable. Cottonwoods and hackberries, on the other hand, were not affected by this turnover. They interacted only with temperature changes and with the stream bank habitat, which after the Ice Age essentially reverted to its previous condition. For horses and grasses, coevolution followed by drastic alteration of essential parts of their environment resulted in rapid evolution followed by extinction, although members of the spear grass and rice grass groups lived on in a few favored spots. For cottonwoods and hackberries, the much milder changes in their environment caused them to evolve little or not at all.

Less complete evidence from other regions indicates that this scenario is not unusual but normal for temperate regions. Woody plants that interact chiefly with physical factors of the climate persist without evolving, whereas smaller plants and animals that interact with each other evolve much more rapidly.

After a major adaptive shift, when a population and its descendants enter a new adaptive zone that offers possibilities for many strategies, evolution becomes unusually rapid in terms of both diversification or branching and the appearance of new adaptive gene combinations. This rapid change is the result of the ability to expand into a newly available environment or to exploit in many different ways a new major adaptive strategy. As the new environment becomes saturated or the new ways of exploiting it become exhausted, evolution slows down and becomes chiefly a series of variations on well-established themes.

The evolution of animals and plants on oceanic islands such as the Galápagos and the Hawaiian Islands provides an example of this kind of cycle. The Galápagos finches gave Darwin an impressive example of rapid evolution in action, while *Drosophila* and honeycreepers (Figure 4-8) in Hawaii are providing contemporary evolutionists with similarly valuable data.

In many areas that are intermediate between forests and deserts—for example, California and the Mediterranean region—particular kinds of soil occurring in areas a few miles in diameter support plant species unknown elsewhere in the vicinity. Because of their inability to compete with the more vigorous vegetation that surrounds them, these species behave as if they were on

Figure 4-8.

Heads and beaks of honeycreeper birds (Drepanididae) found on the Hawaiian Islands. Among birds of the major continents, this amount of diversity would be enough to place them in different families or orders. Each kind of beak shape evolved in response to the environmental challenge of a different kind of food—insects caught on the wing, nectar of flowers having long curved tubular corollas, and large seeds to be crushed. Each kind of change probably required the establishment of many different mutations and gene combinations and became perfected over many generations. (After W. J. Bock, "Micro-evolutionary Sequences as a Fundamental Concept in Macroevolutionary Models." *Evolution* 24: 704–722, 1970).

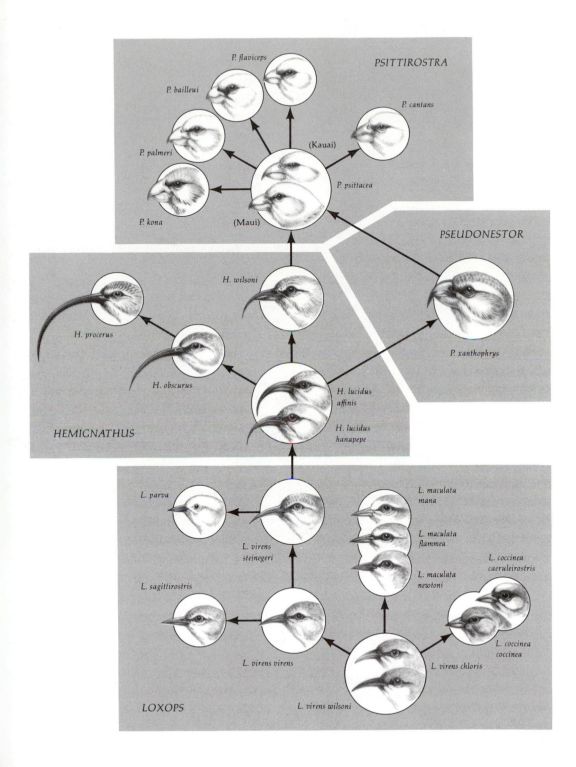

oceanic islands. Where several similar "ecological islands" of this kind occur near each other, they form an "ecological archipelago." Some of the genera that inhabit them—like some genera that inhabit oceanic archipelagoes—have rapidly given rise to clusters of related species that differ from each other with respect to both adaptive and nonadaptive characteristics. Given a change in ecological conditions, escape from these "islands" would be possible for some of these species, which conceivably then could spread into contiguous areas.

The history of the flora of the American Southwest, including California, may have followed such a pattern. After analyzing a large number of rich fossil floras from the western United States, paleobotanist Daniel Axelrod has concluded that the ancestors of the dryland shrubs that form chaparral, as well as ancestors of many smaller plants, became differentiated from forest vegetation in small islands of dry conditons that occurred in the lee or "rain shadow" of mountains that rose from the preexisting plain about 30 million to 40 million years ago. For several million years, these islands formed an ecological archipelago. There, isolation of related populations, accompanied by migration between islands and colonization of new semiarid islands that appeared during a period of increasing drought, stimulated rapid evolution in a fashion similar to that which takes place on oceanic islands. Finally, these "islands" merged to form a "continent" of semiarid and desert conditions over the past 5 million to 10 million years, and the biota that had evolved under insular conditions became converted into the continental desert and semidesert biota that we see today.

MOSAIC EVOLUTION

The evidence presented in this chapter and the preceding chapters shows that rates of evolution are never constant or uniform. While one evolutionary line is racing ahead, another may be nearly or quite static. Among different species of the same genus, some may have been drastically altered during a few thousand years, while others have remained virtually unchanged for millions of years.

Moreover, different parts of the body can and often do evolve at different rates. Legs and feet are transformed into flippers in the evolution of seals, but these animals still retain the facial profile of their doglike ancestors. In evolutionary sequences of fishes, lungs may evolve into swim bladders, but the streamlined fish body is hardly changed at all. Moreover, *mosaic evolution* extends to cells and to macromolecules such as proteins. Some proteins, such as histones, have been virtually constant during hundreds of millions of years, while others, such as those concerned with immune reactions and the "recognition" of transplanted organs, may vary among individuals of the same population.

The modal hypothesis, based on evolution as a multitude of responses to a great diversity of environmental challenges, carries with it mosaic evolution as an expected corollary. Only if evolution were deterministic (its rates and direction having been set by autonomous, internal "evolutionary drives") would we expect rates to be constant.

Nevertheless, evolutionary rates do not vary at random. They are controlled by population–environment interactions. For populations and individual organisms, external factors of the physical and biotic environment are paramount. For protein molecules, however, the internal, cellular environment overshadows everything else. A basic feature of environmental challenges is that they affect some structures and functions of the body more than others. Each response to a particular challenge establishes or modifies a particular adaptive syndrome, consisting of gene-determined structures and functions that contribute to the same adaptive property. Genes that code for structures and molecules that contribute to the same adaptive syndrome evolve at similar rates. These rates may be quite different from prevailing evolutionary rates of structures and molecules that contribute to different adaptive syndromes.

Consider, for instance, three kinds of adaptive syndromes that have been evolving simultaneously during the last 60 million years in primates, the order to which we belong—syndromes associated with circulation of the blood, with diet, and with locomotion. With respect to circulation, the adaptive syndrome that existed already in the earliest primates is adequate for all the successive forms that appeared later; only minor alterations of this system have become established. Much greater selective pressures for change have been exerted on organs associated with food-getting and diet, such as jaws, teeth, stomach, and other internal organs. Selective pressures associated with changes in locomotion have been equally strong and diverse but have been largely independent of those affecting diet. For instance, the shift from running or scrambling along branches to swinging through the forest by grasping the boughs involved great changes in limb and foot or hand structure but affected much less the teeth and internal organs. These latter structures, however, became greatly changed in response to the shift from a diet of insects to one based on fruit-eating, and later in a few groups of monkeys, to grass and other leaves. Still another adaptive shift in some higher primates, from life in the trees to exploiting open savannas, brought about almost simultaneously profound changes in both locomotion and diet.

As is explained more thoroughly in a later chapter, mosaic evolution of different parts of the same body plan has occurred to an even greater degree in plants than in animals. In plants, responses to the inanimate physical environment have been completely different from responses forced upon them by the evolution of new kinds of animals. When evolutionary lines of plants are faced with challenges presented by changes in climate and soil, they respond chiefly by alterations in the number and structure of leaves, stems, and roots. Responses to challenges posed by new kinds of animals—chiefly insects and plant-eating vertebrates—consist of the evolution of insect-repellent bitter or toxic compounds, of flower structures favorable to cross pollination by insects and other animal visitors, and of fruits or seeds that become increasingly efficient for transport by animals. When correlations exist between evolutionary rates of different adaptive syndromes in plants, they are usually due to simultaneous changes in climate and biota.

Of particular interest is the fact that genes coding for cellular proteins often and perhaps always evolve at different rates from those that determine the overall body plan, including anatomical structure. Biochemist Allan Wilson and his associates have shown than, in mammals, cellular proteins are more constant than anatomical structures, while among frogs the reverse is true. Comparing chimpanzees with humans, they found strong resemblances between cellular proteins, in spite of large, obvious differences in external anatomy. Among frogs, pairs of species that are almost identical in overall body plan and anatomy nevertheless are far more different from each other with respect to cellular proteins than are apes from humans.

Why should such differences exist? Is there something about their overall genetic constitution that makes mammals more susceptible to changes in anatomy and frogs more susceptible to changes in cellular proteins? This is Wilson's point of view. An alternative explanation is that a successful response to a challenge can be made relatively quickly, perhaps in a few thousand generations, by anatomical changes. After that, evolution slows down and may proceed very slowly until the population faces another environmental challenge. This hypothesis is in accord with the concept of punctuated equilibria. On the other hand, many environmental challenges may exert only moderate to low selective pressures on cellular proteins. As we know from comparisons between humans and chimpanzees, enzymes having exactly the same molecular structure can function equally well in bodies that differ greatly from each other in outward appearance. Nevertheless, evolutionary changes in these molecules could continue slowly for long periods of time. Possibly, therefore, evolution of anatomical structure and function often proceeds according to the model of punctuated equilibria (Figure 1-3c), while evolution of most cellular enzymes proceeds more gradually (Figure 1-3b). The result of these differences would be a "hare and tortoise" pattern that combines the diagrams of Figure 1-3b and Figure 1-3c. So, one can easily see that, in a young group, newly evolved lines would differ more from each other with respect to anatomy and outward form than with respect to enzymes. In an old group, the reverse would be the case.

This explanation agrees with Wilson's observations. Mammals are relatively young. They diversified rapidly between 50 million and 60 million years ago. Frogs, on the other hand, acquired their present body plan more than 200 million years ago.

IS THERE PROGRESS
IN EVOLUTION?

Many writers of the nineteenth century assumed as a matter of course that the chief trends of evolution consist of progress from lower to higher forms of life. The sequence amoeba → worm → fish → reptile → mammal → ape → human was often said to represent the typical course of evolution. Nobody can deny that such sequences exist, but many of them are hard to recognize as tending to progress. How about amoeba → worm → millipede → cockroach → mayfly → mosquito? Or fish → reptile → tree shrew → mole? A zoologist could mention dozens of others that by human standards would not be called progress. This fact has led to criticism of all concepts of progress in evolution as anthropomorphic; we recognize trends of evolution as progressive to the extent that they resemble the sequence that led to humanity.

This complete denial of progress in evolution seems to many biologists like a swing of the pendulum to absurdity. Isn't the honey bee, which has evolved elaborate societies headed by a queen, a more progressive animal than the lowly cockroach, which may be similar to the common ancestor of winged insects, or the silverfish (*Thysanura*), a primitive, wingless relative of insect ancestors? Isn't the elaborate design of an orchid more progressive than that of the simple tulip? This argument is hard to refute.

During the present century, two evolutionists, G. G. Simpson and Francisco Ayala, have discussed this problem at length. They agree that progress is not a general property of evolution, since many kinds of evolutionary change do not involve progress. But Simpson would recognize as progress any "change toward a particular sort of organism." To him, the evolution of fleas and lice from winged insects or of blind, burrowing moles from active, wide-eyed, shrewlike ancestors would contitute just as much progress as the evolution of bees from less complex insects or of humans from relatively unintelligent, forest-dwelling primates. Ayala, on the other hand, defines progress as "directional change toward the better." This definition leaves us with the troublesome question of whether humans are competent to judge what is "better" in species so unlike our own.

Another possible definition of progress would be "directional change toward more complex organisms." Complexity could be estimated on the basis of anatomy, biochemistry, or behavior. On the basis of anatomy, the evolution of fleas or lice from winged insects or of moles from shrewlike ancestors would be regressive rather than progressive, since the more evolved organisms have lost certain abilities—flight in the case of fleas and lice, and vision in the case of moles—as well as several kinds of cells and organs that are associated with the lost faculties. Orchids would be regarded as more progressive than members of the lily family, even though the latter occupy a greater range of habitats, because orchids have petals that are strongly differentiated from each other and have a great diversity of form among the cells that make up their flowers, whereas this kind of differentiation is minimal in most species that belong to the lily family. Bacteria that live in a medium rich in oxygen and exist efficiently because they have acquired a battery of enzymes capable of converting food into energy in the presence of oxygen are much more complex and efficient from the biochemical viewpoint than are bacteria that live in an atmosphere without oxygen; they could therefore be called biochemically advanced. Finally, based on the criterion of behavioral complexity, honey bees would certainly be regarded as more progressive and advanced than solitary bees, in spite of their narrower ecological range.

If evolutionary progress is defined as trends toward greater complexity, does the definition help us to understand evolutionary phenomena? Using only a few words, we could characterize any trend as toward greater or lesser complexity of form, biochemistry, or behavior, or as toward greater or more restricted ecological success, and thereby avoid the confusion that arises from the use of value-based words such as "better." Perhaps we cannot entirely eliminate from the vocabulary of biology and natural history anthropomorphisms like "progress," "purpose," "sacrifice," and "altruism," but at least we can be aware that, whenever such words are used for nonhuman phenomena, they acquire new and different meanings that must be carefully defined.

An artist's view of the emergence of a qualitative difference: *Liberation* by M. C. Escher. (Copyright © Beeldrecht, Amsterdam/VAGA, New York, 1981. Collection Haags Gemeentemuseum.)

Chapter Five

HOW NEW
QUALITIES EVOLVE

The question of whether natural selection produces truly new categories of animals has bothered evolutionists for many years. Darwin's own approach to the problem was characteristically pragmatic. To him, theoretical deductions were much less important than observable phenomena. He first asked whether artificial selection, practiced by human beings on domestic animals, produced changes comparable to natural changes that would cause a trained zoologist to place an animal in a new category. His observations of domestic breeds of pigeons (Figure 5-1) convinced him that this could happen: "The fantail has thirty or even forty tail feathers, instead of twelve or fourteen—the normal number in all members of the great pigeon family." Later he noted: "Altogether at least a score of pigeons might be chosen, which, if shown to an ornithologist, and he were told that they were wild birds, would certainly be ranked by him as well-defined species. Moreover, I do not believe that any ornithologist would in this case place the English carrier, the short-faced tumbler, the runt, the barb, pouter and fantail in the same genus; more especially as in each of these breeds several truly-inherited sub-breeds or species, as he would call them, could be shown him." He reasoned that any kinds of changes that mankind could produce by artificial

(a)   (b)   (c)

Figure 5-1.

The common rock pigeon (a) and two fancy fantail breeds (b,c). Darwin observed that ornithologists would not classify birds having such large numbers of tail feathers in the family of pigeons. (After Srb and Wallace, from G. Ledyard Stebbins, *Processes of Organic Evolution*, © 1966, pp. 7–8. Reprinted by permission of Prentice-Hall, Inc., Englewood Cliffs, N.J.)

selection could also arise by natural selection, given a great enough adaptive advantage in some habitat.

During the first quarter of this century, many biologists over-looked these observations in favor of the ideas made popular by rediscovery of Mendel's laws and the consequent emphasis on the effects of mutations. Zoologist Richard Goldschmidt observed correctly that the differences between the races of gypsy moth (*Lymantria dispar*), which adapt them to climates as different as those of Scandinavia, Italy, Central Asia, and Japan, are of a nature entirely different from those that set them off from a related species, the nun moth (*L. monacha*). From this he concluded that natural selection of small hereditary variations can produce only the adaptation of similar organisms—such as different races of the same species—to different habitats, but that these races never evolve into new species, genera, or higher categories. He believed that large evolutionary changes depend on the occurrence of "macromutations" or "systemic mutations," which supposedly produce three different kinds of changes in one step: (1) a reshuffling of parts of chromosomes in such a way that the linear order of the genes is completely rearranged; (2) a drastic change in the pattern of development, beginning at a very early age; and (3) the consequent appearance of a "hopeful monster" that could lead a new kind of existence in a new and different habitat.

Like Darwin, Goldschmidt was a pragmatist. The phenomena that he observed were far more important to the formulation of his theories than facts that he gathered from other sources; his firsthand experience, however, was much narrower than Darwin's had been. Moreover, he did not study carefully enough the visible differences between the gypsy and nun moths on the one hand and numerous other tropical species of *Lymantria* on the other. Consequently, a more experienced naturalist and systematist, Ernst Mayr, could point to obvious gaps in Goldschmidt's knowledge that seriously interfered with the correctness of his deductions. Mayr concluded that Goldschmidt had overlooked the importance of geographic variation for speciation. With respect to the macromutations of *Drosophila*, which, along with similar mutations in other organisms, formed the second pillar of Goldschmidt's theory, modern geneticists—for example, Hampton Carson, Francisco Ayala, and the late Theodosius Dobzhansky—have found no indication that mutations similar to those studied by Goldschmidt, which convert balancers (halteres) into wings or front legs into antennae, play a role in the origin of species in this or any other kind of insect. The genetic changes that control differences with respect to outward appearance, behavior, and those physiological differences that cause sterility in interspecific hybrids are not simply adjustments in a small number of "master switches" that in one step turn on a whole new set of characteristics. Instead, each individual gene plays its own definite role in the process of species origin.

Modern discoveries regarding the actual origin of species agree with the theoretical expectations of molecular geneticists and experts in statistical population genetics. Each gene codes for a particular enzyme or other protein that performs one—and only one—specific task in the working machinery of the cell. Products of some genes are single cogs among hundreds that keep each cell in operation and give rise to new and different kinds of cells; products of other genes regulate and integrate complex patterns of development. If a mutation occurs, the gene may produce merely a small change in the way its product performs a single task, or it may alter its product so drastically that the overall cellular machinery becomes completely disorganized. Constructive changes usually come about through individually small alterations that finally produce harmonious changes of the entire gene system. Large changes tend to be destructive to this harmony.

In 1929, R. A. Fisher noted that a large number of observed mutations producing drastic effects were definitely inferior in the laboratory and probably deleterious in the natural state. This is what one would expect, given the view of naturalists then and now that organisms are intricately adapted, both in their internal mechanisms and in their relation to their environment. Another pioneer of modern evolutionary biology, Sewall Wright, put the matter even more forcefully. He wrote in 1977: "It is evident that Goldschmidt had never grasped the creative effect of natural selection, operating generation after generation on small accidental genetic differences and thereby building up large adaptive changes that would be inconceivably improbable as single steps. In rejecting such trial and error processes at the level of small random changes, he was forced to accept a determinative role of catastrophic events of which the adaptive consequences can only be regarded as miraculous."

Thus the consensus among modern evolutionists is not that there are two kinds of evolution—microevolution, which brings about the origin of races and species, and macroevolution, which is responsible for the progression from amoeba to mankind—but that evolution is unitary and operates at different levels. At the level of races and, to a certain extent, species, evolution consists of a complex web of quantitative changes in the frequency of genes that are constantly interacting with each other. The qualitative differences that we perceive at this level as being new are the outcome of these quantitative changes and interactions. At higher levels—such as the evolution of amoebae from bacteria, of jellyfishes and worms from amoebae, of fishes from wormlike animals, of amphibia, reptiles, and mammals from fishes, and of mankind from apelike mammals—the processes are essentially the same.

New kinds of higher-level organisms arise by continuation of the same kinds of processes that bring about new kinds of bacteria from other bacteria, and so on. Here we may need to review the three principles discussed in the last chapter, epigenesis, autocatalysis, and coevolution.

All differences between organisms that we recognize as *qualitative* or "novel" are the outcome of complex networks of quantitative changes. Even the most strikingly new characteristics or attributes that evolutionary change has produced are the result of combinations of genetically based differences, such as those that geneticists observe and analyze in contemporary populations. Each individual change has a relatively simple and *quantitative* effect on the complex, highly integrated system that is the body of a higher animal or plant. Extrapolations from changes at lower levels of the hierarchy of living beings may seem difficult to make, but they become clear in the light of experience in observing living organisms, including the intimate details of their structure and organization, and in thinking logically about them.

A simple example to illustrate the perception of *quality* as a result of complex *quantitative* differences is shown in the three windows of Figure 5-2. The difference between (a) and (b) is obviously quantitative; (a) has twice as many panes as (b). The differences between (a), (b), and (c) appear to be qualitative, particularly if the shading shown on each of the panes represents different colors of stained glass. Nevertheless, this "qualitative" difference is easily reduced to quantitative components. There are sixteen panes in (c) as compared with eight in (b) and four in (a). Each pane in (c) has three sides rather than four; two sides are equal in size to each other and to all of those in (a) and (b), but the third side in (c) is much shorter; and, finally, the impression of different colors in each of the panes of (c) is produced by the transmission of light waves of different lengths. The frontispiece of this chapter illustrates the same principle in a more artistic fashion.

Complexity itself tends to blur the distinction between quantitative and qualitative differences by making quantitative analyses more difficult. To see how this works for inanimate matter as well as for living systems, let us think about water and various sub-

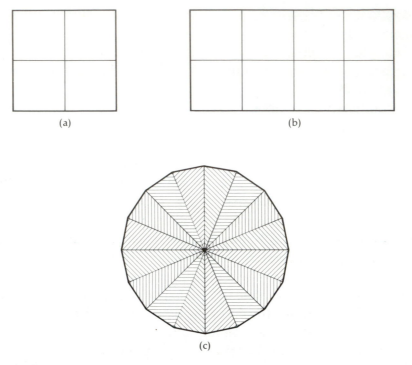

Figure 5-2.

Three different kinds of windows, showing how apparently qualitative differences can arise as the result of several superimposed quantitative differences.

stances that can be put into it. One of the simplest of substances, water exists in three qualitatively distinct states—ice, liquid, and steam. The change from one of these states or qualities to another is abrupt; there is no quantitative transition. Physicists know, however, that the difference between ice, water, and steam depends entirely on quantitative differences in the strength of attraction between atoms and ions.

We can make the water system slightly more complex by dissolving in it, up to saturation, small molecules such as those of salt (NaCl). This will change the temperature at which the change in states occurs, and the change from water to ice will be more gradual, through an intermediate stage of slush. If instead of salt, which has only two atoms per molecule, we add sugar, whose molecules contain 24 atoms and posses a ringlike structure, the distinction between solid and liquid becomes less easy to pinpoint. Depending on the concentration of the sugar solution and on its temperature, it will exhibit every conceivable transition from liquid to solid—thin syrup, thick syrup, molasses, and rock candy. The next stage of complexity could be illustrated by adding several different substances at once, including some with large molecules, such as proteins that contain about 5,000 to 10,000 atoms per molecule. The result would be the formation of a jellylike substance that chemists call a colloid. A familiar example of a colloid is mayonnaise. The Russian biologist Sergei Oparin, an authority on the origin of life, has developed the highly plausible theory that, for thousands or millions of years before the first true cells appeared, the prebiotic organic system was a "coacervate" with colloidal properties. Even the simplest living organism is, however, in terms of complexity, what the mayonnaise factory would be to the mayonnaise itself.

Human perceptions of qualitative differences are highly subjective. The better acquainted we are with objects or people, the more easily we can recognize their different qualities. Transferring this principle to evolutionary studies, anyone who is not particularly interested in insects may think of all flies as essentially alike. A specialist in the genetics of *Drosophila*, however, can easily see qualitative differences between "garbage-loving" species like *D. melanogaster*, inhabitants of mountain forests like *D. pseudoobscura*, denizens of tropical jungles like *D. willistoni*, and cactus-fruit eaters like *D. mulleri*. Moreover—believe it or not—the flies themselves, even with their tiny brains, have a far greater ability to perceive qualitative differences between individuals of their own and of other *Drosophila* species than the human observer does.

We must regard from this basis the evolution of major qualitative differences between organisms. To humans the qualitative difference between mankind and all other animals appears tremendous. Many of us, therefore, have been shocked by recent discoveries indicating that, with respect to the basic body proteins and even genetic coding, human beings are hardly more different from chimpanzees than horses are from donkeys. Evidently we must look for other definitions of "humanness." Now that the apes have been taught the rudiments of speech and can apparently pass on this ability to other apes, while monkeys have been observed to acquire new customs when faced with new environmental situations, distinctively "human" qualities are becoming harder and harder to recognize.

THE ORIGIN OF NOVELTY

Can biological novelty—that is, qualitative difference—arise as a result of novel interactions between populations and their environment, mediated by natural selection, as do the less spectacular, clearly quantitative differences between populations and species? A few examples at the levels of both visible structures and biochemical macromolecules will serve to indicate that this is not only possible but also by far the most plausible way of explaining how novelties of form and function arise.

One of the most novel structures in the plant kingdom is an ear of corn (Figure 5-3). The ear consists of the stiff cob, out of which arise the kernels, each in its individual scaly cup; the husks surrounding the cob, which form a graded series enough like leaves so that we can understand their evolutionary origin but which nevertheless differ greatly from the leaves produced on the cornstalk; and the silks—slender threads through which pollen tubes grow several inches to reach and fertilize the prospective grain. To the casual observer and the botanist alike, the ear and tassel of corn appear to differ as greatly as orchid and iris, daisy and honeysuckle, or larkspur and buttercup. Yet anatomical studies, combined with a wealth of fossil evidence obtained from the remains of the settlements of early mankind in America, have shown clearly that the ear of corn has been derived by artificial selection from a structure borne by an ancestral wild grass, a form of teosinte. It was not unlike a modern, relatively unbranched corn tassel, with seeds on its lower half and stamens containing pollen on its upper part. Teosinte evolved from an ancestor in which pollen and seeds developed from the same tiny flower.

This artificial selection by the early ancestors of the Indians resembles typical examples of natural selection in that it brought about a complex of new adaptations. In the ancestors of corn, the kernels broke off and fell from the plant as soon as they were ripe, and so they were hard to gather. Once firmly attached to the cob, however, they could be harvested whenever the cultivator needed them. The kernels of the original form were edible if cooked, but too hard to be ground easily into meal. Cultivators selected varieties with softer kernels, which proved as succulent to numerous insects as to humans; consequently, adaptiveness required better protection for the ripening grain, achieved by transformation of leaves into husks. As early Indian cultivators were perfecting these structures by artificial selection, genes that caused the silks to become greatly elongated were inadvertently selected, since only in this fashion could the developing kernels be pollinated and continue their growth. Thus the ear of corn represents the adaptive complex of one of the world's finest food plants, which evolved over about 5,000 years by a process of artificial selection that was in every way comparable to natural selection.

By comparative morphology and anatomy, botanists have shown that all the differences between corn and the ancestral teosinte can be reduced to specific quantitative changes. Geneticists have crossed corn with modern counterparts of the ancestral form and have shown that the genes responsible for the differences between the forms are in every way comparable to those responsible for differences between varieties of corn itself.

Darwin pointed out that any characteristics obtainable through artificial selection can also be established by natural selection, given their adaptiveness and the right kinds of selection pressures. Let us now look at a second example, that of jaws and teeth. The earliest fishlike ancestors of vertebrates had no jaws. Their mouths were always open, and, as they swam through the water or wriggled along the bottom of the ocean, they sucked up bits of food just as a vacuum cleaner sucks up dust. Gills along the sides of their heads served to aid in absorbing both the food and oxygen from the water. The supports for the gills were pairs of bones arranged in a V-shaped fashion, just as jaws are. The front pairs

(a)

Figure 5-3.

(a) Part of a pollen-bearing tassel of corn and kernel-bearing ear from the same plant. Both of these structures have evolved from an inflorescence like that of the grass genus *Eulalia* (b) via a complex series of quantitative changes, plus the loss of kernels from the corn tassel and of stamens from the structure that evolved into the ear. Intermediate stages support the hypothesis that this drastic transformation took place step by step, via selection of small differences.

(b)

may well have moved back and forth to aid the sucking process, as our jaws do when we suck liquid through a straw. It is possible that some population of "pre-fish" entered an environment where small particles of food were scarce and large particles, including small animals, were abundant. This would have favored genetic changes exaggerating the movement of the front gill arches and eventually converting them into jaws, leaving the posterior ones to perform the original function of aeration. Such a change is well supported by evidence found in fossils of early fishes.

The addition of teeth to the newly formed jaws would have enabled the fish to impale actively moving prey. The first teeth were much like the hard, pointed scales that covered the body of the early jawless and jawed fishes (Figure 5-4), and they may have evolved by the simple process of repatterning development so that scalelike structures appeared on the jaws as well as on other parts of the skin. These first teeth were implanted only into the skin. Larger teeth, set into sockets of the jaw (as those of humans are), evolved through a successsion of quantitative changes and were not perfected until a much later period of vertebrate evolution. The scales of early fishes evolved into those that cover the bodies of modern fishes, becoming broader, thinner, and more homogeneous in substance, so that in modern fishes like codfish, perch, and trout, scales are so unlike teeth that virtually no resemblance between the two structures can be detected.

The evolution of other structures of the more advanced animals was much more complex and is less well supported by fossil evidence, but it differed only quantitatively from the evolution of jaws and teeth. Examples are the evolution of bird feathers from reptilian scales and the evolution of lungs and ears.

With respect to enzymes and other molecules that perform the biochemical tasks that keep cells and organisms alive, a model for explaining novelty as a succession of quantitative adaptive changes is provided by the evolution of hemoglobin, the substance that makes our blood red and transports oxygen through our bodies. Detailed analyses of hemoglobin and related protein molecules have shown that the molecule ancestral to hemoglobin was myoglobin, which still exists in muscles and provides them with the

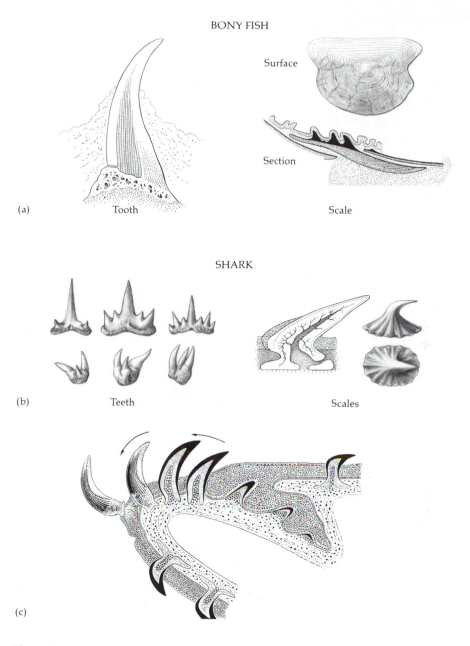

Figure 5-4.

Evolution of teeth and scales of fishes. (a) A tooth (*left*) and a scale (*right*) of a modern bony fish. The difference between them appears to be qualitative. (b) Teeth and scales of a modern shark, which resemble each other more closely. (c) Cross-section through the jaw of a shark, illustrating transitions between teeth and scales and their origin from similar tissues. The jaw of primitive fishes was much like that of modern sharks. (After A. S. Romer, *The Vertebrate Body*, Fourth Ed. New York: Holt, Rinehart and Winston, 1970.)

oxygen they use in contracting. Although the structure of myoglobin is not fully known in primitive animals such as jellyfishes and worms, there is evidence to suggest that molecules ancestral to vertebrate myoglobin—and therefore genes that coded for them—existed in the earliest animals of these kinds. As the bodies of early fishes became larger, the adaptive value of circulatory fluid increased. This adaptation seems to have been achieved by two kinds of changes in the DNA that codes for myoglobin. First, a duplication of a part of the chromosome took place, so that two similar genes, both of them coding for myoglobin, existed side by side on the same chromosome, or perhaps later on different chromosomes. Because one of these genes could provide the cell with all the myoglobin needed for full muscular activity, the other was free to mutate to an inactive state without interfering with normal functions. In this way, it could have become "retooled" by means of changes in molecular structure—for example, the substitution of one amino acid for another at certain critical positions. The new molecule retained the ability to capture and release oxygen and could do this in a fluid passing through blood vessels rather than embedded in muscular tissues. In the surviving descendants of the jawless fishes, the lamprey (*Petromyzon*) and the hagfish (*Myxine*), hemoglobin consists of a single kind of protein chain that has an amino acid sequence similar to the myoglobin in the muscles of the same animal. During the evolution of jawed fishes, a second duplication of the gene for hemoglobin took place, permitting the differentiation into two kinds of protein chains, which unite to make the kind of hemoglobin found in all jawed fishes, amphibians, reptiles, birds, and mammals. Through additional duplication–differentiation cycles later on, mammals acquired two kinds of hemoglobin molecules—one confined to the fetus and the other found in individuals after birth.

In the globin example, the old protein and its new derivative have very similar functions, so that a retooling process that involves a relatively small part of the amino acid sequence is a reasonable explanation of the change. In other examples, however, the change in function is much more drastic. For instance, comparison of amino acid sequences suggests that lactalbumin, a protein found in milk, is derived from lysozyme, an enzyme that digests the cell walls of bacteria. Nevertheless, the basic chemical requirements and reactions between the molecules necessary for these two functions are similar. In other instances, new functions have evolved as a result of two proteins uniting to form a new, more complex protein. For instance, beta galactosidase, a bacterial enzyme that breaks down certain sugar molecules, has acquired more complex activity by adding to its molecule an entire amino acid sequence derived from a completely different enzyme, difolate reductase.

After reviewing the entire field of comparative structure of protein molecules, biologist Emile Zuckerkandl has concluded that, since the earliest cellular organisms appeared billions of years ago, all new enzyme functions have evolved by means of structural and functional modification of proteins already present and functioning. Just as all life has evolved from preexisting life, so all the functioning macromolecules that make up living systems have evolved from preexisting functional macromolecules. Biological novelty at the molecular level can probably be reduced to complex changes and reorganizations of the sequences of nucleotides that make up DNA and RNA, and reorganizations of the amino acids that make up proteins.

Zuckerkandl bases his conclusions on comparison between protein molecules and analogies with the gross anatomy of individual organisms such as crabs and lobsters. These animals have several different kinds of appendages—jaws, claws, and feet—with very different forms and functions, but all are constructed from the same kinds of proteins arranged in different patterns (Figure 5-5). Such changes in pattern are based on differences in the stage of development when particular gene complexes are stimulated to transcribe RNA and thus to make protein synthesis possible, on rates of synthetic processes, and on the position in the embryo where cells are located that will give rise to a particular organ. The difference between the simple form of a sowbug and the intricate and elaborate form of a lobster is comparable not to the difference between a piano recital and a symphony orchestra concert but to two different compositions, such as "Mary Had a Little Lamb" and

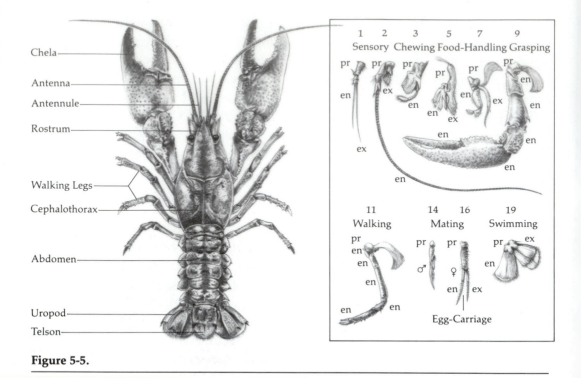

Figure 5-5.

The different appendages of a crayfish, showing a great diversity of organizational patterns among the different parts of the same organism. Since the cells and tissues that make up these structures are very similar among all of them, the apparently novel or qualitative differences between them are entirely at the level of the visible pattern of organization.

Beethoven's "Appassionata Sonata," played on similar pianos. One of the biggest gaps in modern biological knowledge is our almost total ignorance of the mechanisms by which disparate programs of differential gene action are brought about.

The great difficulties involved in the evolution of totally new kinds of proteins are evident in the ways many animals have evolved systems for digesting certain kinds of food. One of the most marvelous digestive systems in the animal kingdom is that of the ruminant herbivores (cows and other animals that chew their cud). They can use plants containing a high proportion of fiber (consisting of complex carbohydrates such as cellulose and lignin), breaking these compounds down to simpler ones that provide them with energy, just as sugar and starch do for us. But these herbivores have not evolved new enzymes for this function over time. Rather, they have evolved an enormously rich and well-balanced biota of microorganisms, consisting of 25 or more species (and billions of individuals) of bacteria and protozoa that digest the food for them. The microorganisms use some of this food to nourish and reproduce themselves, but there is plenty left over for the cow. Just as, in the evolution of the contracting business, general contractors have found that enlisting the services of specialized subcontractors is by far the most efficient method of getting certain jobs done, so in the evolution of herbivores and other organisms that rely on cellulose and lignin for nourishment, delegating the task of digestion to microorganisms is by far the quickest and most efficient pathway of change (Figure 5-6).

This path has been followed also by termites, whose guts contain an extraordinary fauna of elaborately designed single-celled microorganisms. Because of these tiny animalcules, termites can

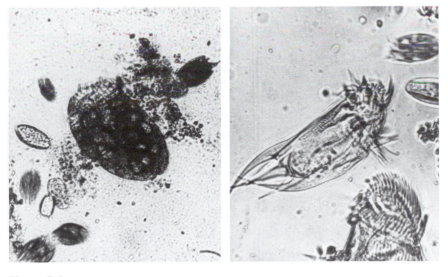

Figure 5-6.

Microorganisms (ciliate protozoa) that live in the stomachs of cattle. (From R. E. Hungate, *The Rumen and Its Microbes*, Academic Press, 1966.)

feed on wood, a substance that is indigestible for most animals. Many other animals may posses biota of digestive organisms, but such ecosystems are poorly known. Their future exploration could easily develop into a fascinating and highly useful field of research—internal ecology. From it we might learn that animals usually evolve the ability to digest previously unused kinds of substances by acquiring microbes in their digestive tracts rather than by evolving new kinds of enzymes as a result of gene mutations.

Novelty does not arise because of unique mutations or other genetic changes that appear spontaneously and randomly in populations, regardless of their environment. Selection pressure for it is generated by the appearance of novel challenges presented by the environment and by the ability of certain populations to meet such challenges. Most often the successful populations are "preadapted" to meet the challenge. Preadaptation, however, does not imply any form of anticipation on the part of the successful population. The term refers only to the fact that a few populations possess particular kinds of adaptations to an old environment that with a relatively small number of genetic changes can adapt them to a completely new way of life. The group of fishes that were ancestral to amphibia possessed adaptations to a particular kind of aquatic habitat that could be converted with relative ease to adaptation for life on land. The reptiles that were ancestral to mammals were subjected to particularly strong selective pressures favoring the acquisition of mammalian characteristics because of their small size and unusual mode of existence. As G. G. Simpson remarked in 1949, "Orientation in evolution is not determined solely by some characteristics within the evolving organisms or solely by factors in their environment, but by both and by interplay between the two." Both novelty and continuity in evolution are based on epigenetic sucessions of responses to environmental challenges, made possible by natural selection, autocatalysis, and coevolution.

If it is correct to assume that the qualitative differences producing evolutionary novelty become established only when the presence of a novel environmental challenge coincides with that of a population prepared to meet the challenge, then one would expect to find many challenges not met and many ecological niches unfilled. Simpson provides an example: "Opportunities for flying insect eaters existed for millions of years before there were any pterodactyls, birds, or bats to exploit them. The opportunity could not be seized until there were organisms that had the required mutations for development of wings and for the many other related necessities of this way of life. These immediately required mutations had to follow after innumerable others for the development of limbs that could become wings, of jaws that could be adapted to seizing insects, and so on through an almost incredibly complex series of changes."

THE CONCEPT
OF TRANSCENDENCE

In sexually reproducing species, every individual contains, to some extent, unique qualities. Consequently, in tracing the course of evolution from amoeba to humanity, we are interested not in novelty itself but in those novel characteristics that produced worldwide major shifts in organism–environment interactions. Can these steps be analyzed or explained in terms of combinations of quantitative differences? Have known sequences of selection— either natural or artificial—brought about comparable changes in populations or species?

A useful concept for identifying the most significant changes in quality is that of *transcendence*, as developed by Theodosius Dobzhansky in *The Biology of Ultimate Concern*. He points out that evolution has taken place at several different levels, of which the easiest for us to recognize are the *cosmic*, up to the appearance of life; the *biological*, from the earliest cell to the appearance of mankind; and the *human*, which concerns primarily our own species. Each of these three levels is subject to particular laws. At the cosmic level, the laws of gravity, thermodynamics, and attractions between particles are paramount. These laws are equally valid at the biological level, but they cannot by themselves explain differences such as those between bacteria, plants, amoeba, insects,and mankind. Biological evolution can be understood only by means of new laws or generalizations, such as Mendel's laws of heredity, information theory as applied to linear codes, and laws of inter-action between complex macromolecules and molecular systems. The most important of these laws include a dimension of time. In the development of an animal or plant, each successive pattern of organization is a modification of the ordered pattern that precedes it. In a similar fashion, the kind of adaptation that a population evolves to meet a new environmental challenge is usually deter-mined by the minimum amount of genetic change that will serve to meet the challenge.

Dobzhansky wrote: "The attainment of a new level of dimension is, however, a critical event in evolutionary history. I propose to call it evolutionary transcendence. The word 'transcendence' is obviously not used here in the sense of philosophical transcen-dentalism; to transcend is to go beyond the limits of, or to sur-pass the ordinary, accustomed, previously utilized well-trodden possibilities of a system."

The two examples that he discusses are the origin of life and the origin of humanity. Both the origin of life and the evolution of humans from apelike ancestors came about through innumerable biochemical and genetic changes that were little or no different individually from the smaller amount of changes that gave rise to all the other millions of species of living organisms. Are we there-fore justified in using a term such as *transcendence* for these and perhaps a few other evolutionary changes? I believe that we are, for at least three reasons. The first is the nature of human percep-tion. Most people regard living beings as entirely different from any inanimate substance and believe that human beings are much more than just glorified animals. Second, both the origin of life and the evolution of mankind are distinctive in their universality. Once living cells evolved, they spread to all corners of the earth and gave rise to ecosystems based on complex interactions between different kinds of organisms; the human species alone has the ability to thrive on all parts of the earth's land surface that can support any form of life. Third, the origin of life and of mankind gave rise to phenomena that were totally unpredictable on the basis of preexisting phenomena. If a chemist knows the formula and properties of a few representative compounds, he can predict with some assurance the nature and properties of other compounds that have similar formulas. However, no inorganic chemist, familiar only with compounds that arise independently of the activity of living cells, could predict the intricacies of the genetic code or the properties of macromolecules such as nucleic acids and proteins that are related to it. Similarly, a well-trained zoologist, familiar with one species of great ape, would have little difficulty in imagining the anatomy and behavior of related species of apes. If, however, such a zoologist should arrive on earth from outer space and see only nonhuman animals, including apes, he might be hard put to predict the existence of an animal species capable of building motor cars or creating artistic and architectural marvels.

On the basis of the criteria of unpredictability and universality, then, what other events of evolution can be said to be transcendent? There appear to be at least six others. The first is the origin of distinct cell nuclei, which gave rise to *eukaryotes* (for example, protozoa and flagellate unicellular algae). It made possible and was closely connected with the origin of both mitotic cell division and the sexual cycle, in which meiosis alternates with the union of entire cells at fertilization. With the appearance of eukaryotes, beauty and diversity of form entered the world of life (see Figure 5-7).

Since their discovery by van Leeuwenhoek in 1676, protozoa have become the means by which thousands of students have learned to admire the exotic beauty of living organisms invisible to the naked eye. Generations of young biologists have been fascinated by the sight of a *Paramecium* cruising through the water like a stately ocean liner, pushing aside the smaller whiplash-bearing flagellates that cross its path; at the trumpet-shaped *Stentor*, gracefully swinging back and forth its arc of whirling cilia; at the ever-changing form of an amoeba as it crawls over the bottom of the water and engulfs the bacteria that form its food; or at the elaborately lovely shells of marine protozoa such as Radiolarians.

Nuclei and mitotic cell division greatly increased the amount of genetic information that could be stored by a single organism. The genes present in the single chromosome of bacteria are much fewer than those borne by the numerous chromosomes of eukaryotes. Genetic segregation during meiosis and recombination at fertilization make possible many more hybrid gene combinations. Moreover, in the chromosomes of eukaryotes, the informational DNA is combined with a variety of proteins lacking in bacteria. Some of these proteins play essential roles in regulating complex patterns of development, and without them the diversity of cellular form necessary to a higher plant or animal would not be possible.

By comparing the total span of evolution carried out by prokaryotes with that achieved by eukaryotes, one can logically infer that the evolution of the eukaryote cell was an essential step in the evolution of higher animals and plants. Prokaryotes have been evolving actively for 3.5 billion years; eukaryotes, for only 1.2 billion years. The evolution of prokaryotes was not halted by the rise to dominance of eukaryotes; in some prokaryote lines, evolution was greatly stimulated. Nevertheless, the response of prokaryotes to the totally new biotic environment created by eukaryotes was not an increase in diversity of form but rather the evolution of increased physiological and biochemical diversity. The most actively evolving modern bacteria are the pathogens that cause diseases of mammals, particularly humans. When seen through the microscope, these cells are almost indistinguishable from their ancestors that lived 2 billion years ago. Their biochemical activity, however, is totally different. During their entire evolutionary history, which includes billions upon billions of cells exposed to mil-

Figure 5-7.

Various kinds of single-celled nucleate organisms, or eukaryotes, showing the diversity of form and structures that could evolve after the perfection of the nucleus, with its many chromosomes, and division via mitosis. (a) Amoeba. (b,c) Amoeboid and flagellated forms of *Naegleria*. (d,e,f) Three different marine relatives of Amoeba. (g–j) Four different kinds of flagellated protozoa. (k–n) Four different kinds of ciliated protozoa.

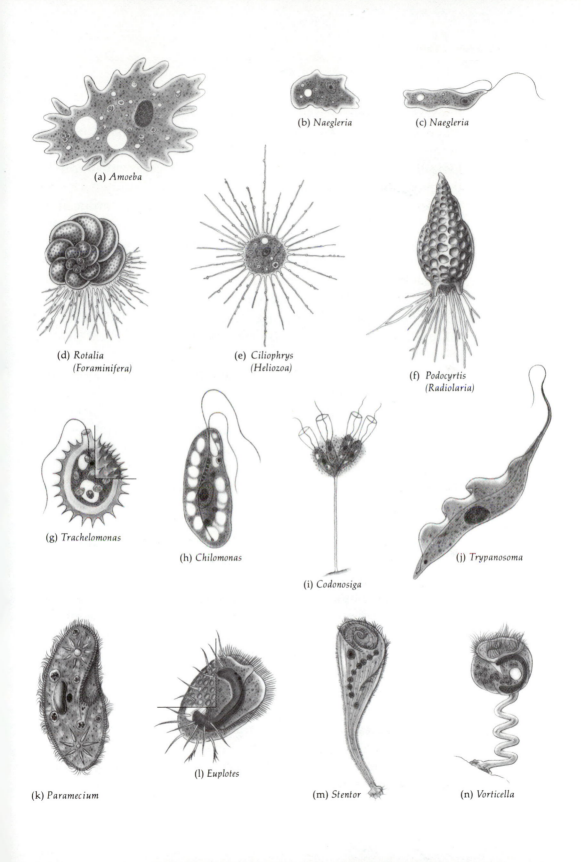

(a) *Amoeba*

(b) *Naegleria*

(c) *Naegleria*

(d) *Rotalia*
(Foraminifera)

(e) *Ciliophrys*
(Heliozoa)

(f) *Podocyrtis*
(Radiolaria)

(g) *Trachelomonas*

(h) *Chilomonas*

(i) *Codonosiga*

(j) *Trypanosoma*

(k) *Paramecium*

(l) *Euplotes*

(m) *Stentor*

(n) *Vorticella*

lions of different kinds of habitats, prokaryotes have always responded to new environments by evolving diversity of physiology and biochemical activity rather than of form. Thus it can logically be concluded that they have never become preadapted to the evolution of great morphological diversity.

Another transcendent event was the evolution of multicellular animals that develop according to a complex pattern of activation and deactivation of genes. Only by acquiring many interacting and cooperating cells, or at least many nuclei within a single cellular cavity, could organisms visible to the naked eye have evolved. The beauty and diversity of form displayed by unicellular eukaryotes can be seen only through the microscope. The beauty of multicellular eukaryotes comprises the beauty of nature that we see around us every day.

Two more transcendent events were the origins of those phyla of animals that include the great majority of the creatures familiar to us—the joint-legged animals, or *arthropods*, and the backboned animals, or *vertebrates*. Arthropods filled both land and sea with small, highly active creatures—shrimps, water fleas, crabs, spiders, grasshoppers, flies, beetles, and butterflies. Among the vertebrates, fishes, amphibians, reptiles, birds and mammals—including humans—are the larger animals we are most familiar with.

The origin of warm-blooded birds and mammals might be regarded by zoologists as of transcendent importance. Two events in the evolution of plants can also be regarded as transcendent. One was the evolution of the woody or vascular tissue in the stems of all larger plants—including ferns, small flowering plants, grasses, trees, and shrubs—which made possible the evolution of forests and shrublands containing habitats that support a wide variety of animal life. The second was the evolution of flowering plants, or angiosperms. Their great diversity of form and adaptation has enabled them to colonize most of the earth, including freshwater lakes, streams, and the margins of the oceans. They include all the food plants essential to our existence.

These examples emphasize the fact that transcendence is not an intrinsic property of evolution so much as a word used to emphasize the profound universal significance of a few exceptional evolutionary changes that have been brought about by the operation of the same kinds of processes that were responsible for evolution in general. The difference between transcendental novelty and the kind of novelty that has been part of every event of speciation is quantitative, not qualitative.

Transcendental novelties in biological evolution can be compared directly to the effects of certain epoch-making events in history that completely transformed the cultural world of humanity. The rise and spread of agriculture, the cultural flowering of Attic Greece, the origin and spread of the great world religions and philosophies such as Buddhism, Christianity, Islam, and Marxism, the discovery of America, and the Industrial Revolution all changed indelibly the entire course of humanity, just as the six or eight transcendent events mentioned above changed entirely the course of biological evolution. Both of these kinds of extraordinary changes were only exceptionally influential examples of essentially similar but less significant kinds of changes that have taken place much more often. The recognition of transcendent events as being different in their influence but similar to ordinary changes with respect to basic processes leads us to an understanding of why extrapolations from changes that can be recognized and analyzed experimentally at the levels of populations and species are justified as a way of interpreting evolution. Such extrapolations are in fact the only logical way of understanding evolution as a whole.

REDUCTIONISM AND HOLISM

In living systems, organization is more important than substance. Newly organized arrangements of preexisting molecules, cells, or tissues can give rise to emergent or transcendent properties that often become the most important attributes of the system. Scientists who recognize this realize also that two opposite approaches to the study of living systems have equal and complementary value. These approaches are *reductionism* and *holism*.

An extreme reductionist uses the "nothing but" approach, maintaining that, since DNA and RNA are molecules containing only common and familiar atoms, a biochemist can find out everything he needs to know about them by isolating the molecules and analyzing their structure, their method of synthesis and breakdown, and their relationships to other molecules. Another reductionist, trained primarily in physiology and cell biology, would conclude that the cell, as the unit of structure, provides the key to understanding the structure and function of the entire organism. A behavioral biologist with a reductionist slant might believe that, if humans are simply one kind of animal, humans can be fully understood by analyzing the genetic and behavioral characteristics of various other kinds of animals.

The extreme holist takes the opposite point of view. A holistic zoologist would maintain that the animal body is an integrated whole, so that nothing is gained by studying its component molecules and cells in isolation. A holistic anthropologist or sociologist might conclude that humans are so obviously different from other animals, and a human from one culture is so different from a human raised in another, that attempts to apply concepts of genetics and animal behavior to an understanding of human nature and society are doomed to failure.

Both the extreme reductionistic and extreme holistic attitudes are misleading; however, the complementary nature of reasonable reductionism and holism can be recognized by experiences in our daily lives. For instance, you may find one morning that your car will not run or that it emits a strange odor. You will amplify this discovery by means of a reductionistic approach, carried out by your reliable garage mechanic, who will analyze the machinery to find the trouble and correct it. He will then use a holistic technique by test-driving the vehicle and asking you to do the same. The job will not be complete until this alternation from holism to reductionism and back to holism has run its full cycle.

The success of evolutionary science owes much to a continuing alternation between holism and reductionism. Darwin—a holist— recognized that visible differences in form could be inherited and that, as a result of natural selection for various traits, populations could evolve in various directions. Mendel, Morgan, and other reductionist geneticists showed that many differences in form are due to differences with respect to particular genes. Fisher, Wright, Haldane, and other population geneticists, by returning to a holistic approach, devised principles and models by means of which the rates and directions of change could be predicted in a general way. Further exploitation of the reductionist approach identified DNA and RNA as the replicating molecules of genes and showed how their chemical organization made possible both their precise replication and their capacity for storing genetic information. During the late 1960s and 1970s, two holistic approaches drastically modified previous concepts of the role of DNA in evolution. Analyzing the entire DNA content of an organism, R. J. Britten and others showed that, in higher animals and plants, very little of the DNA consists of genes that code for protein structure. At the same time, biochemical evolutionists such as A. C. Wilson have determined that, in different groups of organisms, three kinds of changes proceed at different rates relative to each other—changes in protein structure, in the organization of chromosomes, and in the anatomical structure of the body. Reductionist biologists are now analyzing the DNA content of various cell nuclei bit by bit in order to find out why these discrepancies exist. When the analyses have progressed enough so that a return to a holistic approach is practicable, we can expect a major breakthrough in mankind's understanding of biological evolution.

To me, the excitement of science lies not only in discovering particular new facts and principles but even more in scientific dialogue that culminates in fitting these facts and principles into a comprehensible pattern. To understand the diversity of phenomena that comprise biological evolution, one must recognize the complementary value of reductionism and holism.

Part Two

THE COURSE OF EVOLUTION: FROM MOLECULES TO PRIMATES

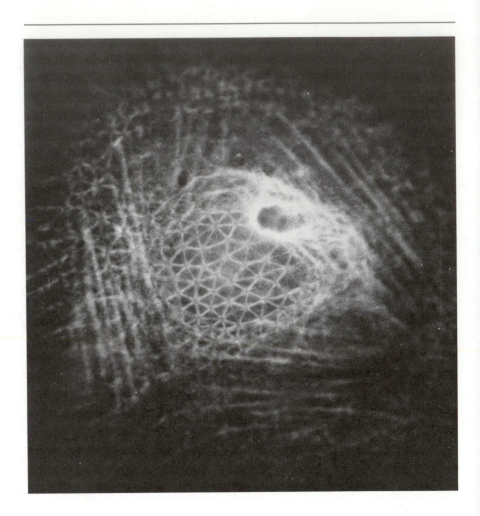

The elaborate cytoskeleton of a eukaryotic cell, organized around the nucleus and responsible for many of the cell's distinctive properties. (Courtesy of E. Lazarides.)

Chapter Six

FROM MOLECULES TO SINGLE-CELLED ORGANISMS

Have scientists actually discovered universal principles that govern the evolution of all kinds of life? Can we believe that the origin of humans proceeded along the same lines as the formation of the first living cells out of the primeval organic soup? Did all major evolutionary advances take place as a result of challenges presented by a new and different environment? Did the successful responses of a favored few individuals with unique characteristics lead to new life forms? Has natural selection always been the mediating factor that has made new ways of life possible?

These questions may never be answered as conclusively as Priestley's demonstration that fire consumes oxygen or Pasteur's proof that all life is derived from preexisting life. But the evolutionist can provide a combination of clearly demonstrated facts, highly probable sequences of events, and the most plausible hypotheses to explain them. He can also make predictions based on these hypotheses and set up experiments to test them.

Certain discoveries and theories serve as guideposts marking the successive changes that are either known or believed to have taken place since carbon-containing organic molecules first appeared on the earth. The reader should bear in mind that only a small portion of evolutionary knowledge has been used to construct this sequence, and only selected major episodes will be reviewed here.

THE ORIGIN OF LIFE

Science has offered only three explanations for the origin of life on our planet. The first of these—*transport of living cells from another planet*—begs the question, because it merely calls for an explanation of life's origin elsewhere, farther back in time. Moreover, the discovery of destructive radiation in outer space mars this hypothesis, and numerous examinations of chemical molecules found on meteorites have turned up no new evidence to support it. At this time, then, we must presume that life as we know it originated on earth.

A second hypothesis is that *the first life arose suddenly, by a chance assembly of organic carbon-containing molecules in an ordered pattern.* Such a system would have to be able to replicate itself, be able to break down other molecules to release energy, and be able to use this energy to synthesize new macromolecules. The possibility that such a complex system could arise by chance alone is so slight that most modern evolutionists have rejected it.

The third hypothesis—*that life evolved slowly and gradually, either through continuous change or by a series of quantum jumps*—is the most plausible thus far. According to this hypothesis, during millions of years before the first living cells appeared, subcellular systems of macromolecules had been evolving in the direction of life, producing new variants that were capable of replicating previously acquired patterns by converting energy from the sun, from the earth's heat sources, or from chemicals. Many biochemists and biologists believe that *variation* as well as a kind of *selection* similar to Darwinian natural selection went on for millions of years before the appearance of what we call life.

Since it left no fossil record, this *prebiotic phase of evolution* cannot be determined by the same methods that scientists use to determine evolutionary sequences of organisms. The prebiotic evolutionist must be part chemist, part physicist, part geologist, and part astronomer in order to reconstruct as accurately as possible the chemical and physical conditions that prevailed when the earth's surface first became a congenial home for living systems. He must then experiment with the modern counterparts of the molecules that were probably there and construct a replica of a kind of system that could have evolved into the first living cell. Only the first steps of this reconstruction have been carried out, but in the past 30 years encouraging progress has been made toward a better understanding of prebiotic systems.

The scenario is reasonably well defined. The earth originated about 4.6 billion years ago, and by about 4 billion years ago its surface had become cool enough to make life possible. By 3.8 billion years ago, rivers were eroding mountains and carrying sediment to valleys and flood plains, forming the first sedimentary rocks. These strata are completely devoid of life. The oldest known fossil cells are from the Fig Tree Chert of South Africa and are about 3.5 billion years old. These data permit us to assume that evolution from the earliest organic molecules to the first cells could have lasted 500 million years—a time span equal to that from the appearance of the first multicellular organisms (primitive jellyfishes) to the age when the first amphibians ventured onto the land.

During the prebiotic stage, the earth's atmosphere contained little or no oxygen. (Our present oxygen supply comes from the photosynthetic activity of green plants, which had not yet evolved.) An atmosphere without oxygen would have allowed ultraviolet radiation from the sun to penetrate to the earth's crust in far greater quantity than at present, a condition that promoted mutational change but could also have destroyed many prebiotic systems and primitive organisms.

The first stage of prebiotic evolution was probably the origin of organic molecules containing carbon, hydrogen, oxygen, and nitrogen, with sulfur, phosphorus, and a few metallic ions such as iron, calcium, potassium, and magnesium. This was indicated in 1953 by Stanley Miller and later elaborated upon by other scientists. Miller supposed that electrical storms first supplied the energy necessary for these syntheses, but Leslie Orgel has calculated that ultraviolet rays from the sun could have supplied continuously all the energy that was needed. During a few million years, these rays, descending to the surface of the ocean and smaller pools, would have produced carbon-containing compounds that turned the water into an organic soup. Some of these compounds could easily have penetrated microscopic fissures in rocks or clefts between soil particles, thereby becoming protected from destructive ultraviolet radiation. The almost infinite numbers of sugar, nucleotide, amino acid, and other molecules thus synthesized could have formed the building blocks from which macromolecules such as proteins and nucleic acids were constructed.

Laboratory production of these molecular building blocks, however, is as far from the construction of a living system as the production of bricks is from the building of a cathedral. Even when amino acids have been strung together to form proteinlike chains, as biochemist Sidney Fox and others have done, a profound gap remains. The millions of artificial proteinlike chains can combine to form tiny spheres as large as bacterial cells, to which they have a superficial resemblance (Figure 6-1), but the majority of protenoid spheres show no chemical activity at all, and only a few have a weak enzymatic activity. Most important, they cannot replicate a particular ordered sequence of amino acids, as the protein–nucleic acid systems of organisms do. Each proteinlike sphere is a separate unit, doomed to exist for a short time and perish without leaving a trace.

Life could not evolve until the appearance of protein chains capable of replication. The special biochemical machinery for this function is found in the colinear relationship between proteins and nucleic acids that is provided by the genetic code. How did this colinear relationship evolve? Three answers appear possible. Either nucleic acids came first, or functional proteins were the first

Figure 6-1.

Tiny proteinlike spheres called coacervates produced by Sidney W. Fox, who heated dry mixtures of amino acids to moderate temperatures. These tiny spheres, under suitable conditions, will grow slowly and eventually form buds. Nevertheless, their activity resembles only slightly that of true enzymatic proteins. (Photo courtesy of S. W. Fox.)

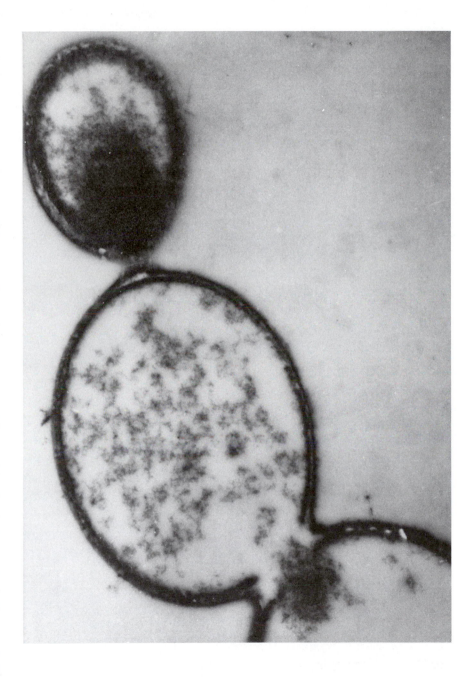

to appear, or the two kinds of chains developed and proliferated in close relationship with each other.

Several geneticists have suggested that nucleic acids with informational content appeared first and that proteins were built around them. This is unlikely, however, because the possibility is low that a nucleic acid consisting of randomly arranged nucleotide units would code for a protein chain having a definite function. Moreover, the relationship between triplets of nucleotides and single amino acids that makes the present genetic code work is highly complex and is almost certainly the end result of a long course of evolution. More direct relationships between nucleotides and amino acids that are sufficiently precise for information to be transferred have been suggested, but the structural relationships between these two kinds of molecules are such that a direct functional relationship is highly improbable. Finally, DNA molecules in modern living systems cannot replicate themselves without the aid of a battery of enzymes, and simpler methods of replication are difficult to imagine. Few biochemists believe that the original macromolecule was DNA or RNA.

The hypothesis that functional proteins appeared first is in many ways more attractive. First, experiments have shown that amino acids can be hooked up into chains (polymerized) more easily than nucleotides can. Second, several of the amino acids found in proteins were probably much more common in the primeval organic soup than nucleotides were. Third, among the millions of short protein chains that were formed, there must have been several kinds that could function as enzymes. We still do not know how the functional proteins became associated with or could have built around themselves nucleic acids that could serve as templates for replicating their particular amino acid sequence.

The third hypothesis, that amino acids and nucleotides evolved their respective polymerized chains—proteins and nucleic acids—in association with each other, is tenable if one can assume that short chains of perhaps five to ten amino acids could have had some kind of enzymatic function in the primeval soup. Such functioning can be imagined, but the nature of the chemical association between the nucleotide and amino acid units poses the same problem for this hypothesis as it does for the hypothesis of a primordial nucleic acid. All three hypotheses, therefore, appear to have a very low degree of probability. For this reason, many biochemists have dismissed the problem of the origin of macromolecules as unsolvable because hard facts that might lead us to a solution probably will never be obtained.

Two biochemists who have not done so are J. D. Bernal and A. G. Cairns–Smith. They point out that proteins can be accepted as the first functional macromolecules if they could have been replicated by totally different kinds of templates with simpler relationships to protein chains and a simpler, though less efficient, method of replication than the nucleic acids have. Bernal and Cairns–Smith suggest that these templates may have been microscopic crystals of various kinds of clay that contain small amounts of metallic ions.

Anybody who has cooked eggs or meat is quite familiar with the tenacity with which tiny pieces of proteins can cling to the surface of a metal frying pan or a clay dish. Bernal and Cairns–Smith suggest that this *capacity for adsorption* may have caused free amino acids or short *polymers* in the primeval *organic* soup to cling to the tiny clay particles at the bottom of rain pools or in the mud of river deltas. After a detailed analysis of crystal structure, Cairns–Smith concluded that crystals containing certain kinds of imperfections and scattered molecules of a different mineral could grow and reproduce faithfully the patterns of the crystal imperfections and the alien minerals. He called these crystals "primordial genes" (in our terms, templates). Cairns–Smith points out that the adsorption of amino acids or protein chains to the surface of a particular mineral template will be specific, depending on the configuration of the organic molecule.

To emphasize his belief that aggregations of clay particles and protein chains could have specific properties and could react differentially to environmental changes, Cairns–Smith invented nicknames for them and suggested how each kind would react to a particular change in the environment. A rather loose aggregation—"sloppy"—could accumulate easily under dry conditions but would be swept away in a rainstorm. A very close association—

"tough"—would replicate so slowly under any conditions that it would always be outdistanced. One containing a high concentration of sugar molecules—"sticky"—would replicate slowly under any conditions. Finally, "lumpy," containing many irregularities, "like badly made custard," would be particularly likely to respond to change by producing more replicates of itself. Thus Cairns–Smith's model postulates that a crude form of selection existed as early as the first formation of functional macromolecules and aided the evolution of the first associations between proteins and nucleic acids.

Depending on the nature and arrangement of the metallic crystals they contain, clay particles can attract specific combinations of amino acids that can perform two functions—chemical selection and physicochemical replication. As they replicate their structure, however imperfectly, they can concentrate in a space of microscopic dimensions thousands of similar aggregates of organic macromolecules.

Cairns–Smith speculates that thousands of millions of such inorganic–organic replicating systems could have been formed over the earth's surface. The vast majority of them remained inert, but here and there a combination of adsorbed amino acids could attack and break down parts of the random, nonfunctional proteinoid sequences in the organic soup, thus obtaining a rich supply of amino acids that could be adsorbed onto their clay templates. This earliest enzymatic activity, though slow and inefficient, would allow the systems to reproduce more rapidly than their nonenzymatic neighbors and thus to dominate estuarine muds or pools, taking over the populations of crystalline–organic template systems just as mutants of bacteria can take over an entire test tube or even a large vat. A stage would arrive during which most of the thousands of millions of crystalline–organic template systems on the earth would have some kind of enzymatic activity.

The next step—the substitution of nucleic acids for microscopic crystalline templates—can be inferred on the basis of observations made by Bernal. He notes that, in modern living systems, nearly all enzyme activity is aided by smaller molecules called coenzymes. The best known of these closely resemble or are identical to RNA nucleotides. These nucleotides consist of three units—a nitrogen-containing base, sugar, and one or more phosphates (Figure 6-2). The chemical bonds between the phosphates and the sugar are called energy-rich, and they efficiently transmit the chemical energy necessary for forging or breaking chemical bonds of other kinds. If such nucleotides became adsorbed either to the crystal surface next to the amino acids (or short polypeptides) or to the organic molecules themselves, their capacity for energy transfer would greatly increase the system's efficiency in breaking down foreign polypeptides, and so would speed up reproduction.

The hypothesis of crystalline–organic particles bound together by adsorption of organic molecules onto surfaces of mineral clay has recently received a great boost from another direction through the experiments of James Lawless, a space scientist whose first interest was the chemical composition of meteorites. Samples of amino acids extracted from meteorites always contain indiscriminate mixtures of isomers (closely related molecules) that are dissimilar, whereas the protein chains of living proteins are always strings of similar isomers. Lawless discovered that amino acids adsorbed onto the surfaces of certain kinds of clay particles are all the same isomer—the one that is found in living systems. In other words, these clays can select from an indiscriminate mixture exactly those amino acids that can combine to form biologically active proteins. Moreover, metal ions such as nickel—present as impurities in the crystals—cause them to adsorb certain amino acids selectively. These metallic ions aid in combining amino acids to form short chains, a process that is also aided by alternation between hot–dry and cool–moist conditions. Perhaps the crystalline–organic systems that had the greatest chance of evolving in the direction of life occurred at or near the seashore between the levels of high and low tide, where they would have been subjected to similar alternating environmental conditions.

Several teams of chemists are actively pursuing this direction of research. In 1980, M. Rao, D. G. Odom, and J. Oró concluded that clays "may have served as very crude analogs of enzymes. As such, clays may have played a role in the very early stages of prebiological evolution: the stages of monomer synthesis and simple monomer polymerization reactions."

Thus experiments apparently show that prebiotic evolution was possible, that at its very beginning it already depended on interactions between evolving systems and their environment, and that various kinds of selection played a dominant role in making possible the sequence of events that led to life. We can expect that further investigations will reveal more about the nature of replication of prebiotic systems.

Once nucleic acids and proteins had become tied together by colinearity and some kind of template code, their self-replicating ordered systems could have continued to evolve toward the much greater complexity needed to form primitive cells in a manner similar to the later evolution of the cells themselves. Molecular biologist Sol Spiegelman and his associates put together a replicating system that in a test tube behaves much like bacterial cells in a culture. Its nucleic acid, RNA, is derived from a small virus, and the enzymes necessary for reproduction are purposely added. The experimenters exposed this system to an environmental challenge by adding a small amount of ethidium bromide, a chemical that acts as a poison. This stopped replication for a while, but in a few minutes it began again. A mutation—one of the many that are constantly occurring in a population of DNA molecules—had enabled the system surrounding it to meet the challenge and to

Figure 6-2.

Structural formulas of the molecular units of RNA and DNA, illustrating the great similarity between coenzymes that transfer energy for chemical syntheses and breakdown and the structure of the nucleic acids RNA and DNA. (a) Adenosine monophosphate, which serves as one of the four units of the RNA helix. (b) By losing one phosphate group, adenosine triphosphate (ATP) can become adenosine diphosphate (ADP) and deliver energy from the energy-rich bond that ties together two phosphate groups. (c) Four nucleotides of DNA, showing their close resemblance to (a) and (b). (From R. E. Dickerson, "Chemical Evolution and the Origin of Life." Copyright © 1978 by Scientific American, Inc. All rights reserved.)

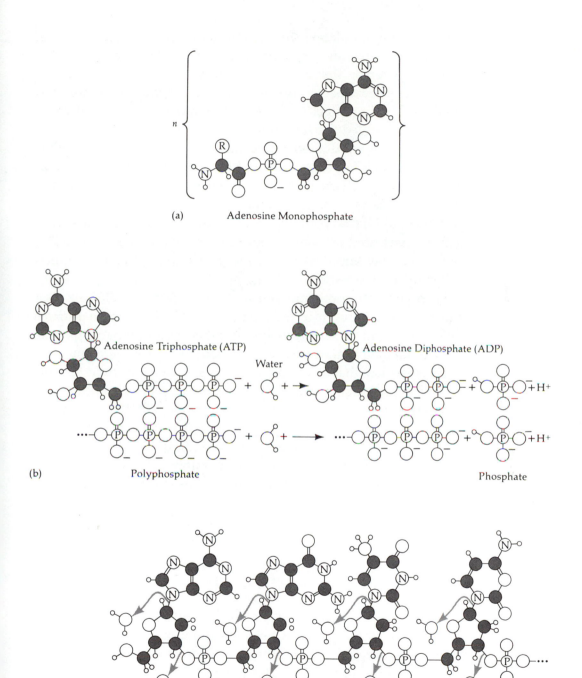

(a)   Adenosine Monophosphate

(b)   Adenosine Triphosphate (ATP)   Water   Adenosine Diphosphate (ADP)

Polyphosphate   Phosphate

(c)   Deoxyribonucleic Acid (DNA)

take over the population, just as resistant cells take over bacterial populations that are exposed to antibiotics. By analyzing the mutant systems, the investigators learned that the successful mutations were those occurring at the exact spot where ethidium bromide became attached to the DNA molecule and that these mutations enabled the molecule to repel the poison.

This experiment demonstrates that, once a prebiotic system has acquired the ability to synthesize functional enzymes and to replicate its ordered structure, its further evolution to form cells of various kinds is possible. Before life on earth originated, this process could have taken place via population–environment interactions, mediated by natural selection, in a manner similar to that which we now know takes place regularly in populations of one-celled microorganisms.

We do not yet know how self-replicating systems retained their integrity and were protected from destruction by external forces before they had evolved the capacity to form the complex cell walls or external membranes we see in all modern microorganisms. One might speculate that abiotically formed, inactive proteinoid spheres existed in enormous numbers in the primitive soup and that self-replicating systems may have used them as protective "homes," much as a hermit crab makes use of a snail shell. Biochemist Sidney Fox has shown that under certain conditions such spheres can be produced in the laboratory in large numbers (Figure 6-1). This, however, is only one of several possibilities still to be explored.

FURTHER EVOLUTION OF PRIMITIVE CELLS

For 2 billion years after the first cells evolved—a period almost three times as long as that between the appearance of the first many-celled organisms and the rise of humanity—evolution was confined to the production of a great variety of simple cells like those of modern bacteria (Figure 6-3). A steady progression of evolutionary change during this period is almost impossible to

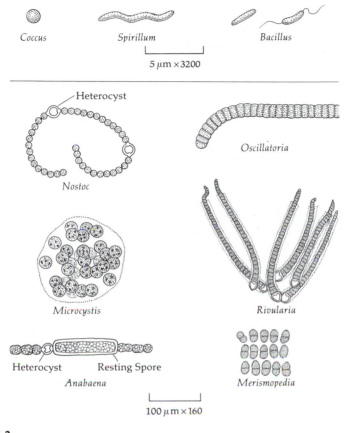

Figure 6-3.

Bacterial cells, showing the diversity of form among various types. (From *Biology*, Third Ed., by G. Hardin and C. Bajema. W. H. Freeman and Company. Copyright © 1978.)

imagine. A stable state with little or no evolution must have lasted for millions of years, with rare episodes of more rapid evolution. Such stability is to be expected on the basis of the interaction concept. Most of the interactions must have been between cells and their immediate physical environment, rather than between different kinds of organisms, because interactions between different kinds of free-living bacteria are extremely rare, and any challenges posed by viruses could have been met by relatively few mutations. Nevertheless, a few important trends must have occurred during this period. The first was the gradual disappearance of the organic soup and its replacement by microorganisms.

Biologists are convinced that all contemporary forms of life are descended from a single original population of cells, each with proteins and nucleotides made up of the same molecular building blocks, the same genetic code, and similar basic enzyme and coenzyme systems. The possibility that such a complex system could have been duplicated by two separate and independent lines of evolution is so remote that it cannot be seriously considered. Nevertheless, it is equally improbable that during the millions of years of prebiotic evolution only one line was evolving in the direction of a cellular system. It is more likely that the prebiotic soup contained thousands or millions of independently evolving lines, all but one of which became extinct before reaching the cellular stage, or shortly thereafter.

Reasons for such widespread extinction are not hard to guess. Cellular organisms live within a fortress formed by the cell wall, which is impervious to most noncellular systems and from which they can excrete enzymes that digest the surrounding organic matter. It has been suggested that some viruses known as bacteriophages, which live as parasites on bacteria, are descendants of primeval precellular systems; most molecular biologists and virologists, however, reject this notion. More probably, after several million years of existence, cellular organisms had digested and absorbed all the remaining organic soup, thus destroying and absorbing the organic resource on which their own lives depended. This destruction might have greatly reduced emerging life, if the trend to photosynthesis had not been occurring at the same time.

Modern life is maintained by the continual activity of green plants converting solar energy into organic food that animals can eat and bacteria can digest and absorb. The earliest photosynthesizers were probably certain kinds of bacteria that still can be found in shallow, muddy ponds and in the deep layers of freshwater lakes. Their photosynthesis differs from that of green plants in that it does not require oxygen, a gas that was rare or absent from the early atmosphere. In other respects, particularly the central activity of the highly complex, light-absorbing molecule *chlorophyll*, as well as a series of elaborate electron transfers and activities of various enzymes and other cofactors, bacterial photosynthesis is much like that of higher plants. While most cells were exhausting the supply of organic matter, others were evolving methods for its replacement. All this probably happened gradually, between 2.5 billion and 1.5 billion years ago.

The appearance of oxygen in appreciable amounts was another of the earliest environmental challenges that organisms faced. Too much oxygen poisons and destroys a living system. Many kinds of bacteria were undoubtedly destroyed by its action, while others persisted in deep pools, mud, and other places that oxygen could not penetrate. Still other organisms evolved the complex system of chemical reactions we call chemical respiration or combustion, by which oxygen is used to break down organic molecules and release large amounts of energy. This respiration, called *aerobic*, is far more efficient than any system of chemical reactions that does not use oxygen, so that aerobic organisms grow many times more rapidly than anaerobic ones. As cells used oxygen with increasing efficiency, they proliferated with explosive speed and soon took over all the habitats on the earth in which oxygen was available.

Among photosynthesizers, primitive anaerobic bacteria gave way to cells that use oxygen for many cellular activities and carry out a modern kind of photosynthesis, in which light energy is transformed into chemical energy by the splitting of water (H_2O) molecules into their component hydrogen and oxygen atoms. This process supplies electrons that can drive specific chemical reactions, particularly the formation of compounds such as ATP (*adenosine triphosphate*), which, because they contain phosphate groups, can release a large amount of chemical energy. Botanists have called these organisms blue-green algae, and they are regarded as a kind of aquatic plant; but we now know that, with respect to both the fine structure of their cells and various biochemical properties, they are completely different from other algae and, except for the possession of photosynthesis, almost identical with certain aerobic bacteria. The name *blue-green bacteria* is now gaining favor and will be used for them in this book.

More than 2 billion years ago, such blue-green bacteria evolved the ability to live in colonies, around which they built protective layers of a hard, limey coating. These layered structures, known as *stromatolites*, still house living blue-green bacteria. Their abundance as ancient fossils shows that ancestral blue-greens formed a dominant component of the earth's biota thousands of millions of years ago.

Until recently, all bacteria were thought to be rather closely related, and bacterial evolution was supposed to be a succession of physiological and biochemical radiations adapting bacteria to life in various kinds of substrates under various regimes of temperature and moisture. During the past few years, however, attention has been focused on a small group known as methane-producing bacteria that contains only three genera and a handful of species. These microorganisms live in mud rich in organic matter, in sludge that contains fecal matter, and less commonly in the guts of higher animals. Among the most strictly anaerobic of any organisms, such bacteria are destroyed by contact with oxygen. They subsist on small carbon-containing molecules of the simplest possible sort—carbon dioxide, acetate, and methanol—which they convert into combustible methane gas.

The first inkling that these microorganisms are not ordinary bacteria came when analysis showed that the cell walls were totally different in chemical composition from those of other bacteria (including the blue-green bacteria). Molecular evolutionist C. R. Woese and his associates compared nucleotide sequences of ribosomal (16S) RNA in a great variety of organisms and found that, with respect to the order of nucleotides in their sequences, all forms of life can be neatly separated into three groups. The first contains amoeba and other protozoa as well as multicellular animals and plants; the second contains nearly all bacteria, including blue-greens; and the third contains only the methane-producing bacteria. Spurred on by this discovery, Woese's group identified additional differences in cellular processes and concluded that methane bacteria form a kingdom of organisms by themselves and are completely different from true bacteria, protozoa, animals, plants, and fungi.

What is the evolutionary position of these strange microorganisms? In this connection, their environment is well worth noting. Sludge and muck that is without oxygen and contains partly degraded fecal matter is much like the hypothetical organic soup on which the earliest organisms are believed to have subsisted. The entire course of evolution from organic molecules to the earliest cells may have taken place in clay-bearing mud at the bottom of shallow pools. It may be that methane bacteria are the scarcely altered descendants of the earliest forms of cellular life, which have avoided all environmental challenges and thus over an enormous stretch of time have evolved little or not at all.

Paleontologist Elso Barghoorn has found deposits in the Fig Tree Chert of South Africa containing concentrated masses of thick-walled cells about 3.5 billion years old. Perhaps a method can be found for isolating the walls of these cells, analyzing them chemically, and comparing them with cell walls isolated from bacteria found in somewhat more recent but still ancient rocks laid down in an environment containing oxygen.

The relationships between methane bacteria, true bacteria, protozoa, and higher animals or plants have been deduced by comparing the chemical structure (nucleotide sequence) of a small but highly significant portion of their RNA (16S RNA) that is transcribed from their genes. This comparison shows that methane bacteria are as different from true bacteria as the latter are from higher organisms, which in turn are also very different from methane bacteria. These three modern groups therefore form an equilateral triangle of relationships rather than a linear series. The most acceptable arrangement, therefore, is to regard each of these groups as a separate early branching from primeval cellular organisms. True bacteria are not ancestral to protozoa and multicellular organisms. They form a separate branch of the evolutionary tree that diversified greatly in its own peculiar way and gave rise to a whole host of microorganisms with comparatively simple cellular structures but highly diversified, complex, and specialized physiology and biochemistry. During most of the Age of Microorganisms, which lasted from 3.5 billion to about 800 million years ago, the living world was dominated by true bacteria, including the blue-greens. As we have seen, for much of this time the ancestors of the protozoa must have occupied a subordinate position, perhaps in the recesses of deep pools or in the crevices of mud.

EVOLUTION OF CELL NUCLEI, ORGANELLES, MITOSIS, AND MEIOSIS

Toward the end of the Age of Microorganisms—about 1.5 billion to 1.4 billion years ago—a totally new kind of one-celled animalcule evolved and soon dominated both the oceans and the fresh waters on all continents. These organisms have much larger cells than those of most bacteria and are enormously diverse in outward form.

Living cells are constructed according to two basically different patterns—*prokaryotic* and *eukaryotic*. Their differences are illustrated in the two parts of Figure 6-4. Bacterial cells have the prokaryotic pattern. Those of all other organisms, including amoeba and other protozoa, animals, plants, and fungi, have the much more complex eukaryotic pattern. By themselves, even these profound differences tell only part of the story. Most prokaryotic cells are several times smaller than the average eukaryotic cell. More important, cells of bacterial prokaryotes are much alike in shape and structure, and, when active, each cell performs most of the functions of the organism. By contrast, eukaryotic cells of protozoa and related forms possess an enormous range of structural variability. The different cells of a multicellular higher animal or plant, also eukaryotic, exhibit an even greater amount of variability (Figure 6-5). The evolution of the eukaryotic pattern opened the door not only to great diversification among single-celled microorganisms but also to division of labor among the cells and organs of larger multicellular organisms. The origin of eukaryotic cells, therefore, is one of the four or five major steps of transcendent novelty in the entire course of biological evolution, as was discussed in Chapter Five.

One important characteristic of eukaryotic cells is that within their outer cell walls they contain several semi-independent cell-

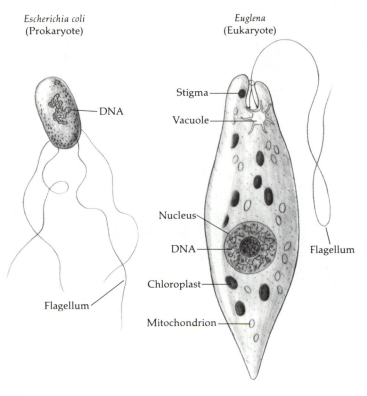

Escherichia coli
(Prokaryote)

Euglena
(Eukaryote)

Stigma

Vacuole

DNA

Nucleus

DNA

Flagellum

Chloroplast

Flagellum

Mitochondrion

Figure 6-4.

Comparison between a prokaryote cell (*Escherichia coli*) and a relatively simple eukaryote cell (*Euglena*). The latter is larger and possesses organelles and details of fine structure that are lacking in prokaryotes. (From J. W. Schopf, "The Evolution of the Earliest Cells." Copyright © 1978 by Scientific American, Inc. All rights reserved.)

like structures known as *organelles*, which can reproduce themselves and multiply within the confines of a single cell. Each organelle performs a particular kind of chemical work essential for maintaining the cell as a whole. Their ability to reproduce is based partly on the fact that each of them has its own particular double helix of DNA, completely separate and containing different information from the bulk of DNA that is enclosed in the nucleus. The most commonly present organelles are *mitochondria*—engines spe-

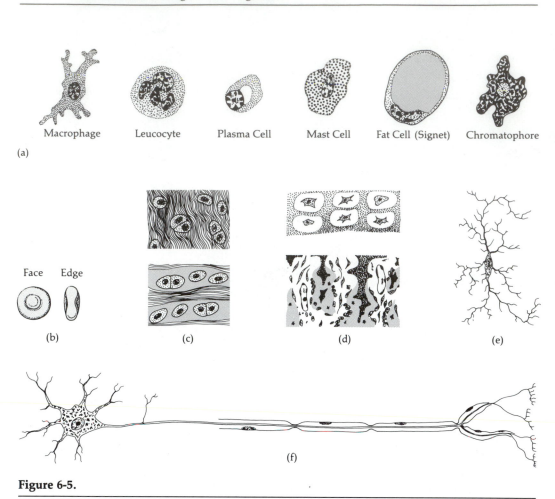

Macrophage   Leucocyte   Plasma Cell   Mast Cell   Fat Cell (Signet)   Chromatophore

(a)

Face   Edge

(b)   (c)   (d)   (e)

(f)

Figure 6-5.

Some of the 250 different kinds of cells found in the human body. (a) Various cells of loose connective tissue. (b) A red blood cell that, when mature, lacks a nucleus. (c) Two kinds of cartilage cells embedded in fibers. (d) Bone cells, which degenerate as the bone becomes hard. (e) A supporting cell from the neuroglia associated with the gray matter of the brain. (f) A nerve cell. Its central part, long and threadlike, is surrounded by nucleated cells of the myelin sheath. The nuclei of all these cells contain the same complement of genes. They become different because in each of them a particular sample of genes is activated to transcribe RNA, which is translated via the genetic code to form the proteins necessary for the construction and activity of that particular kind of cell. The mechanisms that govern this activation are still unknown. (After L. B. Arey, *Human Histology*, Fourth Ed. Philadelphia: W. B. Saunders, 1974.)

cialized for generating the chemical energy necessary for the work of the cell. Photosynthesizing eukaryotes, such as green single-celled protista, algae, and land plants, also contain *plastids*—organelles containing chlorophyll. The cell contains still other organelles such as vacuoles, Golgi bodies, and lysosomes, and many eukaryotic cells also bear on their outer surfaces whiplash *flagella*, organelles constructed on a much more complex pattern than those of bacteria, which make possible the faster and more elaborate movements of many eukaryotic microorganisms.

Even more important in eukaryotic cells is the nucleus, which encloses the bulk of the DNA within a special membrane. This DNA exists in several chromosomes that undergo mitosis, as described in Chapter Two. By contrast, the DNA of bacteria forms a single gigantic molecular double helix that is in direct contact with the rest of the cell and is attached to the inner surface of its surrounding membrane. Its entire course of reproduction consists of duplication, followed by separation of the daughter double helices as the cell wall to which they are attached continues to grow. The nuclear DNA of eukaryotes, much greater in quantity, is bound up in several distinct chromosomes, the number of which varies greatly, even among different single-celled organisms. In addition, the DNA in eukaryote chromosomes is complexed with an array of several different kinds of proteins, each responsible for one of the several functions that chromosomes perform. Finally, eukaryotes evolved the cycle of gene shuffling that is the genetic basis of sexual reproduction.

In bacteria, genetic recombination is accomplished by intercellular transport of small bits of DNA, sometimes by contact between cells but more often through the intermediary of virus particles. In eukaryotes, recombination results from the union of specialized cells known as gametes, the most familiar of which are eggs and sperm. As we saw in Chapter Two, gametes are formed by a modification of mitosis known as meiosis, in which the number of chromosomes in the cell is reduced by half, to be restored when gametes unite in fertilization. The sexual cycle of meiosis plus fertilization must have evolved along with the evolution of

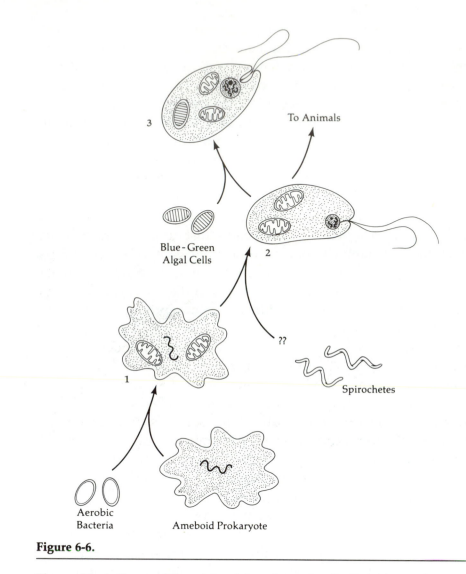

Figure 6-6.

The symbiosis theory of the origin of the eukaryotic cell. (1) Addition of
mitochondria from transformed bacteria. (2) Addition of nucleus and flagella,
possibly from transformed spirochetes. (3) Addition of chloroplasts from
transformed blue-green bacteria. (After Margulis, 1970, from *Evolution*, by T.
Dobzhansky et al. W. H. Freeman and Company. Copyright © 1977.)

mitosis. Recent investigations by protozoologists C. A. Beam and M. Himes of aquatic nuclear microorganisms known as dinoflagellates have shown that, although these organisms have a mitotic sequence less elaborate than that of most eukaryotes, and therefore possibly closer to that of primitive organisms, they nevertheless have a meiotic cycle of mitosis and gametic union as well as typical Mendelian segregation and recombination of genes.

Other investigators have shown that primitive single-celled eukaryotes produce chemical sex attractants that help gametes find each other and unite. As the gametic cells come together, the whiplash flagellae that propel them through the water become entwined in a firm embrace that lasts until the two cells and their contents are fully united.

How and where did this vitally important evolutionary advance take place? Under what conditions would more complex cell structures have had an adaptive advantage over the simpler ones of bacteria? Because details of cell structure are not preserved in fossils, answers to these questions must always be confined to plausible speculations.

There is growing acceptance for the idea that the more complex eukaryotic cells are derived from aggregates of simpler cellular organisms. At least two kinds of organelles—plastids and mitochondria—are derived from entire prokaryotic cells that were originally *ingested* by primitive amoebalike cells and altered as needed to serve the cell in which they found themselves (Figure 6-6). Biochemical evidence for the *symbiotic* origin of plastids is now overwhelmingly strong, and mitochondria are so much like aerobic true bacteria in several features of structure and biochemistry that a similar origin for them is also likely. However, a different origin is more likely for the nuclear membrane, the large numbers of chromosomes found in eukaryotic cells, and various other structures that they possess.

Why should cells swallow other cells and subordinate their victims to their own use? To what environmental challenge would organisms respond by evolving the ability to ingest smaller organisms?

A reasonable answer to the latter question is based on the principle of exploiting new environmental niches and creating new evolutionary strategies. After prokaryotic organisms had existed on the earth for about 2 billion years, they had occupied virtually every habitat available to them, in every imaginable temperature range. A new niche would have to involve a new kind of behavior and evoke a new evolutionary strategy. Only organisms responding to such challenges could replace any of the all-pervasive array of prokaryotes.

The shift from absorption to ingestion of food would have been difficult or impossible in most of the environments that supported early microorganisms. Bacteria, which are surrounded by thick, rigid walls, are incapable of ingesting anything. If these walls are removed, and the bacteria are placed in the usual liquid medium, osmotic conditions within the cell are such that water continually enters it, causing it to swell and burst. This destruction can be prevented only if the surrounding medium contains so many particles of various sorts dissolved or suspended in it that it is nearly as dense as the contents of the cell.

Perhaps, therefore, the ancestors of the eukaryotes lived in the bottoms of tide pools along the seashore or favorable sites on the ocean floor. The sea water already had a higher osmotic value than fresh water and was being constantly enriched by the remains of dead organisms that sank downward from higher water levels. At first, primitive eukaryotes ingested bits of organic matter and small living cells only to obtain food. Once ingestion had become well established as a way of life, some of these amoebalike microorganisms kept within themselves entire cells without digesting them. The imprisoned cells retained their original vital functions, even though their structure became modified in adaptation to life in the new host. If, for example, photosynthetic blue-green bacteria were ingested, their cells became modified into engines for photosynthesis known as plastids, which enabled the host to manufacture food in the presence of light. Ingested bacteria that had been using oxygen to break down and ingest absorbed food were converted into specialized engines that could generate chemical energy by breaking down ingested food. These engines, known as mitochondria, exist in all eukaryotic cells. Some cells contain only one or a few of them, but others contain hundreds.

These newly evolved eukaryotic organisms, which consisted of cells within cells, would have greatly increased their opportunities to occupy new ecological niches. Those that had converted photosynthetic cells into plastids could rise to the surface of the water and carry on photosynthesis on a more efficient and larger scale than could blue-green bacteria. From such organisms evolved various kinds of larger plants. Other primitive eukaryotes that contained mitochondria but no plastids developed more efficient ways of ingesting food, evolving for this purpose new and more efficient ways of moving around. Some of them became larger and evolved *cellular skeletons*—a network of strands or fibrils made of a kind of protein known as *tubulin* (see frontispiece for this chapter and Figure 6-7). Groups of tubulin fibers, arranged into long, thin-walled tubes, are known as *microtubules*. Once their cellular skeleton had been well developed, eukaryotic cells with thin, flexible external cell membranes could leave the osmotically rich bottoms of pools and exploit open salt water or fresh water without danger of bursting. The word *plankton* categorizes the organisms of various shapes and sizes and belonging to several different groups of animals, plants, and micororganisms that float near the surface of the ocean. In modern oceans, plankton is the principal source of food for many larger animals, particularly fishes.

THE DIVERSITY OF UNICELLULAR EUKARYOTES

The illustration of unicellular eukaryotes in Figure 5-7 gives some idea of the great diversity of these microorganisms. The most important differences between them relate to the presence or absence of chlorophyll and their capacity for photosynthesis; in the plastids, the chemical composition of the accessory pigments associated with chlorophyll; and the kind of body plan and external organelles, flagella, or cilia associated with the microorganisms' mobility. Although these differences have been observed

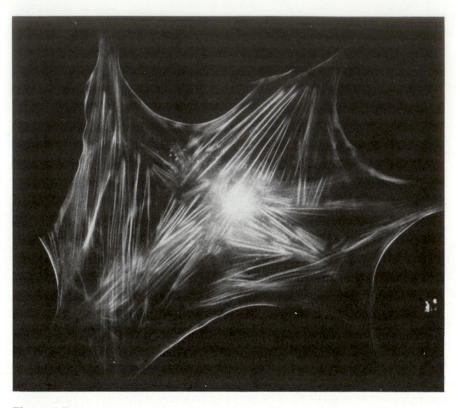

Figure 6-7.

Tubulin fibers that form part of the cytoskeleton of a eukaryotic cell. (Photo by E. Lazarides.)

only in modern forms, there is every reason to believe that counterparts of both modern photosynthetic and nonphotosynthetic unicellular eukaryotes did exist.

After the first eukaryotic cells evolved, about 1.4 billion years ago, the earth was dominated for about 600 million years by single-celled microorganisms. Based on comparisons between different modern single-celled eukaryotes, one can infer that some of these ancient microorganisms possessed chlorophyll and were photosynthetic, others were colorless and absorbed or ingested their food, and still others—like the modern *Euglena* (Figure 6-4)—could either manufacture food by photosynthesis or ingest it. The classification of these modern forms has been difficult. Microorganisms that are *both* photosynthetic and motile are regularly described and classified by botanists as plants (algae) and by zoologists as animals (protozoa).

Three major groups or classes of unicellular eukaryotes can be recognized on the basis of size, form, and nature of motility. The *flagellates* have relatively small cells (Figure 5-7) that move through water by means of two or more whiplash organelles. Several orders of photosynthetic flagellates (sometimes recognized as classes, phyla, or divisions) are distinguished from each other on the basis of the position and mode of action of their flagella as well as the color and chemical composition of the accessory pigments that accompany the chlorophyll in their plastids. Another basis of difference is the chemical composition of their cell walls. Some photosynthetic flagellates are almost identical with motile reproductive cells produced by brown algae or seaweeds; they may be the ancestors of these seaweeds. Other flagellates, green in color, are much like the ancient ancestors of green seaweeds, pond scum, and multicellular land plants. Nonphotosynthetic flagellates include free-living cells that resemble certain flagellated cells found in sponges—the simplest of multicellular animals—and they may well be ancestors of some animal phyla. Flagellates are thus the almost unchanged descendants of the microorganisms that gave rise to many phyla of multicellular animals and plants.

The *ciliates* all lack chlorophyll and feed on smaller live microorganisms or on their dead remains. Most ciliates are much larger than flagellates. Their cell surfaces bear many cilia that closely resemble flagella in their fine structure and are very similar in their rhythmic motion. Their name came from the Latin word for eyelid (*cilium*). Among the larger ciliates, some cilia have been modified to perform different functions—for example, to produce a whirlpool current of water that sucks food through an opening in a cell wall. In some ciliates that live on the bottom of pools, clusters of much-thickened cilia are used as a kind of feet to move the cell about.

The third group or class of unicellular eukaryotes contains the familiar amoeba as well as the marine animalcules whose cells form chalk. The activity of amoeba and its relatives is governed by their flexible cell walls, which enable them to change their shape rapidly and in a regular, functional way. They move over the bottoms of pools by sending out adhesive projections that pull the cell along. If the cell wall comes in contact with a bacterial cell, it is stimulated to form projections on either side of its prey, thereby engulfing or ingesting it. To the naturalists of the eighteenth century who first studied these organisms, these projections appeared to resemble both feet and the roots of plants, and they named this class of animalcules Rhizopods, meaning "root–feet."

A few microorganisms are halfway between flagellates and Rhizopods. One of them *(Naegleria)* alternates amoeboid cells that creep over the bottom of a pond with flagellated cells that can explore new ponds or pools. In addition, a large class of eukaryotic microorganisms includes parasites on larger animals such as the one that causes malaria *(Plasmodium)*.

THE ORIGIN OF SEX

Why did eukaryotic microorganisms evolve the function of meiosis and thereby make necessary the equally complex and hazardous union of gametic cells called fertilization, while others, which failed to evolve sexual reproduction, retained their original adaptations to tidepools or other habitats and persisted with little change?

Biologist G. C. Williams has postulated that sex first originated as an adaptation for exploring new and somewhat different habitats. He points out that primitive forms like coral polyps and sponges possess life cycles that adapt them to both stability and change. The same is true of intermediate Rhizopod flagellates such as *Naegleria*. Their existence on a reef or in a pool can be prolonged indefinitely by asexual reproduction, in which cell proliferation by mitotic division gives rise to individuals having exactly the same genes as the original colonizer. Periodically, the corals and sponges undergo meiosis, forming gametes with genetic constitutions different from each other and from those of the parental cells. Unions of these gametes can provide the fertilized egg with genetic equipment for colonizing a reef or pool that has an environment somewhat different from the parental one. Such exploration is extremely wasteful because there is little likelihood that new gene combinations will find a new and congenial habitat. Nevertheless, it succeeds in increasing and spreading the parental genes because the parents can generate thousands or millions of gametes with little cost to themselves. As environments change more drastically or as new and more different environments open up, asexual reproduction in an established habitat plus the evolutionary opportunities provided by exploring gametes and fertilized eggs would be the most likely sources of further evolutionary change and diversification. The environmental challenge that originally made sexual reproduction highly adaptive may well have been the near saturation of the world's oceans, lakes, and rivers with microorganisms. This saturation greatly increased the adaptive value of exploring pioneers that could find new and different habitats to conquer.

GENETICS AND THE ORIGIN OF CELLULAR DIVERSITY

As compared to prokaryotes, eukaryotes have a larger content of DNA and much larger, more complex cells that require more genes to code for necessary proteins. Nevertheless, at any given stage of its development or activity, a eukaryotic cell uses only a part of its genes. Elaborate structures like the organelles found in the larger protozoa are built up through a succession of gene activations and repressions. Although geneticists do not know how many such steps are needed to construct a protozoan cell, some idea is obtained by extrapolating from the developmental genetics of well-known multicellular organisms, particularly flies. Many fly cells possess giant chromosomes in which the activation of particular genes can be followed by radioactive labeling of the RNA as it is being reproduced. Geneticist D. Ribbert found that, during the development of a single large bristle cell in the fly genus *Calliphora*,

101 different genes are successively activated and repressed. Stage-specific division of labor is characteristic of gene activity in both unicellular and multicellular eukaryotes. The regulatory mechanisms that determine the actual program of activation are still largely unknown, but strong evidence indicates that they are somehow associated with particular proteins among the large number that envelope the chromosomal DNA. In multicellular eukaryotes, and probably also in many unicellular forms, hormones and similar growth-regulating substances interact with these proteins to organize the developmental program.

One of the greatest tasks for evolutionists of the future is to describe and explain evolutionary change in terms of alterations in patterns of development. We recognize that genes differ with respect to their DNA structure, the kinds of mutations that can alter them, and the relative value of these different mutations. Recent research on the structure of DNA within the nuclei of eukaryotic cells has shown that it is far more complex than the DNA of bacteria and viruses. In early eukaryotes, both DNA and the structures for which it codes evolved a far greater degree of complexity than exists in any prokaryotic cell.

Once this dual complexity had evolved, the way was open for the evolution of organisms having many different kinds of cells that are specialized for performing an equal number of different functions. This is the distinctive feature of larger plants and animals, which are discussed in the next three chapters.

A redwood forest on the California coast, Humboldt Redwoods State Park. Here the integrated biotic community depends largely on the photosynthesis and the protection of the giant trees that overshadow it. The tallest trees may be as much as 2500 years old, but they in turn have sprung from sprouts from more ancient stumps that are essentially immortal. (Photograph by David Swanlund. Courtesy of Save-the-Redwoods League.)

Chapter Seven

PLANT EVOLUTION

Plants form the most significant part of the environment in which animals live and on which they depend. Because plants can convert solar energy into food by photosynthesis, they are the ultimate providers of all the energy that animals need for growth and activity. In the ocean, seaweeds (algae) and the most primitive phyla of animals coevolved. By the time animals took the first steps toward inhabiting the land, its surface was already covered with plant life. Reptiles, insects, and spiders—the first animals that lived a completely terrestrial life—evolved with the first forests. The most active of the larger animals—the warm-blooded birds and mammals—did not become common and dominant until the most modern and advanced plants, the flowering plants (or angiosperms), were already widespread and well diversified. Consequently, throughout the evolution of plants and animals, which has lasted for hundreds of millions of years, plants led the procession and animals followed.

Knowledge of plants and their evolution gives us a new outlook on some of the most fundamental life processes. For example, people who are aware only of the life cycle of animals tend to think of death as the necessary fate of every living individual. A well-trained botanist realizes that for some plants death is not inevitable.

Most people are familiar with trees that are much older than any animals; both the California Big Tree and the bristlecone pines that grow on the desert mountains of California and Nevada are often called "the oldest living things." But some smaller plants have even longer lives. One example is the buffalo grass that grows on the dry western prairies of the United States. When a seed of this grass sprouts, it forms first a small clump and later spreads by means of stems that creep horizontally along the ground. A rough estimate of the age of a patch that has grown from a single seed can be made by measuring the size of the patch and determining how far the horizontal stems grow in a single season. According to botanist Agnes Chase, some patches of buffalo grass are probably single plants that are at least 10,000 years old, grown from seeds that sprouted at the end of the Ice Age, when the climate of the plains first became favorable for grasses that require warm summers.

Sod-forming plants, as well as bulb-forming plants like daffodils and tulips, can only die as a result of diseases caused by viruses, bacteria, and molds or because of their inability to adapt to new climatic conditions. It is conceivable that plants of this kind, living in disease-free tropical regions, could have been alive as genetic individuals for hundreds of thousands or even millions of years. (I admit that this is pure speculation—there is no way of proving or disproving such a statement.)

Why is relatively early death inevitable for animals? The answer lies in the growth process. An animal starts as a fertilized egg and then becomes an embryo, in which all cells are dividing and differentiating almost simultaneously into different kinds of tissues. As the embryo grows into a juvenile and finally an adult, the differentiation of tissues and organs becomes complete, and no body cells remain that have the capacity to differentiate further. When the individual has reached middle age, or even before, the capacity for replacing worn-out cells and tissues declines; the body's organ reserve declines, and old age, then death, are inevitable. This process is often hastened by the inability of the blood to provide tissues with sufficient nourishment. Finally, complex living systems are to some extent victims of the struggle for survival. When cells are removed from an animal body and cultured as tissues in the laboratory, they live many years longer than the animal from which they were taken. Old age and death are the prices that animals and humans pay for the vitality, integrated activity, and mobility they enjoy in life.

Plants grow in an entirely different fashion. As soon as a seedling emerges from the parental seed, some of its cells cease dividing and differentiate into functional tissues that either perform photosynthesis or support and protect the emerging plantlet. As the seedling grows into an adult, embryonic tissue reamins at the tips of its stems and continually differentiates into leaf, stem, and finally reproductive tissue (Figure 7-1). Even the oldest of the giant sequoias, bristlecone pines, and sod grasses carry at the tips of their stems hundred of bits of embryonic tissue that can differentiate into new branches or even new plants. Fruit growers, vintners, and other cultivators take advantage of this capacity of plants to regenerate, using young plant stem tips as grafts to improve other plants. Thus many plants are perpetuated almost indefinitely.

Why should plants and animals grow in such completely different fashions? Each pattern of growth and development is the most efficient one for the adult life of the organism. In order to detect, approach, and ingest their food, animals must be active, aware, and responsive to their immediate environment. Natural selection acts continually to establish and maintain gene complexes that favor these capacities. As animals become larger and more active, organs and organ systems that maintain these capacities necessarily become more complex. But whether large or small, animals function efficiently only if their entire bodies are compact and precisely symmetrical. Thus, their pattern of embryonic growth and differentiation is the one most efficient for acquiring these capacities.

Plants, however, get their food by using solar energy to convert simple chemical compounds—water and carbon dioxide—into complex sugar, starch, and fat (lipid) molecules to serve as food. For this function, a sedentary life is most efficient. As the plant gets larger, surface area increases in proportion to size, and grow-

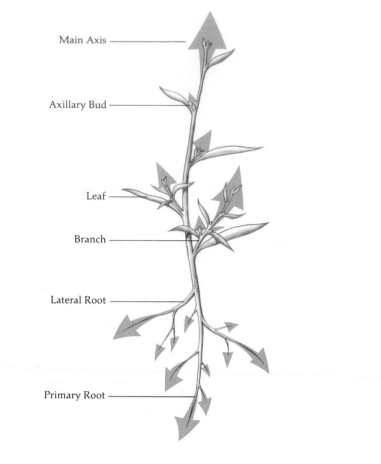

Main Axis ─────────

Axillary Bud ─────────

Leaf ─────

Branch ─────────

Lateral Root ─────────

Primary Root ─────────

Figure 7-1.

Growing regions of a plant. (From *Comparative Morphology of Vascular Plants*, by A. S. Foster and E. M. Gifford. W. H. Freeman and Company. Copyright © 1959.)

ing leaf surface increases the area over which light for photosynthesis strikes the plant. Longer stems elevate the leaves above lower growth and closer to the light. Increased root surface enlarges the area of soil from which water can be extracted. As the plant's surface area grows, its overall symmetry becomes irregular, and precise symmetry becomes confined to individual leaves, stems, and reproductive organs.

An exception to these rules is the coral, so sedentary an animal that it resembles a plant. Coral polyps form the great reefs so familiar to those who visit tropical seas. Anchored to the rock that forms the base of their reef, they wait for ocean waves to transport to their mouths millions of tiny animalcules that float or swim in the warm ocean. Their success depends on the ability of the coral colony to cover as large a surface of the reef as possible. Corals do this by growing horizontally and by retaining within each polyp embryonic cells that can develop into new polyps beside the original one, just as sod grasses spread by means of horizontal stems. Every coral reef is made up of patches or groups consisting of thousands or millions of polyps that are descended from a single fertilized egg. The colony grows by continually budding off new polyps. Theoretically, the genetic individual of a coral animal is as immortal as is that of a sod-forming grass. Corals have lived and evolved for hundreds of millions of years without acquiring the integrated movement and vitality that other animals possess, and so they have not been committed to paying the ultimate price of inevitable decay and death. Their relatives, the free-floating jellyfishes, are as subject to death and decay as any other kind of animal.

Another difference between plants and animals is the greater simplicity of tissues and organs in plants and the greater similarity in basic structure between functionally different parts of the same plant. Most of the organs, tissues, and cells found in the head of an animal are constructed quite differently from those found, for example, in the body or the feet. But the twigs at the top of a tree are quite similar in construction to the main trunk, and leaves are essentially alike no matter where they are located. Even the roots contain tissues essentially like those of the trunk, differing only in the arrangement of the different kinds of cells.

In animals, the most complex pattern of differentiation is that which forms the various parts of its head. The organs thus formed are essential for perceiving and for ingesting food. Evolutionary advancement in animals affects the head region more than any other part of the body. In plants, however, evolutionary advancement is measured chiefly in terms of greater complexity of the reproductive structures. After the initial evolution of woody supporting and conducting tissues that enabled plants to become large while living in terrestrial habitats, the most conspicuous new features were seeds and flowers of increasing complexity. These structures are not essential for the survival of an individual plant but are essential for efficient reproduction. In animals, increases in complexity of reproductive systems are overshadowed by far greater increases in complexity of the entire body. In even the most complex plants, such as orchids and daisies, the gene-controlled pattern of development is far less complex than in most animals. This degree of complexity can be roughly measured by determining the number of different kinds of cells that are differentiated during development from a single fertilized egg. In the simplest threadlike water plants—the green algae—this number can be as low as four or five, including the sex cells. In simple animals (*Hydra*) related to corals, it is about 25, being comparable to those formed by spore-bearing ferns. The largest number of different kinds of cells that any plant embryo can differentiate is about 50, being comparable to the capacity for differentiation by the embryo of a simple worm such as the flatworm (*Planaria*). Earthworms consist of about 66 different kinds of cells; insects, between 100 and 150; and mammals such as ourselves consist of about 250 different kinds of cells. This means that developmental sequences must be two to three times as complex in insects as in plants, and four to five times as complex in mammals.

PLANT EVOLUTION AT THE LEVEL OF POPULATIONS AND SPECIES

Divergence of populations and species from each other is much the same in plants as in animals. Based on data from enzyme differences, the estimated amount of genetic variability present in populations of outcrossing plants is about the same as that in land snails, greater than that in most insects, and more than twice as much as that in birds and mammals. This diversity is maintained, however, only if the plant's genetic properties enforce mating or pollination between genetically different individuals. In many plants, however, male and female organs (*stamens* and *ovaries*) are found on the same flower, and in some of these, pollen can fertilize eggs belonging to the same individual plant. In populations of plants that are regularly self-fertilizing, genetic variability within populations may be reduced to a quantity lower than that found in any sexually reproducing animal species. Among plants as a whole, therefore, a spectrum of variabilities within populations exists that is quite comparable to that found in animals. Consequently, plants are capable of evolutionary responses to challenges of new environments to much the same degree as animals are.

This similarity is reflected in the patterns of racial differentiation found in animal and plant species distributed through comparable ranges of habitats. Among the animals of North America, such as foxes, deer, forest mice, song sparrows, and swallowtail butterflies, the most widespread species are differentiated into similar series of races, each race being adapted to a different geographical region and habitat. The same is true of pine trees, wild cherry bushes, herbs such as the sticky cinquefoil, and grasses such as the prairie bluestems. A difference exists, however, with respect to the distinctness of species. In most animal groups, physiological characteristics, patterns of behavior, and biochemical nature clearly indicate whether an individual belongs to one or another of the recognized species. This is true even in genera like *Drosophila*, which contains more than a thousand species. As we have noted, in places where two related species occur together, they usually maintain their identity with great fidelity. Hybrids between them are relatively rare and are highly sterile. Backcrosses between hybrids and individuals belonging to either of the parental populations are so uncommon that they have little effect on the pattern of variability.

In some plant groups, the situation is similar. Plant genera that contain clearly defined species are most often found among annuals that live for only a single season and die after they have completed one cycle of reproduction. In longer-lived plants—for example, oaks and willows—boundaries between species are less well defined. Related species such as the yellow-barked and scarlet oaks (*Quercus velutina* and *Q. coccinea*) in the eastern United States, as well as coastal and inland live oaks (*Q. agrifolia* and *Q. Wislizenii*) in California, have widely overlapping geographical ranges and occur together in many places. In some places their populations exist side by side and remain completely distinct from each other, whereas elsewhere the limits of the species are difficult to determine because many individuals are intermediate hybrids.

Two other characteristics related to mobility set animals apart from plants. These are the elaborate courtship patterns in many animal species, impossible in sedentary plants, and the more complex, integrated patterns of development found in most animals, particularly joint-legged animals (arthropods) and vertebrates.

Imperfectly developed reproductive barriers between species are most common in trees and shrubs with large, heavy seeds that are unlikely to be transported for long distances. Here, the chances are minimal that a new population will become established far from the parental population and subjected to strong divergent selective pressures—in other words, to the conditions most likely to bring about the origin of new species. Fast-growing annual plants are likely to migrate passively through seed transport by wind or animals. Most animals move around in search of food but remain in a restricted territory for reproduction, while most annual plants can establish themselves in new habitats with relative ease. It is not surprising, therefore, that in plants of this kind, species become nearly or quite as distinct as they are in animals.

Reproductive isolation between species of animals is often strongly reinforced by complex species-specific behavior patterns. Consciously or instinctively, a female chooses between several males that are courting her, generally selects one that belongs to her species, and performs whatever courtship rite is appropriate. Plants, however, lack anything that resembles a nervous system and so are incapable of discriminating between individual neighboring plants that might provide pollen for them. It is true that in some of the most complex flowering plants like orchids, flowers belonging to different species are so constructed that a visiting pollinator such as a bee or fly either avoids visiting successively neighboring plants that belong to different but related species or, if it does visit them, is unable to deposit pollen grains in the right position to deliver the male nuclei to the position of the egg cell. As a result, one often finds in the same locality several different but related species of orchids growing side by side but remaining sharply distinct from each other. This kind of reinforcement of the reproductive isolating barriers between plant species is, however, relatively uncommon.

The greater complexity of animal development is responsible for the fact that animal hybrids are much more likely to die as embryos or juveniles than are plant hybrids. Insect geneticist C. G. Oliver has recently found that, within a single species of butterfly, the Crescent (*Phyciodes tharos*), races exist that are indistinguishable in outward appearance and even with respect to electrophoretic mobility of their enzymes. Nevertheless, when different races are intercrossed in either direction, many of the fertilized eggs produced by backcrossing a hybrid to one of its parents produce embryos that die during early stages of development. These deaths are attributed to clashes between the action of regulatory genes that control different patterns of development. Situations like this do exist among hybrids between plant species, but they are much less common than in animals.

Although plant populations tend not to become differentiated into sharply defined species, evolutionary lines above the species level can and do diverge. But the simpler construction of plants restricts the ways in which they can adjust to the challenge of a new environment. For this reason, many plants with different origins and relationships bear a superficial resemblance to each other. A zoologist can analyze carefully the structure and development of a single organ, such as the antenna of an insect, and determine the evolutionary relationships of the species under study. In plants, this is much more difficult. As a result, during the nineteenth century, botanists concluded that the evolution of the *corolla*—a bell-shaped, saucer-shaped, or tubular structure of the showy part of the flower—occurred but rarely and that therefore plants with similar corollas must be related to each other. Recent comparisons of genera having flowers with united petals have cast considerable doubt on this conclusion, and scientists now know that the common possession of this or any other superficial characteristic is by itself a poor indicator of evolutionary relationships. In short, *parallel and convergent evolution* have blurred the picture.

THE EARLY DIFFERENTIATION OF MULTICELLULAR PLANTS

Unfortunately, the fossil record of plant evolution is incomplete and hard to interpret, mainly because only three kinds of plant tissue are often preserved—the wood that forms the central part of stems, branches, and roots; the surfaces of leaves; and the surrounding walls of pollen or spores. From the earliest seed plants, seeds and their coverings are well preserved, as are the woody cones of fossil pines and other conifers. In other groups, however, well-preserved reproductive parts are rarely found. Generally speaking, botanists can understand the fossil record of plants only by constant reference to other fossils and to living forms.

In the present account, fossil evidence is used as much as possible, but necessarily many inferences are made on the basis of comparisons between contemporary forms.

Multicellular plants evolved first in water, giving rise to seaweeds (marine algae) and pond scum (freshwater algae). The term *algae* is used by botanists for all aquatic organisms, unicellular or multicellular, that are photosynthetic and do not have the more elaborate reproductive structures found in mosslike plants (Bryophyta), fernlike plants (Pteridophyta or vascular cryptogams), and seed-bearing plants. In this chapter we use the term *algae* only for multicellular forms that are sedentary except for their sex cells.

The evolution of multicellular algae displays the parallelism characteristic of plant evolution (Figure 7-2). This conclusion is inferred from comparisons between various multicellular algae and their nearest unicellular photosynthetic relatives. Three different kinds of characteristics form the principal basis of these comparisons. Most important are the biochemical nature and ultrastructure of their cell contents. Based on the color and chem-

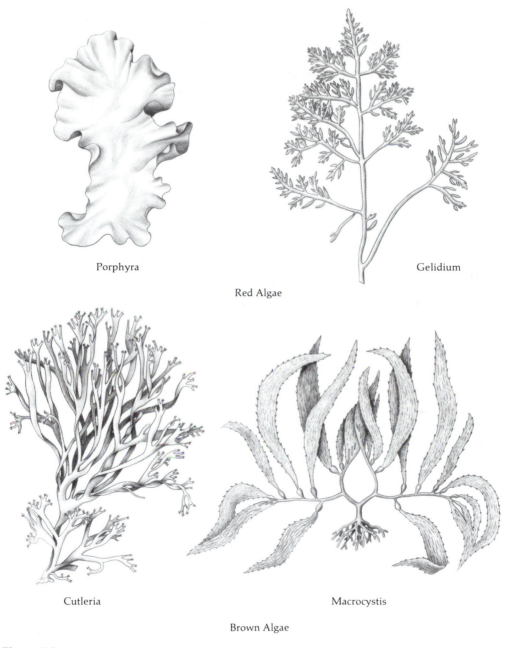

Porphyra

Gelidium

Red Algae

Cutleria

Macrocystis

Brown Algae

Figure 7-2.

Some of the larger marine algae, or seaweeds. (From *Biology*, Third Ed., by G. Hardin and C. Bajema. W. H. Freeman and Company. Copyright © 1978.)

ical nature of the pigments associated with their chlorophyll, three separate divisions or phyla of multicellular algae are recognized—red, brown, and green. These phyla can also be distinguished on the basis of two other characteristics—the chemical composition of their cell walls and the nature and motility of their reproductive cells. Each phylum was derived from a different group of unicellular eukaryotes.

Red algae, chiefly marine, are most common and diversified in warmer oceans. Most are small or medium-sized, thin, and gelatinous in texture, with a delicate, branched structure. Others resemble dwarf corals or corallike incrustations on the rocks of tide pools along the shore. Red algae are distinctive also in the chemistry of their cell walls and in their nonmotile reproductive cells. They completely lack the whiplash flagalla found in the majority of unicellular eukaryotes, which are characteristic of the reproductive cells produced by other kinds of algae.

Brown algae are the commonest seaweeds in relatively cold oceans. Such forms as rockweeds (*Fucus, Ascophyllum*), kelps (*Laminaria, Macrocystis*) and the sea palm (*Postelsia*) of maritime California are familiar sights along the coasts of North America. They are the largest marine plants, sometimes reaching 600 feet in length. Their stems may become differentiated into different kinds of tissues, but they never form woody tissues like those found in the stems of land plants. On the basis of the plastid pigments, cell-wall chemistry, and structure and motility of reproductive cells, brown algae resemble a particular group of unicellular flagellates (Chrysophyceae) so closely that an independent origin from this group can be inferred. Like the red algae, they diversified along their own particular evolutionary lines and did not give rise to any other kinds of plants.

The green algae resemble land plants in having green chlorophyll in their plastids, accompanied by accessory pigments—*carotene* and *xanthophyll*—that are closely similar to the accessory pigments found in the plastids of mosses, ferns, and seed-bearing plants. With respect to cell-wall chemistry, the nature of their reproductive cells, and the fine structure of their mitotic apparatus, green algae are much more diverse than reds and browns. Furthermore, some green algae have cells that each contain only a single nucleus, while others consist of much larger cells, each of which contains a large number of nuclei. The extreme expression of this condition occurs in a group of seaweeds called Siphonales, in which the entire plant consists of a single enormous branching "cell" known as a *coenocyte*. Because of this diversity, some botanists have recently concluded that seven different evolutionary lines of green algae, each of which is homogeneous within itself with respect to the characteristics mentioned above, have evolved separately from unicellular green flagellates, all of which possess carotene and xanthophyll in their plastids. One of these lines is particularly important because it gave rise to land plants.

Green algae differ also from reds and browns in that they are more diversified in fresh water than in salt water. Most of the green seaweeds, like the sea lettuce (*Ulva*), are best developed in tidepools near the level of high tide. Of the seven multicellular lines mentioned above, three (Cladophorales, Siphonales, and Ulotrichales) contain both freshwater and marine forms, while the other four (Chlorococcales, Volvocales, Conjugales, and Oedogoniales) are confined to fresh water.

In addition to the sedentary and photosynthetic multicellular algae, botanists usually accept as plants the fungi, organisms that are sedentary but lack chlorophyll and absorb food either by enzymatic breakdown and absorption of dead organic matter or by becoming parasites of other organisms. Mildews, molds, smuts, rusts, mushrooms, puffballs, and many other colorless plants make up this group. They are not, however, the only colorless plants.

Creeping over other kinds of vegetation, particularly in salt marshes, one often sees slender yellow or orange threadlike stems that fasten onto other kinds of plants and parasitize them by sucking their juice. They belong to the dodder, a parasite that, when ready to reproduce, forms tiny white flowers that resemble in form those of a morning glory. Other kinds of flowering plants, such as the mottled white and black Indian pipes that grow in deep shade of cool northern forests, the bright scarlet snow plant and red-striped sugar stick that are found in the mountain forests of California, and the spectacular ghost orchid found in the forests of the Pacific coast, subsist on decaying leaf mold. A parasite on tree roots found in tropical rainforests of Indonesia, known as *Rafflesia*, has dish-shaped flowers several feet in diameter, the largest flowers known.

Fungi have recently been found to possess chemical characteristics that set them apart from all phyla of algae and higher plants. Most of them have cell walls composed of *chitin*, a substance unknown in plants but characteristic of the skins of insects. With respect to the amino acid sequence of one of their proteins (*cytochrome c*), fungi resemble animals more than they do green plants. The simplest of them are much like colorless unicellular eukaryotes and may have evolved from them. Their independence from the ancestral line and from other plants illustrates parallel evolution in a clear and recognizable form.

HOW PLANTS CONQUERED THE LAND

For plants that spend their lives under water—particularly the relatively calm water of freshwater lakes and slowly running streams—the environmental challenges for adaptation are relatively slight. The slender threads of green algae can become several feet long without evolving cellular structures different from those found in single-celled green microorganisms. Marine algae, most of which live anchored to rocks near the seashore, will survive only if they can withstand the buffeting of the waves. This kind of adaptation is best evolved by acquiring photosynthesizing blades or fronds that have a leathery texture. The larger forms have evolved stems with enough tissue differentiation to make them exceedingly tough and pliant rather then relatively stiff and brittle, as are the branches of most land plants. For all kinds of algae, the problems involved in reproduction and dissemination are far simpler than they are in all but the smallest of land plants. Most of them reproduce by means of motile, flagellated cells known as *zoospores* that are almost identical with the motile swimming cells of the unicellular eukaryotes from which they evolved. For a photosynthetic and aquatic plant, egg cells that store large amounts of food do not provide the same advantage that they do in many-celled animals, in which the young cannot feed until they have become large enough to ingest their food. Consequently, only a few groups of algae have evolved egg cells much larger than the male sex cells or sperms. Fusion of sex cells or fertilization in the water can be achieved by the evolution of powerful chemical attractants that the sex cells detect over long distances. Algae, therefore, need relatively few specialized adaptations for either asexual or sexual reproduction.

For small plants that inhabit the land, environmental challenges are equally modest. They quickly become greater as plants acquire even moderate size, and are tremendous for all kinds of trees. In all except the driest of climates, plants that are thin in texture and hug the ground closely can take advantage of rainy periods when the soil is saturated with moisture to grow and reproduce. As the soil dries up, they can form single cells with highly resistant walls and dense colloidal internal structures that enable them to lose almost all their water and still remain alive. In this form, they can survive for months in soil that is dry as dust.

An example is the ball-fruit liverwort (*Sphaerocarpos*, Figure 7-3). These tiny plants exist by the hundreds on bare, moist soil in regions with seasonal rainfall. They consist of a green disk of tissue, microscopically thin and about one-half inch (4–8 mm) in diameter. When the rains come, each plant grows from a resistant cell or spore and in a few days or weeks forms tiny plantlets anchored to the ground by microscopic rootlets (rhizoids), each of which consists of a single cell. The adult plants form reproductive structures that consist of male and female plants, differentiated from each other by the same kind of chromosomal apparatus that genetically determines males and females in higher animals, including humans. The males produce flask-shaped containers holding differentiated sperm cells. When they emerge, the sperm swim through thin films of rainwater to neighboring female plants, where they enter bottlelike structures that contain eggs. After fertilization, the microscopic fertilized egg grows into a round sphere, within which are differentiated, highly drought-resistant, dustlike spores. After they have escaped, the spores can be blown

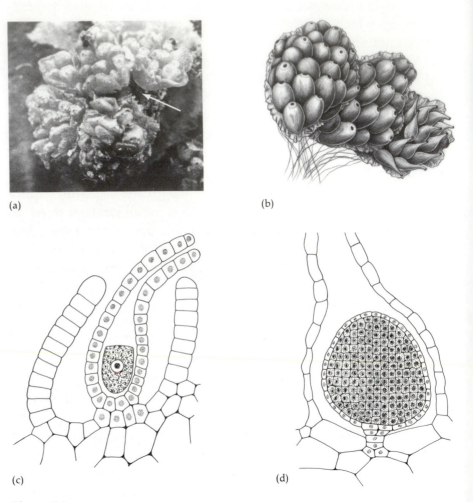

(a)   (b)

(c)   (d)

Figure 7-3.

(a) A group of male (*arrow*) and female ball-fruited liverwort (*Sphaerocarpos texanus*) plants on soil (× 3). Note the inflated involucres each surrounding an archegonium. (b) Male *Sphaerocarpos texanus* plants (× 35). Note the flask-shaped involucres each containing an antheridium. (From H. C. Bold, *Morphology of Plants*, Third Ed., Harper & Row, 1973.) (c) The egg cell and (d) sperm cell containing the archegonium and antheridium, as seen through the microscope × 325. (After G. M. Smith, *Cryptogamic Botany*. New York: McGraw-Hill, 1955.)

or carried over long distances and can survive months of complete drought, until the next season's rains enable them to grow into a new generation of plants. Thus if a plant is very small and close to the ground, it can evolve a fairly complex life cycle, including the formation of three kinds of reproductive cells plus sexual differentiation, while remaining simple in overall structure.

Larger land plants can meet new environmental challenges only by acquiring more complex structures both for survival and for reproduction. The entire course of land-plant evolution is best understood in terms of adaptive responses to these challenges. A tree can stand up under the force of high winds only if it possesses a mechanically efficient trunk. This efficiency is acquired by the differentiation of specialized woody cells, arranged first as hollow cylinders and in older trees as solid heavy columns. In order to supply their leaves with water, large trees must lift several tons of water from their roots to their tops every few days. The faster the leaves evaporate or transpire water vapor into the air, the more rapid and efficient is their photosynthesis. Nevertheless, if the tree loses water too rapidly, it cannot take in enough through its roots to maintain a proper water balance. Consequently, physiological integration is essential for larger plants. Plants living in seasonal climates must be well integrated and able to withstand long periods of cold and drought. The evolution of land plants has been dominated by the adaptive value of increasingly efficient responses to these environmental challenges.

To sedentary land plants, increasingly efficient structures for reproduction, dissemination, and establishment in new localities have equal adaptive value. Botanists have found that the course of plant evolution can best be traced by making comparisons between different kinds of reproductive structures—egg sacs (archegonia), sperm cases (antheridia), airborne reproductive cells (spores), spore cases (sporangia), seeds, stalked pollen sacs (stamens), and coverings for developing seeds (carpels, ovaries).

What was the nature of the first land plants, and how did they evolve from their single-celled ancestors? Botanists have long accepted the hypothesis that populations of green freshwater algae with threadlike branches consisting of single rows of cells changed their overall shape and acquired more drought-resistant cell walls as well as the capacity to produce airborne spores. In this way, they evolved into flat, ground-hugging (thalloid) plants.

Recently, however, this hypothesis has been questioned. In the first place, research with the electron microscope has shown that nearly all many-celled green algae, including the threadlike ones (*Fritschiella*) that were the leading candidates for land-plant ancestors, differ widely from all land plants with respect to important details of mitotic cell division, as well as the chemical nature of important enzymes for photosynthesis (glycolate oxidase) and submicroscopic fine structure of motile cells. Differences with respect to three totally different kinds of characteristics render highly improbable the hypothesis that any modern many-celled freshwater green algae are related to the ancestors of many-celled land plants, except for a few highly specialized forms (*Conjugales, Coleochaete, Charales*) that may well be secondarily aquatic descendants from primitive terrestrial ancestors.

Recent explorations of the biota of moist soils in many parts of the world have revealed a wealth of single-celled green eukaryotic microorganisms. Some of these cells (Prasinophytes, *Klebsormidium*) resemble those of typical land plants such as mosses and ferns with respect to both fine structure and mitotic apparatus. Why couldn't these single-celled forms have existed by the trillions, permeating the entire land surface of the globe before it became covered with many-celled land plants? If this condition existed, the probability that some of these populations of cells evolved directly into flat, tissuelike, multicellular land plants is much greater than the likelihood that aquatic threadlike algae evolved into land plants by means of radical alterations of both their cell walls and their patterns of growth.

Early land plants could not spread and become diversified until they had acquired two methods of reproduction. The first is migration over land, achieved by the evolution of asexually produced cells known as spores. They are microscopic in size, light in weight, and can be carried almost everywhere by air currents. They penetrate even into the deepest caverns, far under the earth's surface. Spores have nearly waterproof walls and dense, solid internal contents that are highly drought-resistant and can remain alive in a dormant condition for weeks or months.

The second method of reproduction is sexual. Adult plants that grow from spores produce differentiated structures like the sperm-containing flasks (*antheridia*) and egg-containing bottles (*archegonia*) illustrated in Figure 7-3. A thin film of water is required for fertilization. All spore-bearing land plants such as mosses, ferns, and their relatives go through a stage when they produce actively moving sperm cells—the only truly motile cells that higher plants possess. In all spore-bearing plants, the cycle of alternating chromosome numbers that results from meiosis and fertilization is timed in such a way that two kinds of many-celled plants or "generations" alternate with each other. Since meiosis occurs just before the differentiation of spores, the latter cells have the reduced chromosome number that in animals exists only in eggs and sperm cells. The plant that grows from the spore retains this number until it is adult and has differentiated egg and sperm cells in their appropriate container, and until fertilization has restored to the egg the unreduced chromosome number. The fertilized egg then grows into a spore-bearing plant (*sporophyte*) having the unreduced chromosome number that in most land plants looks entirely different from the generation (*gametophyte*) that produces the eggs and sperm. Figure 7-4 illustrates this cycle for mosses, in which the spore-bearing plant is relatively small and remains attached to the egg bottle from which it arose, and for ferns, in which it is much larger, independent, and well rooted in the soil.

Spore-bearing plants diverged into two major groups. One of them contains the true mosses and liverworts, so called because they superficially resemble the shape of the human liver. The other contains ferns, club mosses, or ground pines (*Lycopodium*, *Selaginella*), horsetails (*Equisetum*), and quillworts (*Isoetes*). Both primitive, extinct, spore-bearing trees and modern seed plants evolved from extinct plants related to members of the latter group. Because fossils of small plants that lack woody tissue are rarely preserved, and because modern mosses, liverworts, and ferns are separated from each other by wide gaps, the early differentiation of spore-

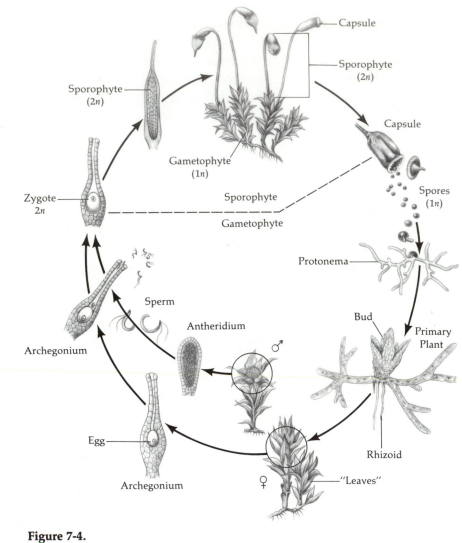

Figure 7-4.

The life cycle of a moss. (From *Biology*, Third Ed., by G. Hardin and C. Bajema. W. H. Freeman and Company. Copyright © 1978.)

bearing plants may never be known. Their common ancestor may have resembled some very simply constructed liverworts such as those illustrated in Figure 7-3, but we cannot be sure of this. We can be reasonably certain, however, that the great majority of mosses, liverworts, and their relatives gave rise only to more greatly diversified plants of their own kinds. Their plant bodies are ideally adapted to growing horizontally over moist soil, wet rocks, and the bark of trees. Some of them have become able to survive weeks or months of complete desiccation in a dormant state. In this way, a few species of mosses have even been able to colonize deserts. But because they have never evolved efficient supporting or conducting tissues, none has been able to grow more than a few inches above the level of the ground.

EVOLUTION OF VASCULAR PLANTS AND SEEDS

Land plants first appeared in the Silurian period, about 400 million years ago. They consisted of simple branching stems a foot or less in height, which in some forms were naked and in others bore small scalelike structures superficially resembling leaves (Figure 7-5). The central part of the stem was a core of woody tissue that served to support the plant and to conduct water from the roots to the top of the plant. Because of the presence of this tissue, they are called *vascular plants*, a term applied to all the larger land plants, including ferns and their allies, coniferous trees, and the large diverse assemblage of flowering plants. The vascular core was in some a solid column, in others a hollow cylinder, and in still others a series of separate strands that often formed a sort of network. Some of the branches were topped by spore cases (*sporangia*) that in the adult plant released hundreds of airborne spores. Plants of this kind dominated the earth's surface for about 50 million years, after which they were replaced by larger, more elaborate plants, most of which bore true leaves or at least leaflike structures. At about this time or shortly afterward, from 340 million to 300 million years ago, vascular plants split into several evolutionary lines, some of which became extinct while others persisted until our time, evolving simultaneously and parallel with each other. The true ferns evolved large leaves or fronds, but most of them have remained small in stature. Large tree ferns evolved in the Devonian period but became extinct during the Carboniferous and Permian, when woody seed-bearing plants became wide-

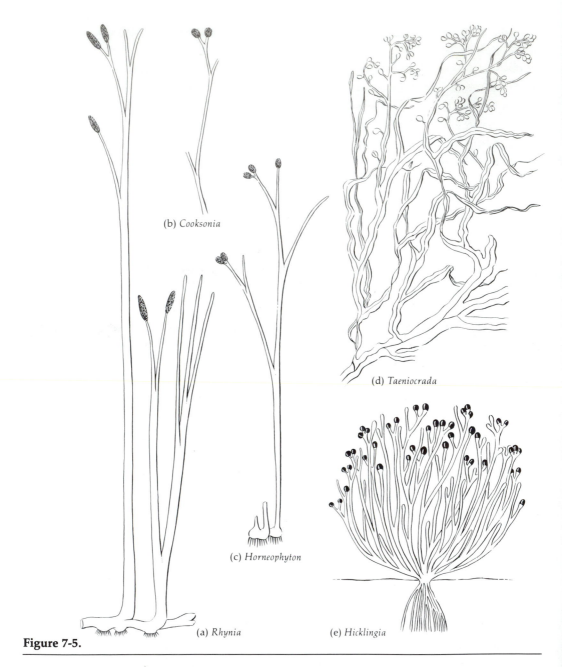

Figure 7-5.

Restoration of five primitive extinct kinds of land plants (*Rhynia, Cooksonia, Horneophyton, Taeniocrada, Hicklingia*). (From *Comparative Morphology of Vascular Plants*, Second Ed., by A. S. Foster and E. M. Gifford. W. H. Freeman and Company. Copyright © 1974.)

spread. The relatively small number of tree ferns that exist today, particularly in the tropics and the south temperate zone, evolved much more recently—probably from smaller ancestors.

The inability of most ferns to compete as trees with seed plants is due probably to the nature of their reproductive cycle, which does not include a stage when the parent plant builds up and transmits to its offspring an ample supply of stored foods. Consequently, young plantlets of ferns cannot compete with the offspring of seed plants that grow beside them.

When a fern spore germinates, it gives rise in a few weeks to a small plantlet (prothallus or gametophyte) thin in texture, less than an inch in diameter, but green and photosynthetic. This prothallus produces bottles, each containing a single egg cell, and sperm sacs that release large numbers of motile sperm cells. Since many sperm sacs are differentiated before egg bottles appear on the plantlet, the sperm usually brings about crossfertilization by swimming to a neighboring prothallus, to which they are guided by a powerful chemical sex attractant. They enter the mouth of the bottle and swim down it to fertilize the egg. The vascular, spore-bearing fern grows as a delicate sporeling from the fertilized egg (Figure 7-6).

Because these plantlets can easily dry up, ferns are best adapted to regions in which the soil is moist during at least some of the warm growing season. Some ferns, however, are highly drought-resistant, growing on the edges of deserts in deep shaded rock crevices. These semidesert ferns are never more than a few inches or a foot in height. Ferns are very abundant and diverse in the moist tropics, but some of them grow in cold regions, almost up to the limit of trees in both subarctic regions and on high mountains. Until about 70 million years ago, when herbaceous seed plants became abundant and diversified, ferns must have formed the bulk of the undergrowth in forests.

Another line of spore-bearing plants are the club mosses or ground pines (*Lycopodium* and *Selaginella*), familiar sights in the deciduous forests of the eastern United States, particularly in winter, when they are green and other low-growing plants are dry and dormant (Figure 7-7). They differ from true mosses in having

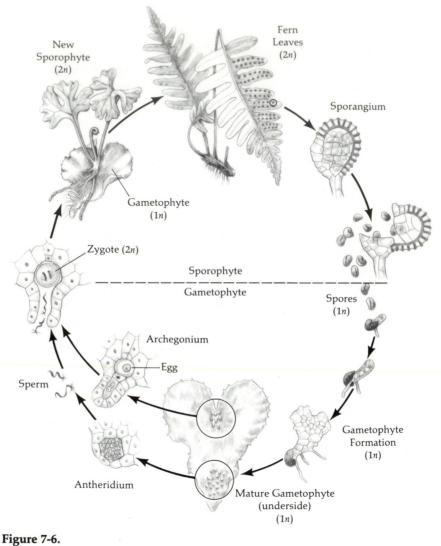

New
Sporophyte
(2*n*)

Fern
Leaves
(2*n*)

Sporangium

Gametophyte
(1*n*)

Zygote (2*n*)

Sporophyte
— — — — — — — — — — — — — — — — — — —
Gametophyte

Spores
(1*n*)

Archegonium

Egg

Sperm

Gametophyte
Formation
(1*n*)

Antheridium

Mature Gametophyte
(underside)
(1*n*)

Figure 7-6.

The life cycle of a fern. (From *Biology*, Third Ed., by G. Hardin and C. Bajema.
W. H. Freeman and Company. Copyright © 1978.)

(a)

(b)

Figure 7-7.

Two species of "club moss" or "ground pine" (*Lycopodium*), a relatively primitive modern vascular plant. (From *Comparative Morphology of Vascular Plants*, Second Ed., by A. S. Foster and E. M. Gifford. W. H. Freeman and Company. Copyright © 1974.)

a core of woody vascular tissues and in their life cycles. Like ferns, the spore-producing generation has the unreduced chromosome number and differentiates airborne spores, while a smaller generation in the typical ground pines is subterranean and lacks green tissues. It produces eggs and sperms like those of ferns. Ground pines have evolved little from their remote ancestors that lived 300 million years ago.

During the Carboniferous (Mississippian and Pennsylvanian) periods, from 340 million to 280 million years ago, club mosses evolved into large trees (*Lepidodendron, Sigillaria*; Figure 7-8). A change in their life cycle helped them to do this. The spore-bearing plants produced spores of two kinds—one large and containing stored food densely packed inside a thick, resistant wall, and the other small and dustlike, able to be blown by the wind from one plant to another. When small spores landed on the surface of a large spore, they released male nuclei that fertilized egg nuclei at the edge of the large spore. The fertilized egg grew into a plantlet that during early stages of growth was nourished by the stored food. In this way, young plants could begin growth rapidly, even in the shade of larger plants. This life cycle is inferred because the tree club mosses resembled in their structures the modern little club mosses (*Selaginella*) in having large and small spores.

Another line of spore-bearing plants (Sphenopsida) flourished during the Carboniferous period but later became extinct except for a single genus, the horsetails (*Equisetum*), which contains several widespread and common modern species. These plants are characterized by jointed stems that are green and photosynthetic and in the modern forms bear dry black or brown scales instead of leaves. Some of their ancestors produced rather delicate leaves. The giant horsetails of the Carboniferous period, like the tree club mosses, produced large and small spores.

The first seed-bearing ferns (Pteridosperms) evolved during the Carboniferous period. Like the large spores of Carboniferous plants, seeds promote the early growth of young plants by storing up food for them but do this more efficiently than spores. The stored food is surrounded not by a single hard wall but by special coverings (integuments) that evolved from flattened branchlets.

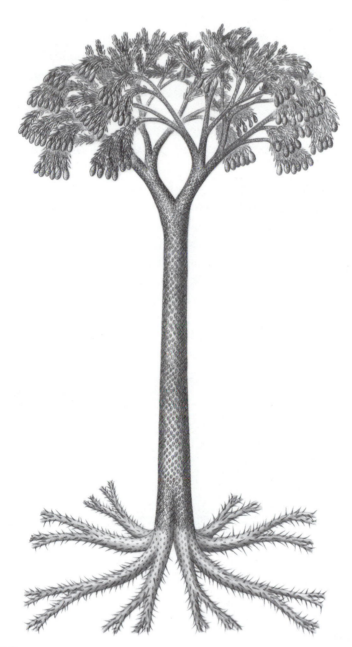

Figure 7-8.

A giant relative of the club mosses (*Lepidodendron*) that flourished during the Carboniferous period, 300 million to 350 million years ago. (Redrawn from *Handbuch der Paläobotanik* by M. Hirmer. R. Oldenbourg, Munich, 1927.)

The earliest seeds were much larger than the large spores of any contemporary tree club mosses or horsetails, and their coverings provided much better protection for the tender young embryo plants that were growing inside them. Almost contemporaneously with the seed ferns, a group of seed plants (*Cordaitales*) appeared with long, strap-shaped leaves and reproductive structures that botanists believe were ancestral to the cones of pines and other conifers.

The forests that occupied thousands of square miles of the earth's surface during the Pennsylvanian period, from 280 million to 325 million years ago, would have appeared strange to us (Figure 7-9). From the swampy ground, tall buttressed trunks of tree club mosses rose into the steamy atmosphere, bearing thousands of scalelike leaves that would have looked unnaturally small. They were accompanied by the gigantic candelabra of the tree horsetails, the graceful fronds of tree ferns and seed ferns, and a few ancestral conifers. The ground was carpeted with mosses and liv-

Figure 7-9.

A reconstruction of the kind of forest that existed in the swamps at the end of the Paleozoic Era. (Courtesy of the Field Museum of Natural History, Chicago.)

erworts, some of which covered the tree trunks, and dotted with smaller ferns. Birds did not evolve until many millions of years later, but the air was filled with giant dragonflies flitting from one tree to another. Worms, scorpions, centipedes, spiders, and other small creatures crawled over the ground and served as food for the long-bodied, short-legged amphibians. A few of the driest spots in the forest were occupied by slow, lumbering, lizardlike creatures, the earliest reptiles. In the absence of animals that could cry, squawk, or even rattle, the prolonged silence would have been oppressive and deadening.

At the end of the Pennsylvanian period, mountain ranges rose up in many parts of the world, and the swampy lowlands that supported the great forests dried up. In the middle of the Permian period, about 250 million years ago, much of the earth's surface in what is now India, South Africa, and other southern continents became covered with glacial ice. The trees of the coal forests could not adjust to these radical changes in climate and so became extinct. They were replaced by trees that would have looked much more familiar to us and were clearly ancestral to modern groups. Among the most conspicuous were conifers ancestral to pines, firs, spruces, cedars, and junipers but looking more like modern conifers that are confined to the Southern Hemisphere, such as the "monkey puzzle tree" and the Norfolk Island pine (*Araucaria* spp.). Their seeds were borne in cones having a much more primitive construction than any modern conifers. Also prominent were stout-trunked, unbranched or sparingly branched trees with thick-textured palmlike leaves. These were the cycads, sometimes known as sago palms (Figure 7-10).

Modern cycads are abundant in many parts of the tropics and the Southern Hemisphere such as Mexico, Cuba, South Africa, and Australia; a single species grows wild in Florida. Although superficially like palms, they are totally different with respect to their reproductive structures. Female cycads produce seeds either on specialized leaflike structures (sporophylls) or hidden within long heavy cones. Male plants produce smaller cones that release masses of dustlike wind-borne pollen. The most remarkable feature of their life cycle is their method of fertilization. When pollen

Figure 7-10.

A large cycad (*Encephalartos*) from Africa. Note the seed-bearing cones, as large as the head of the man standing beside the tree. (Courtesy of R. A. Dyer.)

grains land on a female cone, they germinate to produce a tube that digests its way through the tissue of the cone until its tip is a fraction of an inch away from the egg cell. Up to this stage, the pollen grain and tube behave like those of conifers. In the conifers, the tube grows until it almost touches the egg cell, then releases male nuclei that unite with the egg nuclei in the act of fertilization. In cycads, however, the egg nuclei are near a small open chamber that is filled with water. In the pollen tube are elaborate differentiated sperm cells adorned with rows of cilia that make them look like protozoa (Figures 7-11, 7-12). These sperm swim a tiny dis-

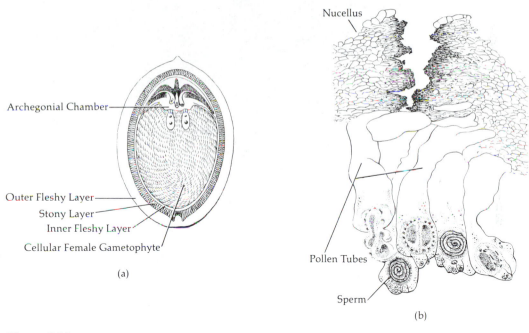

Figure 7-11.

(a) Cross-section of an ovule of a cycad (*Dioon edule*) at the time of fertilization. At the top are two egg-containing "flasks" (archegonia), and above them is the water-containing chamber into which the pollen tubes release their sperm cells. (b) The tissue, pollen tubes, and sperm cells that lie above the pollen chamber, greatly enlarged. From the genus *Cycas*. (From *Comparative Morphology of Vascular Plants*, Second Ed., by A. S. Foster and E. M. Gifford. W. H. Freeman and Company. Copyright © 1974.)

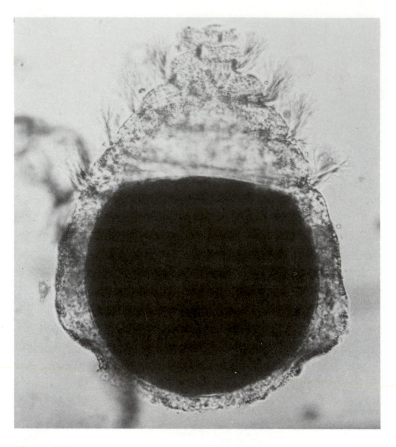

Figure 7-12.

A single sperm cell (× 350) of a primitive seed plant (*Ginkgo*) that resembles the sperm cells of cycads. (Courtesy of K. Norstag.)

tance, hardly more than ten times their own diameter, before releasing nuclei that fertilize the egg cells.

Why should cycads pass through such an elaborate stage of the reproductive cycle when conifers and other modern seed plants are far more successful without it? Comparative studies of the anatomy of all seeds produced by plants of the Carboniferous period reveal the presence of pollen chambers similar to those through which the sperm of cycads swim. The earliest seed ferns may have produced sperm similar to those of modern cycads, which are an elegant example of structures known as vestigial.

The forests of the Permian and Triassic periods contained a number of plants quite unlike modern trees. A few of them resembled the maiden-hair tree (ginkgo) that once grew wild in China but is now almost entirely under cultivation. A much modified descendant of seed ferns having oval, tongue-shaped leaves, *Glossopteris*, had seeds produce in specialized cup-shaped structures. The smaller plants were ferns, club mosses, true mosses, and liverworts. Nothing resembling a flower had yet evolved.

During the beginning of the Age of Dinosaurs—the Jurassic period, which lasted from about 190 million to 136 million years ago—forests consisting of conifers together with ferns of all sizes flourished in most parts of the earth. Cycads, ginkgolike trees, and cup-bearing seed ferns were also found. Another distinctive group of plants looked much like cycads, but in true cycads male and female structures are always on separate plants, whereas in the Jurassic cycadlike plants (Bennettitales) both pollen sacs and seeds were borne on different parts of the same shoot. Botanists believe that these were the first plants that were cross-pollinated by insects. Although superficially similar to the flowering plants that followed them, they did not give rise to any modern plants, becoming completely extinct at the end of the Jurassic.

THE RISE OF MODERN FLOWERING PLANTS

The latter part of the Age of Dinosaurs, the Cretaceous period, witnessed a complete change in the earth's vegetation. Beginning about 130 million years ago as rare intruders into the forests of conifers and ferns, the flowering plants (angiosperms) increased slowly in abundance and diversity for 15 million to 20 million years, then exploded into a great variety of orders and families that dominated the earth's vegetation. By the end of the Cretaceous period, 65 million to 70 million years ago, the earth's forests looked much like those found in tropical regions today.

Why did flowering plants achieve such rapid dominance in most parts of the earth? Several factors contributed to it, all of which are included in the word *versatility*. The most conspicuous factor is the diversity of their reproductive structures.

Botanists use the term *flower* to include a much wider range of structures than common usage of the term implies (Figure 7-13). In addition to conventional flowers like buttercups, roses, poppies, morning glories, petunias, and snapdragons, the pollen-bearing and seed-bearing structures of trees like oaks, birches, and willows are regarded as flowers, as are the seed heads of wheat and the tassels and ears of corn, most of which are clusters of flowers variously arranged on their stalks. This concept is justified because many flowers exist that are intermediate between the conventional forms and those that lack showy petals. The class of flowering plants (Angiosperms) includes all plants that produce flowers in the botanist's sense of the term.

The versatility of form in flowering plants is matched by their versatility in methods of pollination and seed dispersal. For cross-pollination, the pollen of conifers and cycads is borne by the wind—a wasteful method that requires the production of a great excess of pollen. Many kinds of flowering plants are also pollinated by wind, but even more are pollinated by bees, flies, butterflies, beetles, birds, and even bats. The seeds of conifers and cycads are chiefly dispersed by the wind or have no obvious adaptations for dispersal. Those of many flowering plants are also wind-dis-

Figure 7-13.

(a) Staminate (male) flowers or catkins of an oak (*Quercus*). (b) Hanging flower of *Fuchsia* sp. (c) Several flowers of a member of the pea family (*Thermopsis* sp.). (d) An orchid flower (*Odontoglossum* sp.) (e) Two flowers of a larkspur (*Delphinium* sp.). (f) Flowers of the cocoa tree (*Theobroma cacao*). (g) Flower of a Mariposa lily (*Calochortus* sp.), seen from above. (h) Flowers of a species of *Lobelia*. (i) Flower of bush poppy (*Dendromecon rigidum*). (j) Buttercup (*Ranunculus* sp.). (k,l) Pistillate and staminate flowers of water plantain (*Alisma* sp.). (m) Front view of monkey flower (*Mimulus* sp.). (n) Longi-section of beard tongue flower showing position of stamens (*above*) and style (*below*). (o) Section of the flowering head of mule ears (*Wyethia* sp.), a member of the sunflower family.

(a)   (b)   (c)

(d)

(e)

(f)

(g)

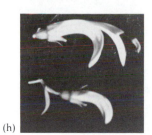

(h)

(i)

(j)

(k)

(l)

(m)

(n)   (o)

persed. An even larger number of flowering plants, however, produce seeds that are spread more widely by clinging to the fur of
animals or to mud attached to the feet of birds or even pass
unharmed through animals' digestive tracts (Figure 7-14).

With respect to seed size, the versatility of flowering plants is
far greater than that of conifers or cycads. A single capsule of an
orchid plant produces millions of seeds as tiny as grains of dust,
barely visible to the naked eye. At the other end of the scale are
huge seeds like that of the coconut. Small seeds have the advantage of being easy to disperse but in many situations are disadvantageous because of the weakness of the seedlings that grow
from them. Large seeds produce more vigorous seedlings but are
costly in energy to produce and transport. The enormous range
of seed sizes that flowering plants have evolved has enabled them
to achieve a great variety of compromises between these advantages and disadvantages. As a result, species of different kinds of
flowering plants can colonize almost every type of habitat found
on the earth.

The principal reason for the versatility of flowering plants with
respect to seed production is the streamlining of seed development. Their pollen grains produce only three nuclei, while those
of other seed plants produce many more. After meiosis in the
female structures, cells of flowering plants having the reduced
number of chromosomes pass through only three mitotic divisions, giving rise to an eight-nucleate structure (embryo sac), of
which one nucleus is that of the egg. Conifers and cycads pass
through many more divisions. Finally, in flowering plants, the
cells that contain stored food for the seedling are produced simultaneously with the growth of the embryo and the seedling itself,
whereas in conifers and cycads most of the food-storage cells are
produced before fertilization. In the flowering plant, there is a
much shorter interval between meiosis and fertilization as well as
between fertilization and the final ripening of the seed. This
streamlining of the reproductive cycle is probably responsible for
the fact that, among flowering plants, short-lived annuals that produce flowers and fruits a few weeks after seed germination have

Figure 7-14.

Diversity of seeds and fruits among flowering plants. *Top row*: Seeds of desert willow (*Chilopsis linearis*); seed of balsa tree (*Ochroma lagopus*); bur of a grass (*Cenchrus ciliaris*); fruit of hawthorn (*Crataegus* sp.); seed of *Magnolia grandiflora*. *Middle row*: Winged fruit of elm tree (*Ulmus* sp.); fruit of thistle (*Crupina vulgaris*), bearing a parachutelike crown of hairs; burlike pod of bur clover (*Medicago polymorpha*). *Bottom row*: Pod and two seeds of loco weed (*Astragalus Hornii*); fruit of blow wives (*Achaerachaena mollis*), bearing a crown of papery scales; appendaged fruit of stork's bill or filaree (*Erodium botrys*); adhesive, burlike fruit of hound's tongue (*Cynoglossum officinale*); fruit of beggar's ticks (*Bidens frondosa*), bearing two barbed projections; acorn of cork oak (*Quercus suber*). The seeds and fruits at the left are light and wind-borne. Those in the middle adhere to the hair and fur of animals as well as to human clothing; berries and red seeds like those of Magnolia are eaten by animals and pass through their digestive tracts unharmed. Acorns are carried about by squirrels and by birds such as jays and woodpeckers; those that are not eaten serve to distribute the species. (Samples provided by State of California, Department of Food and Agriculture, Sacramento.)

evolved in many different families whereas they have never evolved in other seed plants.

The versatility of flowering plants is also expressed in the woody tissues of their stems. In conifers and cycads, all the cells of these tissues resemble each other in size and thickness of cell walls. If the cells are larger, conduction of food and water is easier, but at the expense of the stem's supporting function. Smaller, thick-walled cells provide stronger support but poorer conducting ability. In nearly all flowering plants, the woody tissue contains efficient conduits, relatively large in diameter, for conducting water. They are flanked by narrow, thick-walled fibers that provide good support for even relatively slender stems, such as those of vines. The ability to produce this more versatile tissue is evidence of more complex regulatory systems that control gene action.

Finally, flowering plants are more flexible than conifers and cycads with respect to growth processes themselves, involving both the enlargement of individual cells and the proliferation of cells by mitotic division. In roots, leaves, flowers, and fruits, the contrast between the smallest and the largest mature cells is considerably greater in flowering plants. This capacity for stretching their individual cells enables roots to spread farther through the soil in search of water. It also enables vines to produce prehensile tendrils, as in grapes and pea plants. The quick opening of flowers, the extension of the pollen sacs (anthers) far above the flower itself, and the provision of seeds with feathery hairs or plumes that greatly aid dispersal are all properties that depend on versatility and flexibility of individual cells.

Flexibility with respect to cell proliferation is shown by the presence of embryonic tissue not only at the tips of stems and roots but also in the form of buds at the bases of most of the leaves and in various parts of the plant. The presence of this tissue makes most flowering plants much easier to propagate and to graft than conifers.

Considering all these properties, one need not be surprised to learn that flowering plants have evolved more than 275,000 different species—five times as many as all other kinds of plants put together. Except in forests that are dominated by cone-bearing evergreen trees, in lifeless deserts, and on barren mountain summits, flowering plants give the earth's landscapes their characteristic aspects. The grandeur of tropical jungles, the familiar greens and browns of oak, maple, birch, and other deciduous forests, the velvety rolling prairies, the flocks of olive-green, gray-green, and occasional bright-colored flowers that adorn the scrublands of drier regions, the multicolored tapestry of deserts and prairies in bloom—these and many other kinds of landscapes are shaped by the flowering plants that grow on them. In freshwater lakes and streams, flowering plants such as water lilies, water hyacinths, cattails, bur-reeds, duckweeds, and pondweeds form the bulk of the plant life that they contain. A few kinds of flowering plants, such as eel grass and the tropical mangroves, have even invaded the borders of the oceans.

Unfortunately, the pathways along which these plants evolved are very poorly known. Charles Darwin called their origin "an abominable mystery." More recent discoveries have shed as yet little light upon it. Even compared to other plants, the earliest flowering plants and their immediate ancestors have a particularly spotty fossil record. This defect is easily explained by their nature. Their pollen is well preserved and abundant in both the early and more recent fossil record, and impressions of leaves as well as fragments of wood are commonly found. When these fragments belong to plants that lived relatively recently, less than 40 million years ago, they have such close counterparts among modern forms that the nature of their flowers and fruits can be safely inferred by comparing fossil remains with similar structures that belong to modern species. However, fossil pollen, leaves, and wood of flowering plants that lived from 100 million to 120 million years ago, that is, from the beginning of their fossil record, are so unlike modern forms that the nature of these remote ancestors cannot be reconstructed from fossil remains. The same is even more true of various earlier fragments of plants that some botanists believe belonged to ancestors of flowering plants. Hopefully, future fossil discoveries will eventually clear up at least in part this annoying mystery.

A similar difficulty exists when botanists try to arrange the orders and families of flowering plants into a system that reflects their evolutionary history. Although flowering plant fossils provide nearly or quite as reliable clues to relationships as do fossil insects, the clues offered by comparisons between living forms are far vaguer and less reliable than those that zoologists can use for insects, spiders, and other animal groups that have poor fossil records. This is because of the relatively simple construction of the plant body and the consequent presence of numerous examples of parallel and convergent evolution. In the past, botanists have disagreed widely with respect to relationships among families and orders and even with respect to which modern forms are most like the original, ancestral flowering plants. Although these differences, and the debates that have accompanied them, have diminished in recent years, there is still far less agreement among botanists about evolutionary pathways among flowering plants than among zoologists about the pathways of evolution within any of the major animal phyla. Future research, particularly various approaches based on biochemistry, may clarify the picture. At present, however, the arrangement of different kinds of flowering plants on a hypothetical evolutionary tree is so complex and uncertain that it is omitted from this general volume, but it can be studied in more advanced and specialized botanical works.

A reconstruction of the ocean bottom during the Cretaceous period (the Age of Dinosaurs), 100 million to 150 million years ago. The large ammonoid at right has a diameter of 19 inches; at left are several *Baculites*, having a straight shell except for a small coil at the apex of the cone. (Courtesy of the University of Michigan Exhibit Museum.)

Chapter Eight

THE EARLY EVOLUTION OF MANY-CELLED ANIMALS

The transition from single-celled plants to many-celled plants was gradual and took place independently in many different evolutionary lines. With respect to many-celled animals, the situation was quite different. Among the protozoa, many kinds exist that form colonies, but none is enough like the simplest many-celled animals to be regarded as a link. At least two branches may have evolved separately from single-celled protozoa—one evolving only to sponges, and the other becoming differentiated into jellyfishes, corals and their relatives, and various kinds of worms.

Until recently, the fossil record of early animal evolution was almost blank, but significant clues are now available. The oldest recognizable remains are fossilized worm burrows about 1 billion years old—400 million years younger than the oldest single-celled eukaryotes. Trace fossils of this kind are not common in rocks older than the Ediacara Formation of Australia, which is between 680 million and 580 million years old, where paleontologist M. F. Glaessner found remains of several kinds of worms and jellyfishes (Medusae) but no indications of more complex life. Slightly younger rocks contain minute fragments of the external skeletons of larger multicellular animals, the nature of which cannot be determined. At the beginning of the Cambrian period—about 570 million years ago—external skeletons appear that definitely belong to many living phyla. All phyla that had hard outer parts existed 530 million years ago, and many soft-bodied animals belonging to modern phyla of worms had appeared as well.

Most zoologists believe that the jellyfish phylum (Coelenterata) existed earlier even than worms, which may have evolved from larval stages of some free-living, floating jellyfishes. The principal advantage that both these animals possess relative to unicellular forms is their ability to ingest much larger amounts of food and digest it with the aid of numerous specialized stomach cells. The development of both a front and a hind end was an obvious advantage for worms that sought food by crawling over the bottom of the ocean.

Readers must not interpret the word *worm* to mean only a fishing worm, nightcrawler, or any other creature that resembles familiar modern animals. Linnaeus and other early naturalists used the Latin name *Vermes*, from which our English word *worm* is derived, to describe a highly diverse group of animals. On the basis of their body plans, Vermes are separated into several different phyla. Some are seldom seen in the living state. Among the most familiar are earthworms; tapeworms, which parasitize digestive tracts; nematodes, which attack the roots of crop plants; hookworms, another parasitic group; bristleworms, which live in tidepools; and leeches. Earthworms are among the most specialized of these, having evolved in their own particular direction. They are very different from the probable ancestors of higher animals. Some wormlike animals such as centipedes, millipedes, and caterpillars belong to the phylum of joint-legged animals (Arthropods). Zoologists do not regard them as worms at all.

During the jellyfish–worm stage of animal evolution, the various basic designs of animal bodies evolved. First came the jellyfish (Figure 8-1), a blind sac with only two layers of cells and a single opening serving both as mouth and anus. Jellyfishes have *radial symmetry*; they have no front or hind end. This kind of body plan is adaptive to only two kinds of habitats—floating over the surface of the ocean or attaching in an immobile position to the ocean floor. Jellyfish phyla (Coelenterata, Ctenophora) are dominant in

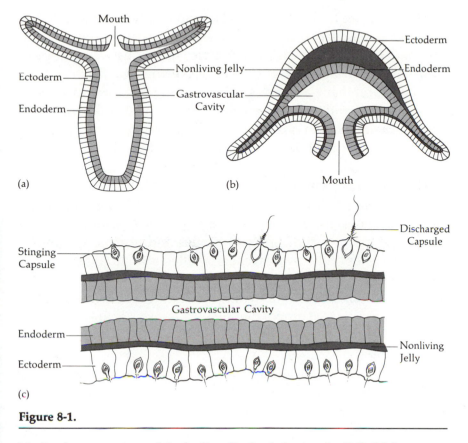

Figure 8-1.

Idealized cross-sections of the bodies of animals having the "jellyfish" (Coelenterate) body plan. (a) A type that is fixed to its substrate (Hydra, coral polyp). (b) A floating form (medusa). (c) Cross-section of the body wall.

both these habitats. Jellyfishes occur in great masses that sometimes cover wide stretches of the ocean with an almost continuous layer of living matter, while their cousins, the sea anemones and corals, line the tidepools of temperate seashores forming massive reefs in warm tropical waters. Although they have been dominant in these habitats for hundreds of millions of years, few of them have evolved species adapted to active, directed movement such as swimming or crawling. Their basic body plan adapts them admirably to certain ecological niches but severely limits their options for colonizing others.

The blind sac body plan has been moderately successful in flat-worms as well. Flatworms have *bilateral symmetry*, a front and hind end, and relatively massive and complex tissues between the outer and inner cell layers (Figure 8-2). Nevertheless, flatworms are nowhere dominant and are successful only as slow crawlers on pool bottoms and as parasites such as the liver fluke and tape-worm. Their ability to crawl comes from having bilateral symmetry and bundles of contractile muscle cells. Jellyfishes also possess both muscle and nerve cells, but these are more diffuse and less well coordinated with respect to the body as a whole.

The remaining three dominant body plans are all modifications of the basic design of a tube within a tube. The animals possess bilateral symmetry, a diversity of tissues and organs, and one or more spaces or body cavities (coelom) filled with fluid between an inner digestive tract and an outer protective fleshy covering of the skin and other tissues (Figure 8-3). A fourth modification of this design, characterized by radial symmetry, is found in spiny-skinned animals (Echinoderms), which include starfishes, sea urchins, and sea cucumbers.

One of these phyla is Mollusca (shellfish), containing among others snails, slugs, clams, oysters, octopi, and squids. Most of these animals are either crawlers or are completely sedentary.

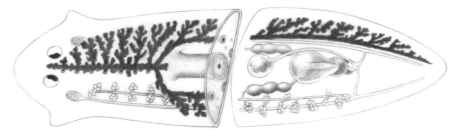

Figure 8-2.

Diagrammatic cross-section through the body of a flatworm, showing several layers of cells as well as a single opening that serves as both mouth and anus. Note the absence of a body cavity. (From *Invertebrate Zoology*, Second Ed., by Paul A. Meglitsch. Copyright © 1967, 1972 by Oxford University Press, Inc. Reprinted by permission.)

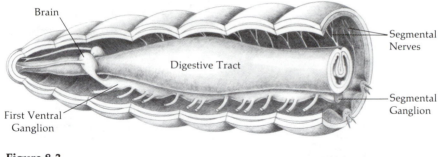

Figure 8-3.

Front end of the body of an earthworm, showing its tube-within-a-tube structure. The outer tube is the body wall. The inner tube is the digestive system, with the mouth at the left and a cross-section of the stomach at the right. The inner tube is surrounded by the nervous system, which sends branches through the body cavity to the body wall. Blood vessels are present but are not shown for the sake of simplicity. (From *Biology*, Third Ed., by G. Hardin and C. Bajema. W. H. Freeman and Company. Copyright © 1978.)

Their proverbial slowness results from having no legs; instead, their principal locomotor muscle takes the form of a single ribbon or foot.

THE EVOLUTION OF CEPHALOPODS

The only group of molluscs that became active swimmers were the class of Cephalopods that includes octopi, squids, and the chambered nautilus. Their evolution illustrates in a dramatic fashion the ways in which modifications of body plan can open up new ecological options and close others, as well as the circuitous route by which new kinds of animals often evolve.

The chambered nautilus has come down to us almost unchanged from the Cambrian period, 500 million years ago. It reveals its connection to other molluscs by its large and roomy coiled shell, which is somewhat like that of some snails (Figure 8-4). Nevertheless, the shell of the nautilus is distinctive in that it consists of a series of chambers with partitions between them (Figure 8-4). As its body and shell grow, the animal constructs a new partition,

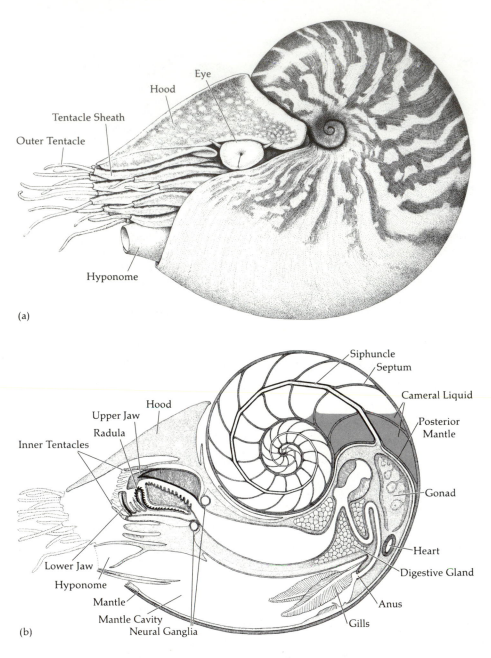

Figure 8-4.

The chambered nautilus. (a) As seen from outside. (b) A section of the shell and the animal, showing the empty chambers previously occupied. (From P. Ward, L. Greenwald, and O. E. Greenwald, "The Buoyancy of the Chambered Nautilus." Copyright © 1980 by Scientific American, Inc. All rights reserved.)

enters a new chamber, and leaves the most recently vacated one behind. The empty chambers, however, are far from being abandoned, useless relics. Running through them is a narrow corridor filled with a slender but highly active thread of living material—an extension of the animal's outer tissue. Its cells secrete enzymes that act as efficient molecular pumps. These pumps drive the water out of the chambers, even against the strong pressure of the surrounding sea, leaving the chambers filled only with gas and thus much lighter than the surrounding water. In this way, the animal attains a specific gravity much like that of its surroundings, and so remains suspended in midocean. The nautilus lives only in the southwest Pacific, and there it is confined to deep water, at least 2000 feet below the surface.

Nautilus has an extraordinary style of movement, which is accomplished by jet propulsion. At the opening of its shell is an open tube like an exhaust pipe, through which water can be ejected with great force, driving the animal backward. Its pace of swimming is leisurely but effective.

The soft body of nautilus is doubled up within its shell, like that of a flexible gymnast. Its head and foot are next to each other, almost forming a single organ; the Greek name *cephalopod* literally means "head–foot" and refers to this peculiarity. The head–foot bears a fringe of about 90 slender tentacles encased in sheaths and bearing sticky surfaces, which grasp the crabs and other crustaceans on which the nautilus feeds and convey them to its mouth, where its strong beak cracks or pierces the victim's shell. After this, the prey is ground up by many small teeth.

Shelled cephalopods similar to nautilus in body plan dominated the class during most of its evolution, appearing at the end of the Cambrian period and persisting until the end of the Cretaceous period, 65 million years ago, when they were replaced by the soft-bodied cephalopods—squids, cuttlefish, and octopi (Figure 8-5). During this long period—almost twice as long as the combined ages of reptiles and of mammals—shelled cephalopods were the most abundant, varied, and successful creatures in the oceans. Until the Devonian period, their rule of the seas was uncontested. For another 250 million years they shared this dominion with fishes and other animals; then their relative number dwindled.

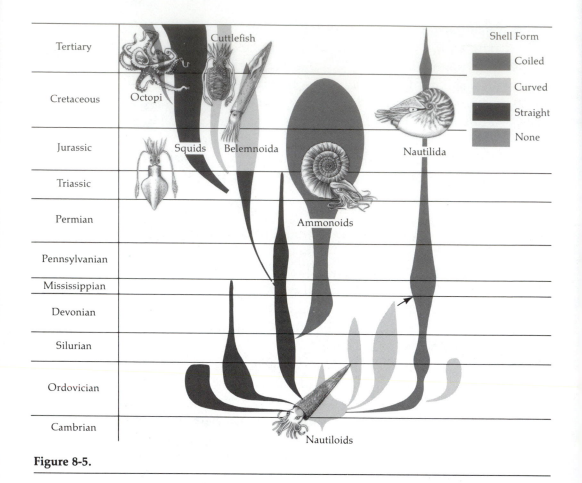

Figure 8-5.

The evolutionary tree of cephalopod molluscs. (From Moore, Lalicker and Fisher, *Invertebrate Fossils*, McGraw-Hill, 1952.)

ADAPTIVE RADIATION OF CEPHALOPODS

My own fascination with cephalopods was largely stimulated by discussions with paleontologist Peter Ward, one of the few scientists who has studied the chambered nautilus in its natural habitat. The following account relies heavily on his research. The family tree of cephalopods is shown in Figure 8-5.

The earliest cephalopods underwent changes in the shape of their shells in response to changes in their mode of life. The earliest shells were loosely coiled and caplike. These soon gave rise in one direction to tightly coiled shells, and in another to those that were completely straight. The latter adapted the animal to much more rapid swimming, both backward and forward. As in modern squids, the cephalopod's jet propulsion vent could be swiveled about. These modifications were associated with exploration of the bottoms of deeper oceans as well as with passive floating on the surface.

A second major adaptive change took place in the late Silurian and Devonian period, 100 million to 150 million years after the cephalopods first appeared. Large and aggressive fishes challenged the cephalopods as rulers of the deep. The response to this challenge was a great elaboration and diversification of their shell size, shape, and internal construction (Figure 8-6). The subclass of ammonoids, in which these features were most strongly developed, could justly be termed the dinosaurs of the deep. The largest ammonoids were giant cartwheels more than six feet in diameter, while the smallest of them were light-shelled, ornate jewels less than an inch across. Through a succession of adaptive radiations, ammonoids evolved 163 families and about 600 genera—more than all the reptiles, including the dinosaurs. Like the dinosaurs, ammonoids became extinct at the end of the Cretaceous period, about 65 million years ago.

Coevolution caused changes in other marine animals. A remarkable correlation exists between the early diversification of nautiloid cephalopods and of trilobites, which may mean that trilobites were the principal prey of the early cephalopods. When nautiloids became able to swim through the water, they could attack trilobites from above as well from the same plane on the ocean bottom. Trilobite survival depended on evolving ways of resisting these attacks, either by hiding, burying themselves in the sand, or evolving more resistant shells. Success for nautiloids then depended on evolving successful ways of overcoming new trilobite defenses. This kind of coevolution continued throughout the Ordovician and Silurian periods and on into the Devonian, when more active,

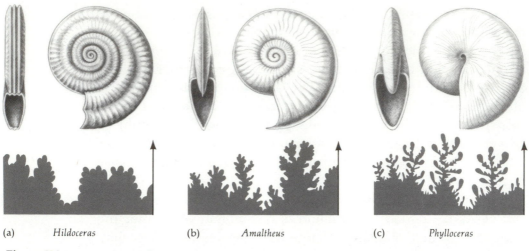

(a)   *Hildoceras*   (b)   *Amaltheus*   (c)   *Phylloceras*

Figure 8-6.

Shells of three genera of extinct Lower Jurassic ammonoids. (a) *Hildoceras;* (b) *Amaltheus;* (c) *Phylloceras.* For each shell, an end view (*top left*) and a side view (*top right*) are shown, along with a portion of the convoluted wall that separates the compartments between the chambers from each other (*bottom*). These complex convolutions are very similar within the genera or species, and differences among them serve to distinguish the species from each other. (From Moore, Lalicker, and Fisher, *Invertebrate Fossils,* McGraw-Hill, 1952.)

larger arthropods such as Eurypterids and particularly large marine fishes entered the environment.

Competition with these new arrivals may have triggered the origin and diversification of the ammonoids. Nevertheless, since neither arthropods nor the earlier fishes had *hydrostatic* mechanisms that would enable them to remain suspended in the water, they either had to swim actively, rest, or crawl over the bottom. That left the shelled cephalopods in command of the midoceanic environment. Only when the advanced bony fishes evolved an inflatable swim bladder that could raise them up when filled with gas and allow them to descend when gas was expelled could they compete with ammonoids and nautiloids and finally become the dominant creatures in all oceanic environments.

During the Mesozoic Era, various kinds of reptiles entered the ocean, becoming abundant and dominant. They probably preyed on the ammonoids.

Evolutionists such as J. B. S. Haldane puzzled over the adaptive significance of the convoluted walls between shell chambers that formed the most diversifying character during ammonoid evolution. Once ichthyosaurs (oceanic reptiles), sharks, and larger bony fishes became abundant, ammonoids could survive only if their shells could resist the crushing jaws of these predators. Convolutions could have been developed as a kind of defense.

CEPHALOPODS AND EPIGENETIC SUCCESSION

Cephalopod evolution also illustrates the great importance of epigenetic succession. Each step of progress toward the highly advanced modern animals was both promoted and restrained by the kind of gene pool that the initial populations inherited from their ancestors. For a mollusc contained within a shell, the evolution of fins or paddles for active swimming was highly improbable; the pathway of least genetic resistance was the evolution of jet propulsion. Once this method of locomotion had been the common property of the class for 200 million years, a shift to the paddle–flipper method would have involved crossing so deep an inadaptive valley that it was essentially impossible, particularly in the face of the immediate challenges posed by animals that had already perfected this method. Consequently, as selective pressure from competitors increased, all the shelled cephalopods succumbed and became extinct.

The way out of this evolutionary dead-end was first taken by a group of straight-shelled nautiloids, the belemnites. During the Permian and Triassic periods, they developed much smaller shells surrounded by body tissues and functioning essentially as supporting props or struts rather than enclosing houses (Figure 8-7). As this change progressed, the body became lighter and more mobile. Faster swimming evolved, along with the ability to swim both forward and backward. Tentacles became longer and more prehensile, with elaborate sucker disks on their surfaces.

Belemnite eyes became more efficient than those of the nautiloids and ammonoids, which had only a pinhole opening but no

MODERN CUTTLEFISH *(SEPIA)*

Internal Shell Nearly as Long as Body
Tentacles Developed

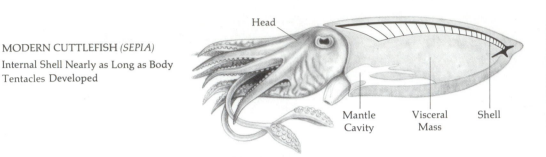

PERMIAN TO JURASSIC BELEMNOIDS

Internal Shell at Base of Body
Tentacles Developed

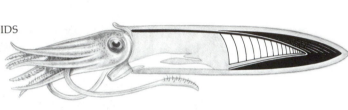

PALEOZOIC NAUTILOID

External Shell Incomplete
Beginning of Tentacle Differentiation

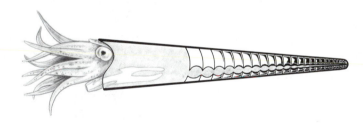

PALEOZOIC NAUTILOID

Well Developed External Shell
No True Tentacles: Front
End as in Nautilus

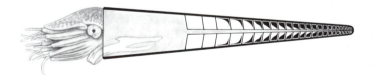

Figure 8-7.

Idealized longitudinal sections of fossil and living cephalopods, showing how the external shell of the earlier forms (Nautiloids, *bottom*) evolved into the internal "skeleton" of modern cuttlefish (*top*). (From Moore, Lalicker, and Fisher, *Invertebrate Fossils*, McGraw-Hill, 1952.)

lens and a poorly developed retina. Squids, which evolved from belemnites, have eyes that in many features such as lens and retina are comparable to those of vertebrates. Thus natural selection based on the belemnite modification of the nautiloid body plan enabled the squids, along with their relatives the cuttlefishes and octopi, to meet the challenges posed by the evolution and adaptive radiation of modern bony fishes and to remain successful up to the present day.

Once cephalopods had evolved bodies with internal skeletons in place of shells, the way was open for new cycles of adaptive radiation. During the past 80 million years, from the Upper Cretaceous through the Tertiary periods, three main branches of advanced cephalopods have developed. The squids—torpedo-shaped rapid swimmers bearing tentacles that could shoot out with lightning speed to grab unwary crustaceans or fishes—succeeded in the open ocean, where they prey on smaller fishes and crustaceans related to lobsters, shrimps, and crayfish, animals that began their own evolution almost contemporaneously with the belemnite–squid line. The cuttlefishes (Sepioideae), relatives of the squid that evolved flattened bodies and became ecologically comparable to sole, flounders, and other flat fishes, can be found in low tide areas along ocean shores.

Though a cephalopod, the octopus has completely divested itself of shell. Moreover, it has lost the two basic adaptive devices that cephalopod ancestors exploited successfully for 400 million years. Octopi do not possess the hydrostatic mechanism that enables other cephalopods to remain suspended in the ocean, and they can move only short distances by jet propulsion. Nevertheless, they are among the most successful, actively evolving marine animals of our time. Their success appears to be based on their nearly perfect adaptation to preying on the crab, an animal with which they coevolved.

Octopi eyes and brains are as well developed as those of fishes and most reptiles and are far superior to those of any other aquatic animals without a backbone. The brain has several lobes and is enclosed in its own brain case. Octopi can be taught to recognize the presence of food-specific shapes of rectangles and other geometrical forms. They can recognize several different colors. Their memories can retain these associations for as long as two to three weeks. The value of these capacities for efficient crab hunting is obvious. It is no accident that the rapid diversification of species and genera in the octopus family during the Tertiary period was contemporaneous with a comparable and even greater diversification of crabs. This diversification took place in the tidal and shallow water coastal zone in which other kinds of marine animals had been evolving for millions of years but which had been closed to cephalopods until the evolution of the cuttlefish and octopus lines.

GENERAL SIGNIFICANCE OF MOLLUSCAN EVOLUTION

The three major classes of molluscs became recognizable at the end of the Cambrian period, and since then they have been following independent paths of evolution. Being principally marine, they have been exposed to similar changes in both the nonbiological environment and the marine biota. Nevertheless, they have evolved very differently with respect to overall form, diversification of genera and species, and habitats. *Gastropods*, which include snails and slugs, are the only animals to succeed both in fresh water and on land. Surprisingly enough, they have deviated little from their original body plan; a modern garden snail looks much like its primitive marine ancestors. Most gastropods feed either on detritus, seaweeds and other algae, or leafy land plants. A few have become parasites, and others have evolved into predators that attack other molluscs by boring into their shells. None of them have become active predators that can hunt, grasp, and devour their prey. Their muscular and nervous equipment for motility was from the start so deeply committed to carrying them about by crawling over a surface that nothing like limbs, fins, paddles, tentacles, or jet propulsion ever evolved.

The pelecypods or bivalves—which include oysters, clams, and scallops—became standardized early on as stationary filter feeders that sift small particles of food through their gills. Their nervous and muscular equipment became committed to opening and closing their shells and to digging themselves deeper into the sand. Scallops have evolved a modest jet propulsion mechanism that enables them to move away from attacking starfish, but this is extremely limited compared to that of cephalopods. Among pelecypods we find the most dramatic examples of long evolutionary stabilization in the molluscan phylum. The prime example is the oyster, which has changed little over 300 million years. Nevertheless, modern pelecypods have differentiated almost twice as many families as have modern cephalopods, and they are comparatively even richer in genera and species. This is due in part to a lower rate of extinction; no group of pelecypods or gastropods has suffered the amount of extinction that overtook the ammonoids during the latter half of the Cretaceous period.

The origin of the molluscan body plan is obscure and somewhat controversial. The pattern of cleavage by means of which a single-celled egg forms two, four, and eight cells of the earliest embryo, as well as the form and structure of the young and floating or swimming larvae, are much alike in molluscs, segmented worms, and joint-legged animals. This suggests a remote common ancestor for the three phyla. On the other hand, the tube-within-a-tube design differs in molluscs from that in the two supposedly related phyla in that the molluscan body cavity is not divided into many separate compartments. In fact, the single uncompartmentalized body cavity found in many arthropods is probably a derived feature.

EVOLUTION OF SEGMENTED WORMS AND JOINT-LEGGED ANIMALS

Segmented worms (annelids) such as earthworms, bristleworms and leeches share a basic body design with joint-legged animals (Arthropods) that include crabs, lobsters and other crustaceans, barnacles, and several classes of land animals such as scorpions, spiders, centipedes, millipedes, and insects. The latter class, which contains 26 orders and between 1 million and 2 million species, is by far the most diverse of all animals (see Figure 8-8).

The modification of the tube within a tube that led to this diversity was an elongate worm whose body cavities between the outer wall and the inner digestive tube, as well as the tissues of the outer tube, are divided into several compartments or segments (see Figure 8-3). This design is ideal for crawling rapidly over the bottoms of marine pools, as bristleworms do. Moreover, a long slender animal can escape predators with relative ease by hiding in crannies and crevices. In the joint-legged animals, which may well have descended from similar ancestors, further protection from predators was achieved through thin armor plates of tough, impermeable, but flexible protein known as chitin.

The elongated, jointed, and compartmentalized design lent itself to the acquisition of chitin-covered limbs with muscles attached to firm plates of the outer skeleton. Because of the larger outer surface to which their skeletal muscles are attached, the more advanced arthropods can move their appendages much faster than any other animals (the exasperating whine of a mosquito is produced by wings that beat back and forth more than a thousand times a second). Another source of arthropod diversity is the great plasticity of the chitin that makes up their external skeletons. It can be molded into organs as different as the rapidly flexing tail of the shrimp, the threadlike feelers or antennae of the shrimp and lobster, the elaborate carapace of the scarab beetle, the swiftly moving paddles of the diving beetles, and the ornate, shimmering wing of the butterfly.

The main restriction that their external skeleton has imposed on arthropods is limitation of size. Their external skeleton grows more slowly than the body inside it, and so must be shed repeatedly. A small animal can easily hide from predators during its highly vulnerable period of acquiring a new external covering, but hiding becomes increasingly difficult as the animal becomes larger. (The internal skeleton of vertebrates grows more slowly than the soft outer surface layers of its body, but it is permanent and keeps up with the growth of the body as a whole.) Moreover, the outer skeleton of an arthropod becomes increasingly heavy as the animal grows larger. Increasing size requires increasingly

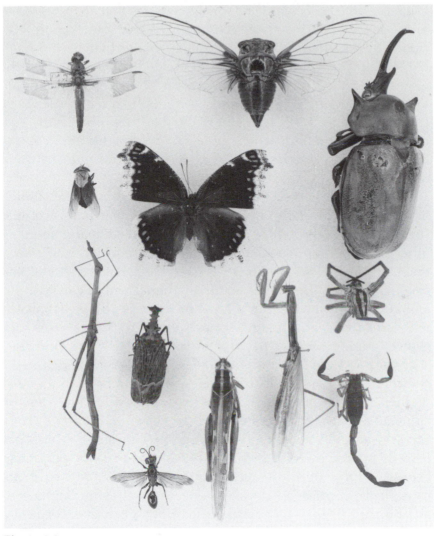

Figure 8-8.

Representative joint-legged animals (Arthropods). *Top row*: (*left*) Dragonfly
(Odonata, Libellulidae); (*right*) cicada (Orthoptera, Cicadidae). *Second row*: (*left*)
Deer fly (Diptera, Tabanidae); (*middle*) mourning cloak butterfly (Lepidoptera,
Nymphalidae); (*right*) scarab beetle (Coleoptera, Scarabeidae). *Third row*: (*left*)
Stick insect (Orthoptera, Phasmatidae); (*middle left*) leaf hopper of rose
(Homoptera, Fulgaridae); (*right center*) praying mantis (Orthoptera, Mantidae);
(*right*) wolf spider (Arachnida, Lycosidae). *Bottom row*: (*left*) Solitary wasp
(Hymenoptera, Sphecidae); (*middle*) grasshopper (Orthoptera, Acrididae);
(*right*) scorpion (Scorpionidae). (Courtesy of Lynn Kimsey, Entomological
Museum, University of California, Davis.)

extensive attachment surfaces for the muscles that are required to move large limbs. Such surfaces are provided much more efficiently by the internal limb bones of vertebrates than by the external skeleton of arthropods. It is no accident, therefore, that the largest arthropod species that are successful in modern oceans are lobsters, weighing 8 to 9 pounds. Although larger species have existed in the past, the largest terrestrial arthropods that are successful in the modern world are a few centipedes about a foot long, and a few species of tropical spiders about four inches in diameter.

This restriction on size of arthropods imposes a second limitation that is particularly important to insects that have acquired social behavior. They are relatively small, even compared to many other arthropods. A small animal cannot have as large a head or brain as a large one. Experiments by zoologists have shown that small brains can sometimes equip an animal with a considerable amount of intelligence, but a good memory is always associated with a relatively large brain. Termites, ants, wasps, and bees have evolved highly complex patterns of behavior. In some instances, such as the "waggle dances" of bees, their behavior requires a considerable amount of learning. Nevertheless, they cannot build on their memories to acquire more knowledge and perception by cumulative learning, as do mammals and birds and to some extent reptiles, amphibians, and fishes. Learning in adult insects has been compared by zoologist E. O. Wilson to the developing of an exposed film with information imprinted on it by innate, gene-controlled processes of development. By contrast, chimpanzees and humans begin their process of learning when their brains are still immature, and their elaborate, flexible, and intelligence-dominated patterns of behavior are possible only because they have already learned simple patterns and can build on them. For this cumulative process, a memory bank of information with a large storage capacity is essential. The basic body plan of insects does not allow for such a memory bank.

The body plan found in vertebrates, including ourselves, is shared by some marine phyla, including the Echinoderms (starfishes, sea urchins, and sea cucumbers). The wormlike ancestor of this group of phyla is unknown, either living or as a fossil, but probably it was a burrowing worm that lived on the bottom of shallow seas and possessed tentacles at its upper end, by means of which it captured and devoured microscopic forms of life. Its body cavity probably consisted of two or three compartments. In Chapter Nine we will discuss the long and complicated evolutionary pathway from such worms to higher vertebrate animals.

DIFFERENTIATION OF ANIMAL PHYLA

Zoologists divide the animal kingdom into about 25 major phyla. These phyla do not form a ladder of increasing advancement or complexity; rather, their origin is the result of a few successive adaptive radiations. Between and following these radiations, several evolutionary lines simultaneously diverged from each other (Figure 8-9). Because the animals involved were all soft-bodied and are now extinct, the pathways of their divergent radiations probably will never be accurately traced. Eminent zoologists such as Jägersten, Beklemishev, and Valentine, on whose speculations my own are based, have been investigating them for many years.

The first truly multicellular animal, ancestor to all the modern phyla except the aberrant sponges, could have been a member of the jellyfish–coral phylum (Coelenterates). It probably evolved from floating colonies of ciliated protozoan cells. As an adult, it had two layers of cells organized according to the blind sac body plan. One of the most important features of this early animal was that it led a dual life. As with the earliest representatives of nearly all the major phyla, its young were many-celled *larvae*, equipped with rows or tufts of protoplasmic hairs or cilia, that floated and drifted through the upper layers of the ocean (Figure 8-10). These larvae subsisted on microscopic one-celled plankton, which they ingested through their rudimentary mouths. Before becoming capable of differentiating eggs and sperms to reproduce their kind, the larvae settled on the bottom of the ocean and became modified into *sessile* (nonmotile) organisms with waving tentacles that could

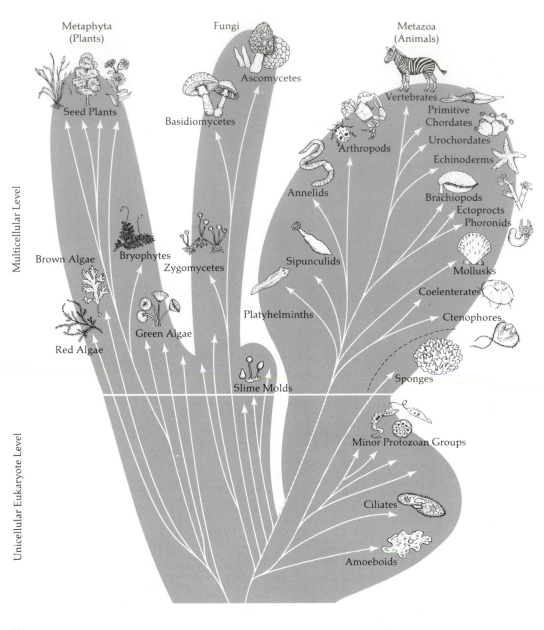

Figure 8-9.

The family tree of nuclear (eukaryotic) organisms. (From J. W. Valentine, "The Evolution of Multicellular Plants and Animals." Copyright © 1978 by Scientific American, Inc. All rights reserved.)

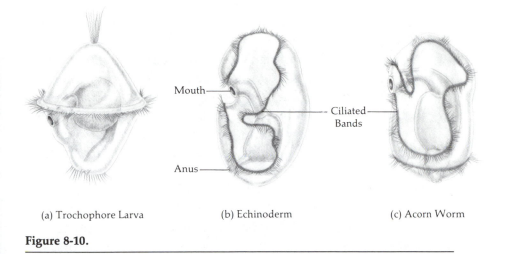

Mouth
Ciliated
Bands
Anus

(a) Trochophore Larva   (b) Echinoderm   (c) Acorn Worm

Figure 8-10.

Three kinds of floating or swimming larvae that develop into sedentary marine animals. (a) Segmented worm or mollusc (trochophore larva). (b) Sea urchin or starfish (Echinoderm). (c) Acorn worm (*Balanoglossus*); the adult has some anatomical structures similar to those of vertebrates.

catch small animals and convey them to their baglike stomachs. Some such type of floating larvae exists in nearly every modern phylum, and about 80 percent of the invertebrate animals now living in or near the sea today have a life cycle that alternates between a floating larval stage and a bottom-dwelling adult stage.

Because of this dual existence and the entirely different organism–environment interactions associated with the two separate stages, selective pressures on the larvae and on the adults would have differed greatly, causing independent divergence of the two stages. This is the best explanation of the fact that molluscs and segmented worms (annelids) have very similar larvae but completely different adult stages. However, the great differences between both the larvae and the earliest embryos of many phyla, particularly the mollusc–annelid–arthropod group as compared to the echinoderm–chordate group, are much more difficult to explain. The tube-within-a-tube body plan may have evolved from the blind sac in at least three and possibly more separate groups. This means that the separation between the two major groups of land animals—joint-legged and vertebrate—goes back to the original jellyfishlike ancestor.

During the late Precambrian to early Cambrian periods, when animal phyla were becoming differentiated from each other, the oceans were becoming filled with animals that alternately preyed upon others and were preyed upon. Biotic communities had not yet reached the relatively stable condition of saturation with different forms of life that they now possess. The autocatalytic nature of animal evolution must have been manifest everywhere. One challenge followed another, extinction and replacement must have been commonplace, and natural selection favored numerous harmonious combinations of genes that made animals capable of exploring new ways of life and successfully eluding predators. The unstable nature of biotic communities in their formative stages may have been the principal cause of the appearance of nearly all the animal phyla during the relatively short period of 150 million years, from about 680 million to 530 million years ago. This time period is certainly long enough for genetic changes to produce the different phyla; we need only assume strong selection pressures on populations with gene pools capable of generating new and adaptive gene combinations.

We can come closer to understanding the environmental challenges that triggered the evolution of new phyla and classes by considering additional aspects of marine ecology, particularly the commonest ways by which animals feed. Many of them—such as wheel animalcules (rotifers), clams, mussels, and various kinds of marine worms—scoop up indiscriminately a great variety of microorganisms that are suspended in water. They do this by means of various structures that surround the mouth and suck in food by creating currents of water. Animals of this kind, known as suspension feeders, have not evolved mechanisms for capturing particular kinds of prey. Many of them are firmly fixed to the ocean floor and move around little or not at all. They are often buried in sand or mud. Their most common specialized adaptations are those that protect them against the battering of the waves or against predators that might dig them up and eat them.

The body cavity and tube-within-a-tube organization probably evolved in animals of this kind. Various modern marine worms live in underground burrows, whose walls they fortify by secreting a cementlike substance. Such burrows are among the oldest trace fossils known from Precambrian times. An efficient way of life in such burrows makes use of a hydrostatic skeleton. When ready to feed, the animal blows itself up like a balloon by taking water into the body cavity; the water pressure causes the mouth in the front part of the body to emerge from the burrow, where it sucks in its food by a suspension-feeding mechanism. When danger appears, or when the animal has finished feeding, the cavity is emptied of water and the animal retreats into its burrow.

The body cavity—that most basic element of the body plan of the higher animals—probably evolved as a response to the challenge of remaining sedentary and being able to feed under the hazardous conditions that prevailed along the seashore. Other seashore animals live on the bottoms of tide pools and forage there for dead organic matter—they are known as detritus feeders. Their biggest problem is how to escape from predators while feeding. Two main methods of protection have evolved. Snails, abalones, and their relatives live in hard shells firmly attached to rocks by a muscular organ known as a foot, which makes them difficult or impossible to pry loose. The segmented worms and joint-legged animals have acquired thin, flexible armor and the ability to swim quickly away from danger. This latter strategy opened the way for differentiation into the large number of modern animal species, particularly those belonging to the class of insects.

Additional adaptive strategies adopted by differentiating phyla involved swimming through the water rather than living on the bottom. The first swimmers, which included a few worms like the arrow worm (*Sagitta*) and the lancelet (*Amphioxus*), were suspension feeders. The importance of their strategy lies in the fact that it led to the evolution of fishes and so to land vertebrates, including humankind.

A tree shrew, native to the tropical rain forests of Malaya. The shrew is related to mammals that existed 80 million years ago; it forms a link between them and the order of Primates, to which apes and humans belong. (Zoological Society of London.)

Chapter Nine

FROM WORMS
TO PRIMATES

Zoologists call the phylum to which humans belong *Chordata*. Included in it are fishes, amphibians, reptiles, birds, mammals and a few rare marine creatures that either possess certain anatomical structures similar to those found in fishes or develop embryo or larval stages similar to those of extinct transitional forms between wormlike animals and fishes.

The roots of the chordate phylum are found in a group of primitive multicellular animals that may have had a common body plan with early echinoderms (the ancestors of modern starfishes, sea urchins, and sea cucumbers). Although the adults of these animals are vastly different, the larvae of the two phyla resemble each other so closely that evolutionary relationships between chordates and echinoderms can hardly be doubted (see Figure 8-10).

The evolutionary pathway from the burrowing, tentacled worm that was the most likely intermediate stage between jellyfishes and the chordate–vertebrate line is not known, but it must have been governed largely by interactions between its adult form and the sea-bottom environment. The free-floating larval stage was reduced and eliminated. Perhaps this was because the animals lived in small tidepools partly separated from the open ocean or in estuaries, where plankton food would have been scarce and continual foraging over the bottom would have had an adaptive advantage.

The first step in the evolution toward fishes was the elimination of the burrowing worm stage. The larvae developed into sessile adults fixed on the ocean floor, sweeping in food with their tentacles and eliminating food and waste through their gill slits. Modern creatures of this kind are tiny rare pterobranchs, which live on the bottom of deep oceans, chiefly in the Southern Hemisphere. They form small colonies and protect themselves by forming simple or branching tubes (see Figure 9-1).

The extinct ancestors of pterobranchs are unknown as fossils. Nevertheless, the modern form, and certain animals like the acorn worm (*Balanoglossus*), which are widespread along sandy seashores, have anatomical characteristics and larval development that strongly suggest their relationship to ancestors of the vertebrates.

When sedentary, filter-feeding, wormlike animals were evolving into free-living fishlike animals, an important shift in the early nutrition of the embryo took place. While growing in its mother's ovary, the egg cell acquired a large amount of fatty yolk, which the embryo could digest before beginning its independent larval life. This shift from filter feeding to dependence on nutrition from the mother and egg occurred independently many times in various lines of animal evolution. It usually accompanied the shift from a dual existence to continuous development either on the sea bottom or in the open ocean. Its occurrence and adaptive importance are the first manifestations of a principle that will become increasingly evident as we follow the evolutionary pathway that led to humanity: Increasing complexity of both form and behavior is greatly promoted by increasingly close relationships between parents and their offspring.

The course of evolution from sessile colonial animals to free-swimming fishes is not documented by the fossil record. Educated guesses can be made, however, as to the probable characteristics of the "missing link." Attention has focused on *Amphioxus*, a small, almost transparent animal a few inches long, shaped like a slender fish (Figure 9-2). As in fishes, the front end of its body is perforated on both sides with a row of gill slits, but it lacks a distinct head and has neither jaws nor eyes. Its mouth is sur-

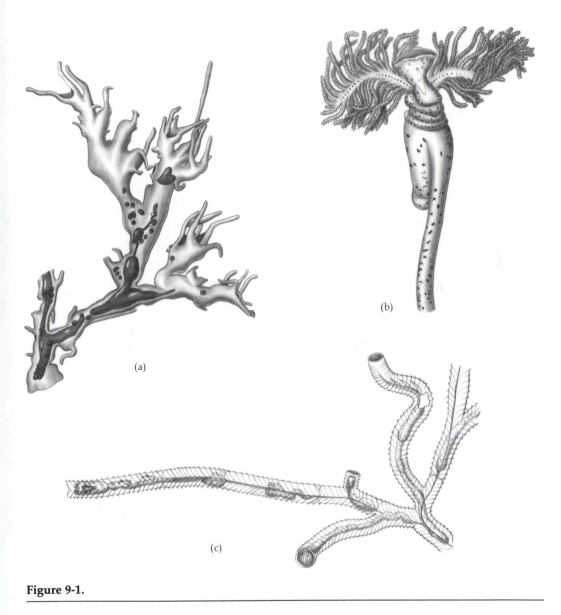

Figure 9-1.

Pterobranchs, colonial wormlike animals that resemble the probable ancestors of *Amphioxus* and vertebrates. (a) Portion of a colony. (b) A single animal. (c) Cross-section of branches. (From E. J. W. Barrington, *The Biology of Hemichordata and Protochordata*, Oliver & Boyd, 1965.)

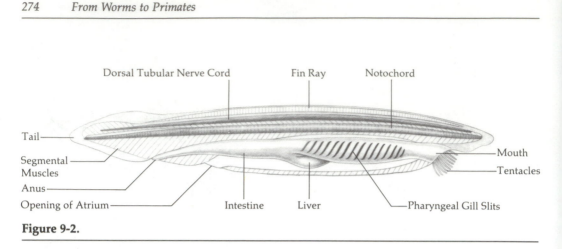

Dorsal Tubular Nerve Cord   Fin Ray   Notochord

Tail

Segmental
Muscles

Anus

Opening of Atrium   Intestine   Liver

Mouth

Tentacles

Pharyngeal Gill Slits

Figure 9-2.

The lancelet (*Amphioxus*), a possible ancestor of vertebrate animals.

rounded by a row of short tentacles, by means of which it sucks in water and the microscopic sea life on which it feeds. Although it can swim, it does so rather poorly, and it spends most of its time embedded in sand, with only its front end protruding. Not surprisingly, this fragile creature has no ancestors or counterparts in the fossil record.

The prevailing opinion among zoologists is that *Amphioxus* was derived from larvae of sessile animals similar to pterobranchs. Perhaps, although less probably, the ancestral adults were like acorn worms or the saclike forms known as sea squirts or tunicates. In any event, the conversion from a sessile adult to a fishlike, swimming animal may have resulted from a shift in the timing of differentiation of the reproductive organs. If they came to be differentiated early on, at a stage that ancestrally had been that of an immature, free-swimming larva, many of the larval characteristics (for example, mobility) could have been preserved. This is the phenomenon of neoteny (first discussed in Chapter Four), probably the final outcome of selection pressure for prolonging larval life. As the sea bottom became crowded with a great variety of different kinds of animals, open niches in which small, defenseless larvae could find a resting place became scarcer and scarcer, so that survival depended on the ability of the larval "scout" to continue its explorations for increasingly longer periods before coming to rest. *Amphioxus* has evolved an admirable solution to this problem, since at almost any developmental stage it can burrow quickly into the protective sand, then move on when conditions permit.

The internal anatomy of *Amphioxus* contains two structures that relate it closely to fishes—a spinal cord and a *notochord*. The first is a bundle of nerve fibers that extends the length of its back; the second is a firm rod that maintains the animal's fishlike shape and enables it to swim without telescoping. The notochord consists of cells that contain discs of muscle surrounded by fluid-filled spaces. Muscular contraction stiffens the rod and the animal; relaxation makes it more flexible.

Animals like *Amphioxus* gave rise to primitive jawless fishes through changes in the head region, body covering, and internal skeleton. They developed eyes, nostrils, a rudimentary brain, and bony plates that formed the beginnings of an internal skull. Behind the head region, their bodies became covered with bony scales that in some forms were not very different from those of many modern fishes (Figure 9-3). The notochord was retained and

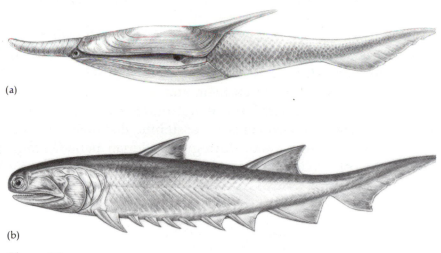

(a)

(b)

Figure 9-3.

(a) A primitive jawless fish, *Pteraspis*, that lived during the Ordovician period.
(b) A primitive jawed fish, belonging to the group of spiny-fins (Acanthodii) and near the ancestral line of modern fishes.

became partly surrounded by the beginnings of vertebrae and a backbone. Their mouths were still jawless, but at least some possessed special bony plates surrounding their mouths, so that they could nibble as well as suck in food. The most conspicuous feature of the earliest fishes, which disappeared in their descendants, was a large bony carapace that covered the head region.

Jawless armored fishes (Ostracoderms) are well preserved, common, and varied in rocks of Silurian and Devonian age, and scraps of their scales and bony armor have been found in Ordovician and Cambrian strata. These fishes overlap in time with the earliest bony fishes, sharks, and lungfishes, all of which were present in the Devonian period. Ostracoderms lacked the side fins acquired by later fishes to help them in directional swimming and turning. Modern jawless fishes include the eellike lamprey and the hagfish (Cyclostomes), which bores into the sides of other fishes and parasitizes them.

Teeth and jaws first appeared in fishes known as spiny-fins (Acanthodii), small minnows with sharklike tails equipped with well-developed lateral fins (Figure 9-3). There can be little doubt that the teeth of spiny-fins were derived from the body-covering scales that existed near their jaws.

The spiny-fins were in every way true fishes, with well-developed bony skeletons as well as jaws, lateral fins, and scales similar in substance and structure to those found in typical bony fishes. Nevertheless, the gap between them and the contemporary jawless armored fishes is large and is not bridged by intermediate forms. This gap is best explained by assuming that their common ancestor lived in fresh water during the Silurian period. Many Silurian deposits contain fragmentary fish scales from both spiny and armored fishes.

FROM FISHES TO
AMPHIBIANS AND REPTILES

During the early Devonian period, the lungfishes (Dipnoi) and the lobe-fins (Crossopterygii)—two related groups of fishes—coexisted. Both had lungs, but more important, they had thick, fleshy lateral fins. This characteristic distinguished them from all contemporary and later fishes.

Lungfishes and lobe-fins had existed as separate groups since at least the beginning of the Devonian period, soon after the appearance of the spiny-fins. The connection between these two kinds of fishes is hidden in the imperfections of the fossil record, but apparently they differed greatly from each other in both diet and locomotion.

Lungfishes were sluggish creatures, large and heavy, with relatively slender side fins that helped anchor them to aquatic vegetation in shallow water. During the Devonian and the first half of the Carboniferous period, they gave rise to several different forms, but after flourishing for about 70 million years, they gradually declined, becoming almost extinct by the end of the Triassic period, 130 million years later. Three species have survived in contemporary lakes and rivers—one in Australia, another in tropical Africa, and a third in South America. They are classic examples of species that have changed little during the last 190 million years.

Lobe-fins had strong, muscular bodies and large side fins that equipped them well for swimming or paddling over the bottom of shallow water. Their numerous sharp teeth could easily have impaled the smaller fishes, shrimps, or other actively swimming animals that were their probable prey. The hind ends of their bodies, where many of the fins were concentrated, would have enabled them to make sudden lunges after swift-moving prey.

During the Devonian period, these kinds of fishes evolved in many directions (Figure 9-4). In one branching line, lungs were converted to a swim bladder filled with fat and connective tissue that served as a hydrostatic organ, enabling the fish to suspend itself motionlessly in the water. This line migrated from fresh water into the ocean, where they evolved into the coelacanth fishes that flourished until the Cretaceous period, then became almost extinct. The discovery during this century of a living coelacanth, *Latimeria*, off the Comoro Islands in the Indian Ocean was one of the most dramatic finds of the century, but its importance must not be overemphasized, since it tells us little about the nature of fishes that gave rise to amphibians. Its ancestors separated from

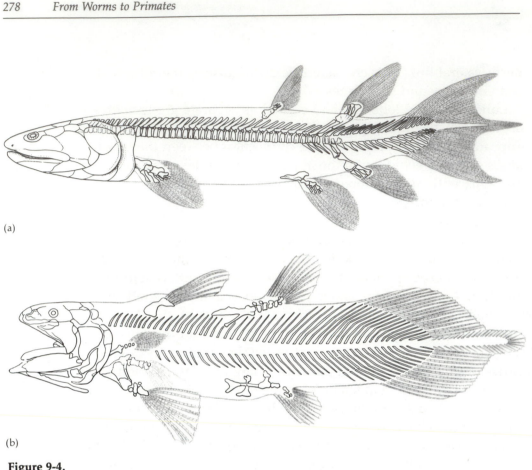

(a)

(b)

Figure 9-4.

Two kinds of ancient lungfishes. (a) A rhipidistian (*Eusthenopteron*), related to the ancestor of the amphibians. (b) A coelacanth (*Laugia*), which may have been the ancestor of the modern genus *Latimeria* that lives in the ocean east of tropical Africa.

the line leading to land animals 375 million years ago and for 250 million years evolved in another direction, toward adaptation to the marine habitat. The conversion of lungs to swim bladders took place not only in the coelacanths but also in the main line that led to bony fishes. These organs are present in all modern fishes except sharks, lungfishes, and the rare bichir of Africa.

During the Devonian period, the main line of lobe-fins evolved into the rhipidistians, which gave rise at the end of this period to amphibians. The evolution of a land animal from an aquatic fish is certainly a profound change, but it did not occur suddenly, even in terms of the geological time scale. Lungs and air-breathing evolved about 70 million years before feet, and even after the amphibians had acquired feet, they were by no means land animals. Feet may well have evolved as an adaptation to crawling over the bottom of shallow lakes and streams, where the water was too shallow to permit these rather large animals to swim through it. Many modern salamanders that spend their entire lives in the water—for example, the mud puppy *(Necturus)* of the central United States—use their feet in exactly this way. Even toads, which live as adults in gardens and similar habitats, must return to the water to lay their eggs. Complete freedom from aquatic life was not attained until the middle of the Carboniferous period, 40 million years after the appearance of the first four-footed amphibian and 90 million years after the appearance of fishes that had lungs as well as bones in their fins that corresponded to the bones of amphibian feet.

THE ORIGIN OF FISHES AND LAND VERTEBRATES

The evolution of early fishes illustrates in a striking fashion several principles that have already been discussed in earlier chapters. The first is that evolution goes much faster and produces much more diversity in organisms that depend for their existence on complex relationships with other organisms (for example, active predators and their equally active prey) than in organisms that interact more directly with their physical environment or in a more general way with other organisms. This principle is well illustrated by the evolution of lungfishes as compared with evolution of the advanced lobe-fins that gave rise to amphibians. Teeth equipped lungfishes for feeding on all sorts of sessile shelled animals, which they could gather without having to chase them. Lungfishes lived in shallow lakes or stagnant streams that finally became oxygen-poor. When this happened, lungfishes gave up breathing through gills and obtained oxygen by rising to the surface, gulping air through their mouths and storing it in their lungs. During particularly prolonged periods of drought, when their shallow pools dried up, lungfishes were not equipped either to return to deeper waters—where they would have been attacked by actively swimming predators—or to venture onto the land. Instead, they encased themselves in cocoons made of mud and remained dormant until the drought was over. Modern lungfishes are all adapted to survive in just this way. The naturalists who first studied the African lungfish in its home on the shores of Lake Victoria were astounded to discover that, if an African lungfish is plunged directly into water while in its mud cocoon, it drowns. Since it cannot breath either through its gills or by coming to the surface, it cannot revive. It must be moistened and awakened gradually while its gills recover their capacity for water-breathing.

This peculiar mode of life is directly connected with the evolutionary stagnation that has overcome the lungfishes during the past 190 million years. After they perfected their peculiar form of existence, they became so highly specialized that they could not give rise to descendants having a different mode of life, and so they persisted only in habitats where the population–environment interactions remained unchanged. This sluggish mode of existence and evolution contrasted sharply with the evolutionary life of rhipidistian fishes. Being active predators, they responded to the drying up of these same pools with more action. They managed to maintain an active life through the critical Devonian droughts either by seeking prey among the joint-legged animals—primitive centipedes that were beginning to populate the land—or by waddling slowly from one pool to another that was less threatened. Consequently, their descendants, which were full-fledged amphibians, flourished in the widespread, teeming swamps that covered much of the earth during the Carboniferous period.

The principle that extensive adaptive radiation is triggered when a group enters into a habitat to which it is better adapted than any of its competitors is illustrated by the expansion and successive adaptive radiations of bony fishes from the Carboniferous period through the Cretaceous period and the failure of either the coelacanths or the sharks to keep up with them. Bony fishes radiated extensively during the Carboniferous and Permian periods contemporaneously with the early amphibians and reptiles, first in fresh water and later in oceans. Nearly all the radiants of this early group (Paleoniscids) became extinct, but during the Triassic and Jurassic periods a few of their descendants that had acquired greater specializations of fins, scales, and head structure (Holostei) underwent a second series of adaptive radiations. Fishes constructed according to this intermediate plan also became extinct, except for a few modern survivors—but before they did so, they gave rise to the still more refined and efficient bony fishes (teleosts) that are the dominant fishes of our own time. This radiation took place during the Cretaceous period, contemporaneously with the decline of the dinosaurs and the rise of modern mammals and birds.

By the time the lobe-fins had radiated into the oceans, they were in direct competition with the more numerous and efficient ray-fins, so that they were successful in only a few niches. They achieved a modest amount of diversity during the Permian and Triassic periods, but after that they declined, surviving only in the deep-sea habitat that is marginal even for ray-fins. Their only living genus, *Latimeria*, has already been mentioned. Fishes that have evolved adaptations for survival in dark, lonely recesses on the bottom of deep oceans face much less competition than those that are adapted to life near the surface and in other aquatic habitats.

Sharks have been highly successful, in a limited way, but have diversified relatively little because of competitive disadvantages in all but a relatively small number of oceanic habitats. Shark skeletons consist of soft cartilage rather than hard bone, and this group lacks the bladder that serves bony fishes as a hydrostatic balancing organ. This restriction does not interfere with the shark's success in open ocean habitats, but it puts sharks at a great disadvantage relative to bony fishes in shore and tidepool habitats. This has been one factor in preventing the larger modern sharks from entering fresh water. The first sharks—some of them freshwater animals—appear in the fossil record somewhat later than bony fishes, and for more than 350 million years they have existed side by side with bony fishes in the oceans, but they have always been inferior in numbers and diversity to the bony types.

A third principle, illustrated by the evolution from lobe-finned fishes to amphibians, is that animals can make an adaptive shift to a new way of life only if existing structures prepare them for the shift. These structures must be such that they can be modified to serve the new function rather easily. In other words, the animals may be regarded as *preadapted* to making the big shift. Preadaptation is usually based on one of two kinds of situations: (1) In the preadapted animal, some vital function, such as breathing, may be carried out by either of two structures, such as gills or lungs. (2) Alternatively, certain structures may be versatile enough to carry out a necessary function in both old and new habitats.

The advanced lobe-fins that were ancestral to amphibians possessed both these kinds of adaptations. In open, well-oxygenated water they were gill-breathers, but in water having little oxygen they rose to the surface and became air-breathers, making use of their lungs. Their thick, firm, muscular fins enabled them to waddle over the bottom of shallow water, thus serving as crude legs even when they were fishlike in appearance. Only these fishes possessed this combination of characteristics, and only they gave rise to terrestrial descendants.

Behavior often acts as the pacemaker of evolution. The three related groups—lungfishes, coelacanths, and lobe-fins—are believed to represent predators with three different kinds of predatory behavior. Lungfishes subsisted on armored, slow-moving prey that they crushed with their massive teeth; consequently they became increasingly sluggish and finally adapted themselves to an extreme habitat by avoiding the most unfavorable season with dormancy. Coelacanths were constantly swimming and capturing active smaller fishes by chasing them. Consequently, their successful descendants acquired swim bladders, and they diversified in open marine habitats. Advanced lobe-fins may have spent much of their time half-buried in mud, but they were always ready to make sudden lunges at actively moving prey. This form of predation would have been equally effective in catching either active fishes or crustaceans and other arthropods. Consequently, as the ancestors of scorpions, spiders, centipedes, and insects started to invade terrestrial habitats, some of the rhipidistian lobe-fins or their amphibian descendants followed them in this direction and became exposed to selective pressures that hastened their evolution toward genuine land animals.

Mosaic evolution is the term used in Chapter Four to describe the way in which different characteristics evolve at different rates in a particular evolutionary line. An animal that is transitional between two major groups is much more likely to be a mixture, or *mosaic*, of old and new characteristics than to be intermediate between two groups with respect to all its characteristics. This principle can be recognized most easily by comparing modern amphibia (salamanders, frogs), lobe-fin fishes, and the earliest amphibians with respect to fins, feet, skull, and skin covering (Figure 9-5). Both kinds of fishes have fins in the positions in which the two kinds of amphibians have paired front and hind feet; the fins end in thin flaps without a skeleton, and the feet end in toes or digits. Although the two kinds of fins are superficially similar and the overall difference in form between fins and feet is great, nevertheless the internal structures of the four kinds of appendages are much more similar. Moveover, the fins of modern bony fishes, as well as those of the remote common ancestors of bony fishes and amphibians, consist almost entirely of a nonskeletal flap with little internal skeleton. The fins of the later rhipidistians had a well-developed skeleton that can be compared almost bone for bone with the legs and feet of amphibians. In short, the rhipidistian fin was a mosaic that combined a fishlike flap with an amphibianlike skeleton.

When skulls are compared (Figure 9-5), the striking similarity between those of ancestral ray-fins, lobe-fins, and early amphibians becomes apparent. By contrast, both modern amphibians (salamanders, frogs) and modern bony fishes are quite different from their respective ancestors. The earliest amphibians, therefore, were a mosaic of rhipidistian fishlike skulls and essentially modern amphibian legs.

The situation with respect to skin covering is somewhat similar. The bellies, flanks, and backs of the earliest amphibians were covered with bony scales much like those that covered the bodies of rhipidistian fishes. By constrast, the scales of modern bony

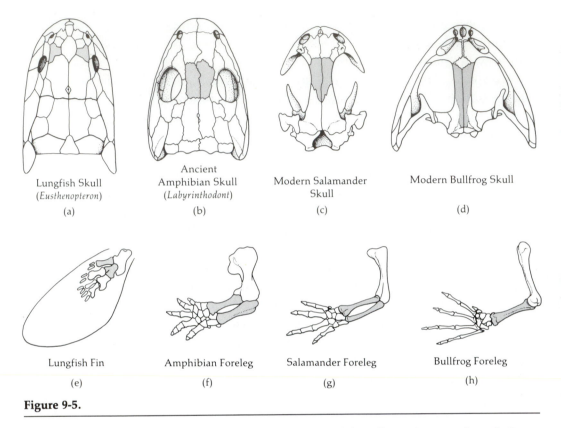

Lungfish Skull
(*Eusthenopteron*)

(a)

Ancient
Amphibian Skull
(*Labyrinthodont*)

(b)

Modern Salamander
Skull

(c)

Modern Bullfrog Skull

(d)

Lungfish Fin

(e)

Amphibian Foreleg

(f)

Salamander Foreleg

(g)

Bullfrog Foreleg

(h)

Figure 9-5.

Skulls and forelimbs from lungfish (*Eusthenopteron*) to amphibia, illustrating mosaic evolution. Note that the *Labyrinthodont* has a skull like that of the lungfish but a forelimb resembling those of modern amphibia.

fishes are much thinner and simpler in construction than those of either rhipidistians or the spiny-fins that were the ancestors of both rhipidistians and bony fishes. Modern amphibians differ from their ancestors by the complete lack of scales. If we consider the structure of the scales of primitive amphibians, we can call them mosaics that combined lungs and feet with fishlike scales. Since, however, some parts of their bodies were already scaleless, we can also regard them as transitional between fishes and modern amphibians with respect to skin coverings.

These few comparisons indicate that we simply cannot understand either the rates or directions of evolution by adopting a strictly taxonomical approach in which named groups are regarded as fundamental, integral units. Instead, groups must be regarded as *assemblages of characteristics* that form adaptive combinations with respect to particular population–environment interactions. The evolution of classes such as amphibians, reptiles, and mammals can best be understood by analyzing each characteristic separately with respect both to visible changes in form and to the selective pressures that brought them about. The final synthesis will be successful only if it is based on careful analysis of the separate characteristics.

The relationship between quantitative and qualitative differences in evolution is illustrated by comparing the fin of an advanced lobe-fin fish and the leg of an amphibian. Most biologists would agree that fins and legs are qualitatively different with respect to both appearance and function. Nevertheless, the totality of obvious differences can be reduced to a series of separate quantitative differences. In the fin, the terminal or distal skin element is longer and flatter than in the toes of the leg. The number of finger or toe bones is greater in the fin than in the leg, and each bone is much longer and narrower. With respect to the proximal part of the appendage—that is, the part nearest to the body—most of the bones present in the leg have their counterparts in the fin, but they differ quantitatively in both shape and relative distance from each other. Each of these quantitative differences could be expressed as a precise mathematical ratio, and they could be compounded so as to express a total quantitative difference that would be equivalent to the qualitative difference between the two kinds of appendages. If the same kind of comparison were made between the rhipidistian fin and that of a modern bony fish, the quantitative aggregate of differences might well turn out to be greater than the total quantitative differences between rhipidistian fin and amphibian leg. Thus the differences that we perceive as differences in kind or quality are usually—perhaps always—assemblages of quantitative differences so numerous and complex that they are not easily analyzed. (This topic is discussed in greater detail in Chapter Five.)

FROM AMPHIBIANS TO
REPTILES AND MAMMALS

From the earliest amphibians to *Homo sapiens*, the evolutionary pedigree is almost continuous. Complete gaps in the fossil record are minor. More common are gaps represented by only a few parts of bones, teeth, or skulls of some extinct animals.

The transition from amphibians to reptiles was a gradual one. A form known as *Seymouria* has in its skeleton a mosaic of amphibian and reptilian characters. According to paleontologist A. S. Romer, *Seymouria* "stands almost exactly on the dividing line between amphibians and reptiles." Several other animals that were *Seymouria's* contemporaries during the Carboniferous and Permian periods are similarly intermediate, but most are regarded as amphibians because of indirect evidence suggesting that they had a gill-bearing, aquatic, larval stage. The earliest or "stem" reptiles (Cotylosaurs) had somewhat narrower and higher skulls than the primitive amphibians, with eardrums on the sides of their heads rather than near the top and a somewhat different brain case.

The characteristic of reptiles that is usually recognized as the distinguishing feature of the class is their relatively large egg, which is enclosed in a tough, water-resistant outer covering and contains both a large amount of yolk and three different membranes. The most important of these membranes—the amnion—encloses the embryo in liquid (Figure 9-6). Since eggs are only rarely preserved in the fossil record, the evolution of this structure cannot be traced. A reasonable guess is that it evolved as part of a response of Carboniferous reptiles to predation of their young. Animals that acquired the characteristic of laying eggs on the still sparsely populated dry land and then evolved eggs that could withstand increasingly long periods of desiccation would have eventually outreproduced those whose eggs were laid in watery areas where aquatic predators abounded.

The transition from reptiles to mammals took longer than that between any other two classes of animals (Figure 9-7). It began during the Upper Carboniferous or Pennsylvanian period with a now-extinct order called pelycosaurs. During the following period—

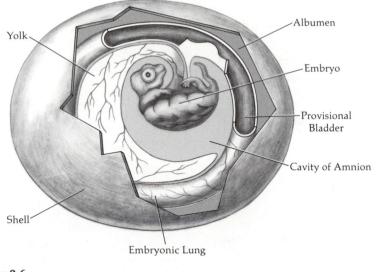

Yolk

Albumen

Embryo

Provisional Bladder

Cavity of Amnion

Shell

Embryonic Lung

Figure 9-6.

Semidiagrammatic view of a turtle's egg, typical of reptiles. The cavity of amnion is filled with fluid, providing the embryo with a water bath similar to that which surrounded the embryos of the reptile's ancestors. The ample food supply allows the young turtle to develop fully before it hatches; the tough, water-resistant outer shell protects it from the drying effects of the atmosphere. (After A. S. Romer, *The Procession of Life.* Cleveland: World Publishing Co., 1968.)

the Permian—they shared with stem reptiles the domination of terrestrial habitats. The most distinctive feature of pelycosaurs, which they handed down to all their descendants, was the pattern of openings in the skull that are needed for the functioning of sense organs. Amphibians have only two pairs of such openings— one for the nostrils and another for the eyes. This condition is retained in stem reptiles and in one modern reptilian order, the turtles. All other modern reptiles, as well as the extinct dinosaurs, characteristically have four pairs of openings—one for the nostrils, one for the eyes, and two in the basal region of the skull for the ears.

The pelycosaurs evolved gradually into another order, the mammallike reptiles called therapsids. The break between the two orders is artificial and is due to the accidents of fossil-bed pres-

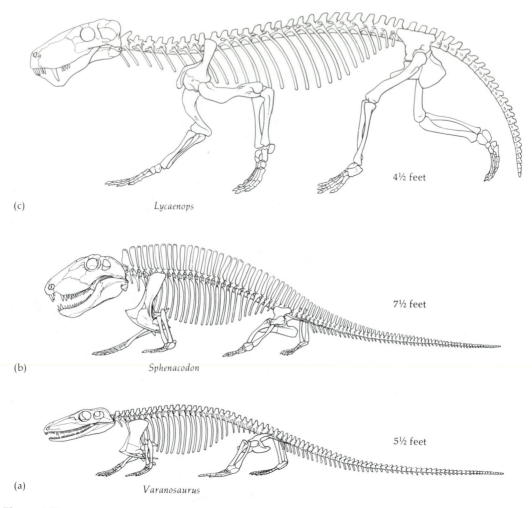

(c)   *Lycaenops*   4½ feet

(b)   *Sphenacodon*   7½ feet

(a)   *Varanosaurus*   5½ feet

Figure 9-7.

Skeletons of three reptiles, illustrating changes in structure preadaptive to the evolution of mammals. (a) *Varanosaurus*, which lived about 270 million years ago (Lower Permian) and was in all respects lizardlike. Its leg bones were relatively small and could not have supported the animal's body if it was raised continuously from the ground. It was about 5½ feet (1.4 m) long. (b) A more advanced pelycosaur, *Sphenacodon*, which lived contemporaneously with *Varanosaurus*. Its limb bones were larger and stronger, although not large enough to enable it to walk erect as mammals do. The beginnings of tooth differentiation can be seen. The animal was about 7½ feet (2.3 m) long. (c) A mammallike reptile, *Lycaenops*, which lived about 240 million years ago. It walked erect as mammals do and had prominent canine teeth. Its other teeth, however, were reptilian and formed the bones of its lower jaw. It was smaller, about 4½ feet long. (Parts a and b are from Romer and Price, Geological Society of America Special Paper No. 28, 1940; part c courtesy of The American Museum of Natural History.)

ervation. Most pelycosaurs have been found in fossil beds of Lower Permian age in the American Southwest and in Europe. Mammallike reptiles have been dug up by the thousands from the Karroo beds of South Africa, which are among the richest of any known fossil deposits. They were laid down in the Upper Permian period, 10 million to 30 million years later than the early Permian beds of Kansas, Oklahoma, and Texas. A skull of the most primitive mammallike reptiles placed in an assemblage of the more advanced North American pelycosaurs would fit well within the range of variation found among these earlier forms.

During the long period of evolution of mammallike reptiles, which lasted for 60 million years and produced the dominant land animals of their time, two major characteristics evolved that anticipated the appearance of mammals. One was posture (Figure 9-7). Neither amphibians nor any modern reptiles—turtles, crocodiles, lizards, or snakes—walk or run in the same way as mammals and birds, with their legs under their bodies, which are held high above the ground. Dinosaurs acquired the ability to support their bodies high above their hind feet, and this characteristic was passed to their descendants, the birds. The ancestors of dinosaurs, however, acquired this ability separately from and later than the mammallike therapsids.

The second advance was the diversification of teeth (Figure 9-8). The teeth of amphibians and most reptiles are alike for the most part, except for minor differences in size. The original mammals, and many of their descendants, including ourselves, have teeth of four kinds. Those in the front of the jaw are cutting teeth, or *incisors*. The back teeth are shearing teeth and grinders—premolars and *molars*. Between them is a single pair of stabbing teeth, or *canines*, on both upper and lower jaws.

The evolution of these four kinds of teeth can be mapped in a gradual series of progressive stages. A slight differentiation among teeth existed in some stem reptiles or cotylosaurs. Differences among the kinds of teeth became increasingly evident among the pelycosaurs and early mammallike reptiles. In their later descendants, the dog-tooth or cynodont, the incisors, canines, and molars are clearly differentiated from each other.

A third advance is in a way the most important, since it is the one used by paleontologists to distinguish reptiles from mammals. The lower jaw of reptiles contains several bones, of which two are important to us. One of these, the dentary, bears the teeth, while the other, the articular, smaller and at the hind end of the jaw, forms part of the hinge between the lower and upper jaw (Figure 9-8). The other part of this hinge is the quadrate, a small bone in the head portion of the skull, or cranium. Immediately behind these two small jaw bones is the middle ear, within which sound waves are amplified and transmitted by a special nerve to the brain. In reptiles, amphibians, and fishes, this amplification is carried out by a single small bone. By contrast, the lower jaw of mammals consists only of the tooth-bearing (dentary) bone, which is hinged to another bone, the squamosal, also in the cranium. The two bones that form the hinge of the reptilian jaw have not disappeared. They are represented in mammals by two small bones in the middle ear connected with the counterpart of the single reptilian ear bone. In reptiles, amplification of sound waves in the middle ear, carried out by a single bone, is relatively inefficient. The three bones in the mammalian ear do this job much more effectively, so that the hearing of mammals is much better than that of reptiles.

Figure 9-8.

A series of skulls showing a few of the numerous transitional forms that, via a series of adaptive radiations, resulted eventually in the origin of modern mammals. (a)–(c): Three typical reptiles. (a) A primitive *Captorhinus* that, like early amphibians and modern turtles, has only one pair of openings in the skull in addition to the nostrils. (b) A primitive ancestor of lizards, *Youngina*. (c) A modern lizard, *Varanus*. (d)–(i): Six reptiles that were on or near the line leading to mammals. (d) and (e) Two pelycosaurs that were typical reptiles but show the beginnings of tooth differentiation. Note that the hindmost bone of the lower jaw (angular, a) is nearly as large as the tooth-bearing bone (dentary, dn). (f) and (g) Two early mammallike reptiles, showing further tooth differentiation, plus reduction in size of the angular bone. (h) and (i) Two later forms of reptiles that, with respect to tooth differentiation and reduction of the angular bone, were much like mammals. *Diarthrognathus* was almost completely intermediate between reptiles and mammals. (j)–(l): Three kinds of mammals. (j) *Sinoconodon*, the earliest of these, still retained a number of reptilian features. (k) A later form, *Deltatheridium*, was very similar to modern shrews. (l) A modern opossum (*Didelphys*). The skulls are drawn at different scales of magnification. Those in the center column are at natural size or somewhat reduced; those in the right column are somewhat magnified.

In order to classify fossil animals neatly and clearly as either reptiles or mammals, most paleontologists and nearly all textbooks classify as reptiles all bony-limbed animals that have a liquid-filled amniotic egg and a jaw hinge formed by the two small bones, articular and quadrate, along with a single ear bone. Mammals differ in having the tooth-bearing (dentary) lower jaw bone

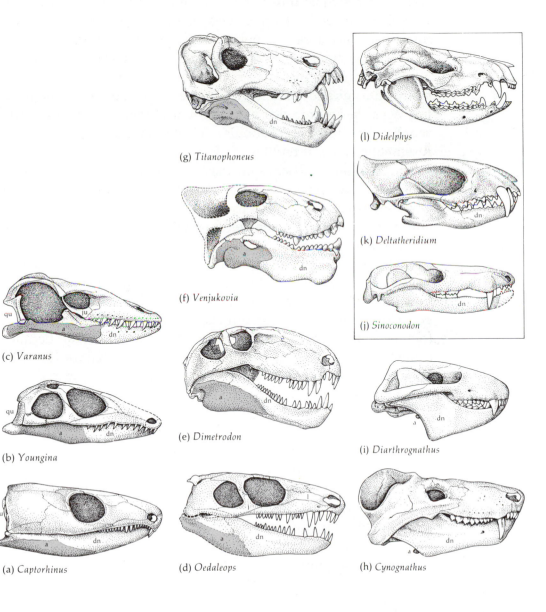

(g) *Titanophoneus*

(l) *Didelphys*

(f) *Venjukovia*

(k) *Deltatheridium*

(c) *Varanus*

(j) *Sinoconodon*

(b) *Youngina*

(e) *Dimetrodon*

(i) *Diarthrognathus*

(a) *Captorhinus*

(d) *Oedaleops*

(h) *Cynognathus*

articulated directly with a bone of the cranium (squamosal), plus three small bones in the middle ear. Tooth structure also helps in classifying them. Nevertheless, an animal that has almost mammalian teeth but a reptilian jaw hinge and middle ear bone is called a reptile. Mammallike reptiles are all classified as reptiles on the basis of this character, even though the advanced dog-tooth has teeth that resemble those of primitive mammals more than they resemble the teeth of the earliest mammallike reptiles or their immediate ancestors, the pelycosaurs. Likewise, the earliest animals having three bones in the middle ear are called mammals, although, like the primitive mammals of modern Australia—the spiny anteater and platypus (monotremes)—they may well have laid eggs, lacked nipples or teats, had skeletons showing some reptilian features such as shoulder girdles, and had chromosomes resembling those of reptiles.

Several recently discovered fossils stand between reptiles and mammals with respect to jaw and ear bones. Their principal lower jaw bone is already jointed to bones of the upper jaw and skull and at the same time is attached to one or both of the small bones that in mammals are located in the middle ear. One of these fossils has been named "double-jaw joint" *(Diarthrognathus)* because it possesses this intermediate characteristic. Some of these transitional forms are believed, on the basis of other characteristics, to be the direct ancestors of mammals; others apparently were ancestors of animals with mammalian characteristics that became extinct. Animals with this transitional feature existed for about 15 million years during the late Triassic period.

The evolution of the soft parts of the mammalian body—particularly hair, mammary glands and teats, and the uterine structure that enables females to bring forth their young alive *(vivipary)*—cannot be deduced from the fossil evidence because such parts are not preserved. Since hair is perfectly developed in egg-laying monotremes as well as in viviparous *marsupials* and placental mammals, and since several kinds of evidence suggest that the line leading to monotremes separated from that leading to viviparous mammals immediately after and perhaps even before mammalian jaw and skull structure had been reached, one might suspect that the dog-tooth therapsids already had hair. On the other hand, monotremes have internal mammary glands but no teats, so that the young lap up their milk rather than suck it from a nipple. Consequently, the evolution of the mammary apparatus may well have begun only with the attainment of mammalian jaws and skulls. By the same reasoning, one could suggest that the earliest mammals were egg-layers and that bearing live young was the last mammalian characteristic to be developed.

The origin of warm-bloodedness, including a constant body temperature, must be inferred from other lines of evidence. Recent analyses of bone structure in reptiles have led some paleontologists to believe that dinosaurs were warm-blooded. The bone structure of pelycosaurs and the early therapsids that evolved into mammals is that of cold-blooded animals, so the characteristic of temperature regulation probably did not appear before the origin of the dog-toothed animals. Nevertheless, monotreme body temperatures are not as well regulated as those of most marsupials and placental mammals. Possibly this mammalian characteristic evolved chiefly during the 80 million to 100 million years that separated the earliest mammals from those that resemble modern marsupials and placentals.

The biggest gaps in the fossil record from fish to primate are found in the 80 million to 100 million years just after mammalian ear bones first appeared. The mammals that existed during this time were small, relatively uncommon, and probably secretive in their habits. They were agile, possibly more intelligent than contemporary reptiles, and probably lived in trees. For all these reasons, they were unlikely to be preserved as fossils. For the most part, they are represented only by teeth and fragments of jaws. Eight orders are recognized, all sharply distinct from each other and from both mammallike reptiles and modern mammals. Six of these orders are contemporaneous with each other in the Upper Jurassic and Lower Cretaceous, but the earliest are separated by about 20 million years from the very early mammals of the Upper Triassic and Lower Jurassic period, while the latest are separated from the later marsupials and placentals by an equal or greater time span, except for an anomalous side line (Multituberculata) that superficially resembled squirrels, rats, and mice. The modern egg-laying monotremes found in Australia have no known fossil record but may be descended from one of the orders that flourished during the Lower Cretaceous—probably a different one from the ancestors of the marsupials and placentals. Figure 9-9 shows the period during which the reptilian–mammalian evolutionary line was represented by transitional forms between typical reptiles and modern mammals. This period was even longer than the span during which modern mammals have dominated the earth.

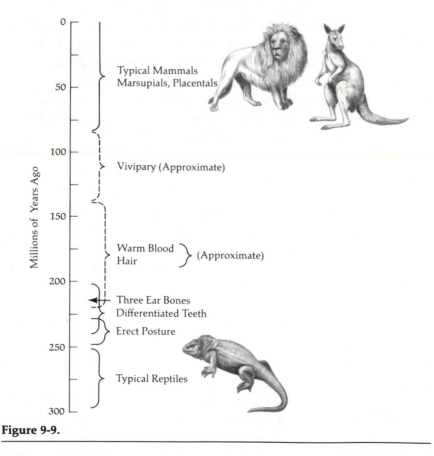

Figure 9-9.

Timing of the appearance of distinctive characteristics during the transition from reptiles to mammals. The times of origin of warm blood, hair, and vivipary (young born alive) are deduced on the basis of other characteristics.

Why did mammals evolve from reptiles? From the earliest amphibians to the dominant mammallike reptiles, each newly evolved amphibian or reptile succeeded if it could exploit more efficiently some available niche on dry land. By the beginning of the Triassic period, the earth was full of reptiles large and small, predators and herbivores, terrestrial, swamp-inhabiting, and aquatic. Some were thick skinned and slow moving; others were slender, lithe, and active. How then could an environmental challenge arise that would be strong enough to trigger the evolution of mammals, a class of animals capable of exploiting an entirely new way of life?

Paradoxically, the answer to this question lies in the diversity of reptiles itself. Complex interactions arose between different kinds of reptiles. Most of the mammallike reptiles were clumsy, slow moving, and unintelligent, easy prey for more active animals and inefficient in protecting their eggs and young. Nevertheless, the larger of them must have preyed on the smaller. At the height of their dominance, other independent species derived from the still-surviving reptilian stem line gave rise to animals with long hind legs and relatively massive tails that supported their bodies when they rose up and ran rapidly on their hind legs. These reptiles were the ancestors of dinosaurs and birds. They literally must have run circles around the larger kinds of mammallike reptiles and probably contributed more than anything else to their extinction during the Jurassic period.

The only mammallike reptiles that could escape from these new enemies were small and equally active. Escape became even easier if they could hide in the trees during the daytime and seek their food at night. This brought about strong selective pressure for regulation of body temperature to keep the animal warm enough during cool evenings so that it could forage actively for food. Equally strong was selective pressure for an insulating coat of hair. Once mammallike reptiles appeared with these new characteristics for keeping active during the night, the smaller ones could survive after their larger relatives perished.

Two other mammalian traits helped these small creatures to keep active during the night—differentiated teeth and larger, more sensitive ears. A cold-blooded animal can afford to gulp its prey whole, then digest it slowly while it is resting in a dormant condition. If an animal is to be active during cool nights, however, it must be constantly provided with fuel in the form of digested food. The changes in jaw and tooth structure that characterized the most advanced mammallike reptiles were the kind that helped them to chew. In most reptiles, jaws can move only up and down; they are excellent for biting but inefficient for chewing. As the tooth-bearing bone of the lower jaw became larger and developed new connections with the palate and upper jaw (including rearrangement of the musculature), the lower jaw acquired much greater freedom of movement. The later mammallike reptiles and early mammals became able to move their jaws sideways, as we do, as well as up and down, thus greatly aiding the chewing process.

Tooth differentiation in early therapsids consisted chiefly of differentiation of cutting incisors and stabbing canines, both of which helped the animal to capture its prey. In later therapsids and early mammals, the changes affected chiefly the back teeth, converting undifferentiated conical teeth into broader, more complex teeth having a shearing action, and finally into shearing premolars and grinding molars. The latest therapsids must have led lives somewhat like those of modern shrews, constantly scurrying around at night catching, chewing, and digesting insects of all sorts and perhaps also seeds and other forms of rich organic matter.

Animals that are active at night can survive only if they have keen night vision and particularly good hearing. This necessity greatly increased selective pressure for enlarging the middle ear and adding the two small bones (hammer and anvil) that amplify the smallest vibrations enough so that they can be perceived by the brain. Increased sensitivity to smell may also have been favored; the nasal opening in the skulls of dog-toothed animals are larger than those in most pelycosaurs and stem reptiles. Finally, an animal is not helped by having more sensitive sense organs unless its brain develops the capacity to integrate the incoming sense impressions and react to them. The increased intelligence of mammals as compared with reptiles is a natural outcome of this entire network of adaptive changes.

The development of mammalian reproduction also had greater adaptive value in association with greater activity and intelligence. Such small animals cannot carry, lay, and incubate large numbers of eggs. Large eggs that contain enough food to nourish the young offspring until they can fend for themselves are also a handicap to survival. The only possible compensation for reduced size and fertility is more intensive care of the young. Early mammals developed egg cells of microscopic size that became embryos within their bodies. They produced young that were smaller at birth and were less able to take care of themselves than young reptiles, then kept them close to the mother's body and fed them with mother's milk until they were able to find food for themselves.

The development of reproduction by live birth was probably the latest characteristic acquired by early mammals; monotremes never acquired it. During the Jurassic or early Cretaceous period, when it probably occurred, predatory dinosaurs dominated the earth, always ready to pounce upon unwary victims. Among the few skeletons that have been found of the feet of early mammals, some have the first toe opposed to the other toes as in modern tree shrews, a modification that enables the foot to grasp small branches of trees. An animal that took refuge in the trees but still laid its eggs on the ground would often be forced to abandon its young. Birds build nests in trees and carry food to the brooding mother and fledglings, but this alternative is not available to animals that cannot fly.

The multituberculates, Cretaceous counterparts of modern squirrels, may have escaped enemies by raising their young either in hollow trees or in underground burrows. Except for the Australian egg-laying monotremes, some of the most primitive of modern animals are entirely arboreal (tree shrews, for example) or climb trees when escaping from enemies (the opossum).

Whatever may have been the reason, the fossil record tells us that, during the height of the Age of Dinosaurs, mammals evolved that had either of two highly important structures for protecting and nourishing their young. One was the marsupial pouch, still found in the opossum and most of the mammals of Australia. The other was the *placenta*, a complex system of tissues by means of which the mother can feed her offspring continuously and gen-

erously while it is still within her uterus. The placenta has provided mammals with the most intimate physiological connection
between mother and offspring that could possibly have been
evolved.

Opinions among specialists are divided as to whether the marsupial pouch preceded the placenta or whether these two ways of
caring for the young evolved separately in different lines. Mammalian biologists disagree also as to whether the placenta is a more
advanced and efficient structure than the pouch. Some authorities
point out that Australian marsupials have been highly successful
for about 60 million years, have conquered a great variety of habitats, and are still flourishing. One thing is certain—during 70
million years, marsupials have not given rise to any lines that in
social behavior and intelligence are comparable with groups of
placentals such as monkeys, apes, and humans. Family relationships are probably the universal and necessary prerequisites to
the formation of complex societies. In placental mammals, the
appearance of an offspring with the ability to develop over a long
period of pregnancy and to be born when it is sufficiently
advanced to respond to the attentions of its mother paved the way
for the evolution of the family and the first mammalian societies.
The marsupial pouch apparently contains no such potential.

Finally, the placenta and the long period of pregnancy also made
easier the evolution of increased intelligence in many placental
mammals, including primates, as compared with marsupials.
Careful comparisons of numerous mammals have revealed a close
correlation between the length of pregnancy that is characteristic
of a species and its average brain size and level of intelligence.
Human mothers who are struggling through the last uncomfort-
able weeks of their pregnancy may be consoled by the fact that
their reward is the gift of high intelligence, reasoning power, and
imagination to their offspring.

At the end of the Cretaceous period, the dinosaurs became
extinct. Why this happened is still somewhat a mystery, although
many paleontologists believe that a cooling of the earth's climate
or radioactive fallout from an exploding nova star was a primary
cause. Placental mammals began one of the most active bursts of

adaptive radiation into different habitats and new ways of life that is known anywhere in the fossil record. One of the earliest orders that appeared as a product of this adaptive radiation was the primates, which eventually gave rise to humans.

THE PATTERN OF
VERTEBRATE EVOLUTION

The course of evolution from fish to primate to humans, as documented by the fossil record, fails to support three kinds of statements that often appear in textbooks of evolution and beginning biology. The first of these is that transitional forms between major groups, such as classes and orders, are lacking or rare. This apparent lack of transitional forms is actually a function of the way fossils are classified. Paleontologists agreed long ago to assign every newly discovered fossil species to a named genus, family, order, and class. They did this by devising previously unnamed genera, families, and orders to accommodate their new finds, but they included the orders within the framework of already existing zoological classes by expanding the definitions of classes. They recognized as the basis of difference certain characteristics of the skull and skeleton that are relatively easy to recognize in fossils. The boundary between fossils classified as amphibians and those called reptiles is vague, and the same forms have been classified differently by different authorities. The boundary between reptiles and mammals was established on the basis of jaw and ear bones; other criteria are used for other classes.

Some transitional steps are in fact represented by smaller populations than those of dominant animals that are typical of their group. The origin of radically new ways of life requires profound transformations of the body with respect to many characteristics. Such transformations require strong selective pressures that will bring about rapid changes. Natural selection is most effective when acting on relatively small populations. Hence, many transitional stages between major groups must have been represented by relatively few individuals that existed for comparatively short periods of time. Moreover, many of the habitats that favored transitional forms were restricted in space and lasted for relatively short periods. If these explanations are valid for the scarcity of many transitional forms, intensive search by paleontologists in certain favored spots should have uncovered "missing links" whose existence had not previously been suspected. Such intensive exploration has, in fact, been rewarding almost beyond belief with respect to mammallike reptiles and the earliest mammals. It has had even more spectacular success in uncovering the immediate ancestors of humans.

A second widespread misconception is that the major classes of animals have evolved in succession; that is, the evolution of a more primitive class is believed to have been nearly or quite complete before a more advanced one originated or became common. Belief in the concept of succession in time is a legacy of the preevolutionary concept of the ladder of nature. This concept was proposed by Greek philosophers such as Aristotle, was widely known to naturalists of the seventeenth and eighteenth centuries, and was incorporated by Lamarck into his theory of evolution. Even after Darwin had demonstrated the importance of natural selection and adaptive radiation, evolutionists such as the German Ernst Haeckel erected ancestral trees that placed each class in a lower or higher position according to supposedly primitive or advanced characteristics. Modern understanding of natural selection and adaptive radiation tells us that a successional placement of lower groups before higher groups would be expected to only a limited degree, and present knowledge of the fossil record confirms this expectation.

The evolution of amphibians and reptiles did not cause the evolution of fishes to cease, even though members of these two new classes remained in or reentered aquatic habitats. One of the greatest bursts of evolution in fishes, which gave rise to the majority of modern forms, took place after the amphibians had been reduced to their present few living forms and the reptiles had begun their decline. The modern orders of amphibians did not become abundant until after reptiles that were derived from much earlier orders of amphibians had passed through their periods of greatest dominance and diversity. Mammals and birds were evolving the efficient specializations that make them the dominant land animals of our time, not as successors to reptiles but during the heyday of reptilian dominance, largely in response to intense competition from the reptiles.

The same kind of picture emerges from analyses of the evolution of invertebrate animals. Molluscs, crustaceans, spiders, and various orders of insects evolved parallel to and contemporaneously with each other, not successively. This principle holds even for the evolution of many groups of unicellular organisms such as protozoa and diatoms. To be sure, rhipidistian fishes and the earlier orders of mammals became extinct largely because of unfavorable competition with descendants that had evolved from them, but this is only one of many reasons why orders and families of animals have become extinct.

Another widespread misconception is that ancestors of new major groups are to be found among the more generalized members of existing groups. This misconception probably arose because of indiscriminate comparisons between modern and extinct forms. Most certainly, amphibians did not arise from modern bony fishes; reptiles did not arise from salamanders, frogs, or amphibians that resembled them; birds and mammals did not arise from turtles, crocodiles, lizards, or snakes. Such origins would have been impossible, since the amphibians appeared long before modern bony fishes began their evolution; reptiles had appeared before salamanders and frogs appeared; and birds and mammals evolved from reptilian lines that had split off from the ancestors of modern reptilian orders long before these orders had given rise to modern reptiles. The only reasonable question to ask is: Were the immediate ancestors of the new classes generalized or specialized as compared with their contemporaries, that is, with animals that belonged to the same class as the ancestors of a new class and lived at the same time but did not evolve in radically new directions?

The answers to this question are unequivocal. Compared with other fishes that flourished during the Devonian period, the rhipidistian ancestors of amphibians were highly specialized as to both their structure and the habitats they occupied. The more generalized Devonian fishes gave rise to many evolutionary descendants, but these descendants all remained fishes. Reptiles diverged from amphibians before amphibians had begun to diversify. The earliest stem reptiles lived contemporaneously with the first diversified families of amphibians and so might be regarded as one of several lines of adaptive radiation to which the earliest amphibians gave rise. Birds arose from highly specialized small dinosaurs, and the ancestors of mammals were the most specialized of mammallike reptiles. If attention is restricted to the animals that were contemporaneous with the ancestors of a new major group, these ancestors are found to be more specialized than their contemporaries and appear generalized only with reference to unrelated members of the ancestral class that evolved much later.

Finally, this story of vertebrate evolution reveals the biological truth of an old Latin proverb, per aspera ad astra: we go through adversity before rising to the stars. The earliest representatives of each new successful class laid the groundwork for its success by evolving adaptations that enabled them to survive adverse conditions so severe as to cause the extinction of most of their contemporaries. Rhipidistians evolved into amphibians in stagnant, deoxygenated water that was drying up, conditions that in modern times have repeatedly caused the death of thousands of bony fishes. The amphibian ancestors of the reptiles evolved a new level of reproductive efficiency by protecting their eggs and offspring from the depredations of contemporary predators. Birds and mammals prepared themselves for taking over the land surfaces of the earth by evolving effective defenses against the tyrannical ruling dinosaurs. To what extent has this principle governed the evolution of humanity? This question is considered in later chapters.

Grey gibbons, the apes that are most similar to the common ancestor of apes and monkeys. (Zoological Society of London.)

Chapter Ten

PRIMATES:
THE BACKGROUND
OF HUMANITY

With respect to the evolution of primates and particularly humans, Charles Darwin, the founder of evolutionary theory, had an insight that was in some ways prophetic. Twelve years after he published one of the great classics of human culture, *The Origin of Species*, Darwin published a second book that was even more controversial than the first. In some ways *The Descent of Man* was more prophetic than *The Origin of Species*. With almost no solid facts to guide him, he deduced a scenario of the evolution of humans from apelike ancestors that in its essentials differs little from that held by most modern evolutionists. We now have a great wealth of evidence from a variety of disciplines—anatomy, biochemistry, chromosome cytology, studies of behavior—and a fossil record that, although still fragmentary, has become for some characteristics almost continuous. A few quotations taken from the second edition of *The Descent of Man* will serve to illustrate the prophetic nature of this work.

Prof. Huxley, in the opinion of most competent judges, has conclusively shown that in every visible character man differs less from the higher apes than these do from the lower members of the same order of Primates (p. 2).

Bischoff, who is a hostile witness, admits that every chief fissure and fold in the brain of man has its analogy in that of the orang; but he adds that at no period of development do their brains perfectly agree; nor could

305

perfect agreement be expected, for otherwise their mental powers would have been the same (p. 6).

Judging from the habits of savages and of the greater number of the Quadrumana (monkeys and apes), primeval man and even their ape-like progenitors probably lived in society. With strictly social animals, natural selection sometimes acts on the individual, through the preservation of variations which are beneficial to the community. A community which includes a large number of well-endowed individuals increases in number, and is victorious over other less favored ones; even though each separate member gains no advantage over the others of the same community (p. 63).

It has often been said that no animal uses any tool; but the chimpanzee in a state of nature cracks a native fruit, somewhat like a walnut, with a stone (p. 82).

Brehm states, on the authority of the well-known traveler Schimper, that in Abyssinia when the baboons belonging to one species (*C. gelada*) descend in troops from the mountains to plunder the fields, they sometimes encounter troops of another species (*C. hamadryas*), and then a fight ensues. The Geladas roll down great stones, which the Hamadryas try to avoid, and then both species, making a great uproar, rush furiously against each other. Animals also render more important service to one another. . . . The Hamadryas baboons turn over stones to find insects, etc.; and when they come to a large one, as many as can stand round, turn it over together and share the booty (p. 82).

These quotations have been carefully selected to outline the principal points that I wish to make in this chapter. They show that Darwin intuitively grasped the relationships and manner of evolution of apes and humans as we understand it today.

Several scientists have recently written important books on primate evolution and human history, among them Bernard Campbell, Clifford Jolly, Richard Leakey, David Pilbeam, F. E. Poirer and Sherwood Washburn. The account of primate evolution presented in this chapter is based on studies of these authors' works and on discussions with some of them. It is meant to serve as a basis for a better understanding of the general principles of evolution discussed in previous chapters. Did these same principles apply to human evolution as they did to the evolution of microbes, plants and animals? To answer this question, we must explore the evolution of our ancestors, the primates. Before we begin, however, let us review some of the principles we have encountered in Parts One and Two and pose some of the questions we shall attempt to answer in Part Three.

Animal evolution has been accelerated both by complex inter-actions between individuals belonging to the same species and by interactions with other species of organisms, including animals, plants and microbes. Has this also been true of human evolution?

When an evolutionary line of animals entered a new habitat or adopted a new way of life, it often triggered a rapid burst of evolution, chiefly through adaptive radiation. Has the human evolutionary line reacted in the same way?

Animal evolution has been aided by preadaptation, a series of gene-determined traits already present in a population that make response to a new environmental challenge easier. Has preadap-tation also been an important factor in human evolution?

In higher animals, particularly those that form societies, changes in gene-determined behavior patterns have enabled animals to enter new habitats, and changes in structure have improved their adaptations to those habitats. Have these changes also regulated human evolution?

The evolution of major classes of animals has been of a mosaic character—certain characteristics of a newly evolving group have appeared early during a transitional period, and other character-istics have appeared later. Was mosaic evolution characteristic of the line that led to humanity?

Some new kinds of organisms—snakes, for example—appear to have arisen rather suddenly in terms of the geological time scale, whereas others—squid, octopi, fishes, mammals—developed more gradually. What is the situation with respect to humans?

The origin of some animal groups—for example, amphibians and early mammals—apparently took place under conditions that presented a maximal environmental challenge. Was this also true at any critical stage of human evolution?

Is there a special transcendent quality—humanness—that dis-tinguishes us from apes and all other animals? If so, can this qual-ity be reduced to quantitative differences?

After reviewing primate and human evolution as paleontologists and particularly anthropologists now understand it, we may be in a better position to find some answers to these intriguing questions.

NONHUMAN PRIMATES

With respect to form and function, the primate is one of the most paradoxical of mammals. The order can be defined as much by the characteristics that its members lack as by those they possess. Primates lack the sharp tearing teeth and piercing claws of carnivores; their legs and feet are not adapted for fast running or springing, like the long legs and variously adapted hoofs of horses, goats, deer, and other herbivores; their teeth are far less powerful than those of rodents; they cannot fly like bats or birds; and they evolved far fewer anatomical specializations than have elephants, armadillos, or anteaters. The chief structural specializations that distinguish most primates are flat fingernails and frontally positioned eyes. Because of this eye position, primates have stereoscopic vision and superior depth perception. Some primates, including humankind, have exceptionally large brains and high intelligence, but most have no larger brains, in comparison to body size, than the majority of mammals that belong to other orders.

One might expect that an order so little specialized would be linked to other orders by transitional forms. Although there is no proof, some authorities believe that tree shrews (*Tupaeoidea*)—small, arboreal ratlike animals of Southeast Asia (page 304)—may be descended from a common ground-living ancestor that took to the trees. We do know that the earliest primates and nearly all their modern descendants were and are adapted to climbing and moving through trees. The exceptions are a few species of Old World monkeys (such as baboons) and chimpanzees, gorillas, and humans.

Did the arboreal life of primate ancestors prepare the earliest humans for the kind of existence they ultimately adopted, including walking or running on the hind legs and fashioning and manipulating tools? Did it also influence them toward and prepare them for their omnivorous diet, for which proteins, starches and roughage in the form of plant leaves and fibers are essential? Most importantly, did arboreal life promote increased intelligence?

The case for an affirmative answer to these questions is good. Active animals can survive in the trees only if their limbs are flexible and free-moving. Arms and hands of this kind are essential for manipulating tools. The primate way of hanging on to branches by grasping them, as opposed to the rodent or squirrel strategy of digging sharp claws into bark, preadapts primates to grasping and manipulating tools. Primate forefeet or hands can also be used to seize food and manipulate it. The majority of primates run or climb through trees on all four feet, but they nevertheless sit on branches when at rest and use their forefeet for grasping food and putting it into their mouths.

The ancestors of primates were insect-eaters accustomed to digesting proteins. Most modern primates are omnivorous, although some—for example, colobus monkeys, gelada baboons and gorillas—subsist entirely on plant materials. Others—for example, chimpanzees—eat smaller mammals as well as insects, fruits, and nuts. The shift from a relatively fixed life in the trees to a highly active, wandering existence in the savanna was easier for those primates that were omnivorous. Human ancestors were such animals.

For a small animal that lives in trees and constantly seeks fruits, nuts, and insects, highly developed sense organs and the intelligence to make use of them are essential. We have seen that mammals have better capacities for hearing, touch, smell, and balance than do the reptiles from which they evolved. These characteristics probably were selected as mammals became adapted to a nocturnal and arboreal existence. Environmental pressures were acting with heightened intensity on early primates. During the Paleocene and Eocene epochs, 65 million to 40 million years ago, small primates were hunted by far more active predators than those to which late Triassic and Jurassic forerunners of mammals were exposed, among them snakes and owls. In addition, during the

Eocene epoch, tree-climbing carnivorous mammals similar to modern civets and weasels were evolving. The primates that successfully eluded these enemies and produced the most young were those whose senses and agility were the best developed. Evolutionary theory based on response to environmental challenge through natural selection is supported by the fact that the evolution of primates, more than that of any other mammalian order, continued the kinds of trends that began with the earliest mammals.

Modern primates can be grouped into five distinct suborders. The bushy-tailed lemurs (Figure 10-1) of Madagascar are related to the lorises of India and to the potto and the bushbaby of Africa. The tarsier (Figure 10-2), which has no close living relatives, is a small, long-tailed, hopping animal with large eyes that inhabits the jungles of Southeast Asia. The New World monkeys are relatively small and slender. Some of them can hang from tree branches by their long prehensile tails. The Old World monkeys (Figure 10-3) are mostly larger, have nonprehensile tails, and either walk along branches or, like the baboons, spend all their time on the ground. Finally, the tailless manlike apes, the largest primates, include gibbons, siamangs, orangutans, gorillas, chimpanzees, and humans.

THE FOSSIL RECORD
OF PRIMATES

The pattern of primate evolution is still imperfectly known because of gaps in the fossil record. This fact should not surprise us. The kinds of land animals that are most easily fossilized are relatively large, live near quicksands, swamps, or rivers where their bones are easily buried, and are relatively unwary, so that they succumb easily to unexpected disasters. Small, quick primates usually meet a different fate. They may be killed by falling from trees, after which their bodies are quickly eaten and their bones broken by scavengers; or they may be killed by stealthy catlike or weasellike predators or by hawks, owls, and other birds of prey. The only

Figure 10-1.

A ringtailed lemur. (San Francisco Zoological Gardens.)

Figure 10-2.

Philippine tarsiers. (Zoological Society of London.)

parts of their bodies that are likely to survive such disasters are their hard enamel teeth set in their tough jawbones. Occasionally pieces of their skulls are preserved, or, more rarely, the entire skull.

In spite of this handicap, an evolutionary tree of primates has been constructed using supplementary biochemical evidence that shows how the modern forms are related to each other. The fossil record begins in the Upper Cretaceous, about 80 million years ago. A few fossil teeth characteristic of tree shrews have been found in North American deposits of late Cretaceous and Paleocene age. Similar remains that resemble modern lemurs have been found in Paleocene and Eocene deposits in western North America and Europe. Neither of these groups is represented by fossils of Oligocene, Miocene, or Pliocene age, but we can be sure that intermediate forms of modern tree shrews and lemurs must have

Figure 10-3.

A Doric langur, an Old World monkey from Southeast Asia. (San Francisco Zoological Gardens.)

existed somewhere on earth throughout these epochs. In many evolutionary lines of animals belonging to other orders, fossils that closely resemble modern forms are found both in recent and in relatively ancient deposits but not in intermediate strata. The prevalence of such gaps leads evolutionists to assume that most gaps in fossil sequences result from the problems of fossil preservation rather than from the nonexistence of intermediate members of these evolutionary lines.

Several kinds of primates related to the tarsier lived in Europe and North America during the Eocene, Oligocene, and early Miocene epochs. In the few forms of which skulls are preserved, the eye sockets are huge and turned forward, as in the modern animal. In addition, Eocene and Oligocene tarsiers had acquired protective bones surrounding their ears, which are absent in tree shrews and lemurs but present in monkeys, apes, and humans.

The forerunners of monkeys and apes appear simultaneously in a single rich fossil bed of early Oligocene age, 30 million to 35 million years old—the Fayum series in Egypt. Some of the monkeys may have been transitional to tarsiers; others may have been generalized ancestors of modern Old World monkeys. The earliest known fossils of New World monkeys are from late Oligocene deposits about 25 million years old in South America.

Tailless apes similar to the gibbon, as well as more primitive forms (*Propliopithecus*) that could have been generalized ancestors of all the apes, occur in the early Oligocene Fayum deposits, along with a more advanced type (*Aegyptopithecus*) that might represent a transition toward the common ancestor of chimpanzees, gorillas and humans.

From the early Oligocene to the middle of the Miocene epoch— an interval of about 10 million to 15 million years—there is practically no record of primate fossils that could have been ancestral to modern Old World monkeys, apes, or humans. Europe, southern Asia, and Africa have yielded Upper Miocene ape fossils about 10 million to 20 million years old, collectively known as the Dryopithecines, which apparently represent common ancestors of gorillas, chimpanzees, and humans. A single African form—*Pro-*

consul—is known from a well-preserved skull. The Proconsul brain case is only about 7 to 10 percent smaller than that of a chimpanzee and of the same general shape, but it lacks the massive brow ridges of that animal, and its face and jaws protrude somewhat less. Proconsul's limbs were shorter than those of chimpanzees, and probably it was less agile at arm-swinging (brachiation) through the trees. The foot structure suggests that Proconsul may have stood on its hind legs more often than modern gorillas and chimpanzees do. In some ways, Proconsul was intermediate between Old World monkeys and apes, but a few characters foreshadow those of hominids and humans. All in all, Proconsul pretty well fulfills expectations for a common ancestor of chimpanzees, gorillas, and humans.

During the latter part of the Miocene epoch—from 7 million to 12 million years ago—Dryopithecines evolved several adaptively radiating lines. Many of the fossils that document this radiation have been discovered only recently, and it is not yet possible to say just when the line leading to humans diverged from the line leading to the modern African apes. A divergence about 7 million years ago is compatible with both paleontological and biochemical evidence, although it is regarded as improbable by most paleontologists, who believe that between 12 million and 15 million years ago is a more probable date. Certain biochemical data based on resemblances with respect to DNA and proteins suggest an even more recent date for the separation—between 4 million and 6 million years ago.

The fossil record of the transition from these common ancestors to modern humans has been greatly enriched during the past forty years by the brilliant discoveries of such researchers as Raymond Dart, Davidson Black, the Leakeys, Elwyn Simons, and David Pilbeam. The meaning of their discoveries can best be understood in the light of equally provocative evidence unearthed by other scientists showing that, from the biochemical, genetic, and psychological points of view, modern gorillas, chimpanzees, and humans are much more alike than had previously been suspected.

INTERPRETATION OF THE
EARLY PRIMATE FOSSIL RECORD

Evolution from tree shrews to lemurs consisted chiefly of adaptations for increasingly efficient arboreal life. Toe claws evolved into toenails. The tail became larger and more useful as a balancing organ—a trend that was followed somewhat later and independently in squirrels. The big toe (or thumb) and the adjoining toe (or finger) acquired greater independence of movement from other digits, thus increasing the animal's ability to manipulate objects. The brain case increased in size and became more rounded and vaulted, indicating that even primitive lemurs were more intelligent than their tree shrew ancestors.

Beginning with lemurs, and continuing through tarsiers to monkeys, revolutionary anatomical changes occurred that greatly altered their lifestyle. Nocturnal ancestors gave way to descendants active during the daytime. The reason for this change is unknown, but pressure from newly evolved nocturnal predators may have been at least partly responsible. The most important anatomical changes were reduced sense of smell and increased power of vision. The long ratlike head of the tree shrew gave rise to an intermediate form in lemurs and the more rounded shape seen in tarsiers and monkeys. Nostrils no longer protruded from above the upper jaw but either were flattened against the middle of the face (as in New World monkeys), descended from a distinct nose (as in humans), or occupied an intermediate position. Eye sockets became much larger, and the position of the eyes shifted more and more toward the front of a flatter face. Changes in the optic nerves produced a single stereoscopic visual field. The higher primates began to perceive depth of field and distance far more accurately than other mammals, and differentiation of cells lining the optic retina into rods and cones gave them color vision. As an adaptation to integrating these new visual stimuli, the midbrain (cerebellum) greatly increased in size along with a lesser increase in size of the associative regions of the forebrain (cerebrum). At the same time the olfactory bulbs, which receive sensations of smell, became much reduced relative to overall brain size.

The adaptive relationship between these anatomical changes and the shift in behavior is easy to recognize. At night, enemies are more easily smelled than seen, and in tropical jungles sources of food can be recognized by subtle odors. Eyes placed on the sides of the head give a wider range of vision than those directed forward. In the dark, escape is easier if this range is wide; it is more important to detect moving shadows from any direction and of any kind than to identify specifically the animal to which they belong. Escape by hiding is easier than running away. By contrast, in daylight, vision is far more important than smell. Edible fruits and insects can be recognized by their shape and color. Escape from predators involves running along narrow branches, leaping from one branch to another, and grasping the safest branch with extended limbs and digits. Stereoscopic vision greatly increases the efficiency of these movements, particularly leaping with great precision from one branch to another, but it is useless unless governed by a larger, more efficient brain.

As arboreal animals become larger, the tail loses its efficiency as a balancer, and balance must be achieved through better-coordinated limbs. The reduction of the tail in Old World monkeys and its conversion to a prehensile, clinging organ in many species of New World monkeys are therefore completely attuned with other facets of the adaptive shift.

Fossil evidence indicates that the shift from nocturnal to diurnal activity did not take place as a single trend but occurred repeatedly in various lines of early primates. In a few lines, the shift was not completed before the animals became extinct; in various Old and New World monkeys, this shift represented a new way of life; in still other primates, it became the basis for further changes that led to the development of apes and humans.

The record suggests that some primates had perfected adaptations to diurnal life toward the end of the Eocene or during the beginning of the Oligocene epoch, a few million years prior to the time when the Fayum fossils were laid down and therefore prior to the time when precursors of Old World monkeys, gibbonlike apes, and possible ancestors of Dryopithecine apes lived side by side. We can only speculate that the perfection of adaptation to the diurnal niche in tropical forests may have triggered adaptive radiations that led to the divergence into the various lines of monkeys, apes, transitional hominids, and humans.

The population–environment interactions that dominated the diversification of primates during the middle of the Tertiary period—and particularly the divergence of apes from humans— were brought about by climatic changes. During the Eocene and the early Oligocene epochs, the earth's climate was warm, moist, and temperate, so that lush tropical forests were widespread. Primates lived in these habitats. From the middle of the Oligocene epoch—35 million years ago—climates all over the earth, particularly in the Northern Hemisphere, became colder and drier. In many regions, unbroken tropical jungles were replaced by a mosaic of forest, open parklike savanna, scattered trees, grassland, and desert. Some populations of primates possessed gene pools that could respond to the challenges of these new environments. The mosaic structure of the mixed environment would have caused the responding populations to break up into many small subpopulations that were partly isolated from each other. As we know, this kind of population structure is highly conducive to rapid evolution.

Probably, ancestors of ground-inhabiting apes, including humans, were not driven out of the trees by climatic changes but entered the savannas as temporary pioneers. Undoubtedly, many of these exploring lines failed to make the necessary adjustments and became extinct. Human ancestors were among the few that succeeded. One might even speculate that the evolution of tailed monkeys to tailless apes and the shift from four-footed walking along branches to arm-swinging were associated with temporary adaptation to ground life followed by the return of descendant evolutionary lines to the trees. There is every reason to believe that the diversity and changeability of terrestrial environments during the past 35 million years has had a great effect on the evolution of all kinds of mammals, including primates and human ancestors.

BIOCHEMICAL SIMILARITIES
BETWEEN APES AND HUMANS

Fossil evidence supports the hypothesis that some of the Miocene epoch apes were the common ancestors of gorillas, chimpanzees, and humans. During the approximately 20 million years of their existence, they occupied forests in Africa, southern Asia, and Europe. With respect to behavior and ecological response, the divergence of humans from our ancestors of 10 million years ago was vastly greater than that of gorillas and chimpanzees from their (and our) common ancestors. Biochemically speaking, it was not. With respect to proteins and nucleic acids, humans and apes are far more alike than are many species of flies belonging to the genus *Drosophila*.

These startling discoveries, made independently during the past few years by both biochemists and geneticists, deserve our special attention. With respect to the two kinds of protein chains found in hemoglobin—alpha and beta—humans and chimpanzees have the same primary structure or sequence of amino acids, while humans and gorillas differ with respect to a single amino acid in each of the two chains. With respect to another protein, cytochrome c, all three forms are identical. Similarities in other proteins for which amino acid sequences are not available have been obtained by immunological methods. While some differences have been found, particularly between an enzyme known as lysozyme in humans and its counterpart in the gorilla, they are small and are more comparable to differences among species of other animals that belong to the same genus (such as dog/coyote, domestic cat/bobcat, and the species of *Drosophila*) than to animals that are classified into different families, as are humans and apes.

The biochemical similarity between apes and humans extends to their genes. A common way of comparing DNAs belonging to different species is to split the double helix of each separate species into two separate single strands and to break up each strand into many smaller pieces. The pieces of single strands derived from an individual belonging to one species are then placed in the same agar medium with single strands of another individual belonging to the same species or with comparable strands of DNA from a different species. When single strands derived from individuals belonging to the same species are combined, perfect pieces of double helix are reconstituted. If the DNA is derived from individuals belonging to completely unrelated species—for example, humans and insects—no double helix is formed. DNA derived from species having moderately close relationships to each other will form new pieces of double helix, but these are likely to be unstable and to come apart when heated. The percentage of new double helices and their heat stability form an approximate measure of the degree of relationship between the two species from which the DNA was derived. Table 10-1 shows a comparison between human DNA and that of various species of animals obtained by this method. According to this estimate, humans and chimpanzees differ from each other with respect to only 2.5 percent of their genes, as compared with 11.2 percent difference between sheep and cows, 30 percent difference between rats and mice, and 75 percent difference between two species (*D. melanogaster* and *D. funebris*) of the genus *Drosophila*.

This low value for the proportional difference between chimpanzees and humans does not mean that we are genetically the same. Humans and chimpanzees differ from each other with respect to about 2,500 gene pairs—a number large enough to account for the observed differences in body size and proportions,

Table 10-1. Percent difference in nucleotide sequences of DNA between selected pairs of animal species.

Human/chimpanzee	2.5
Human/gibbon	5.1
Human/green (Old World) monkey	9.0
Human/capuchin (New World) monkey	15.8
Human/lemur	42.0
Mouse/rat	30.0
Drosophila melanogaster/D. funebris	75.0

head and brain size, and mental ability. Nevertheless, the data on DNA and protein similarities between humans and apes underscore the findings that in every visible characteristic and in every measurable biochemical characteristic—including the genes that are the basis of heredity—humans differ less from the higher apes than these apes differ from the lower members of the same order of primates.

THE BEHAVIORAL BACKGROUND
OF HUMANITY

An employer who is interviewing an applicant for a position is much less interested in the applicant's height, hair color, or other physical traits than he is in the applicant's potential performance. The important question for the employer is: What can the applicant do? Similarly, we cannot understand primates as the background for humanity unless we know something about their behavior. As an aid to understanding the problems that modern human beings face, the study of behavioral evolution is by far the most important part of the science of evolution.

For most species of animals, behavioral patterns are highly conservative. Only a few populations can respond to challenges of a drastically changing environment by evolving new ways of life. Human behavioral patterns can be just as conservative. Our ability to adjust to new environmental challenges is often hampered by powerful bonds of reaction. Nevertheless, during the approximately 100,000 years that *Homo sapiens* has existed, small groups of humans have repeatedly broken these bonds and pioneered new ways of life. Under what circumstances is such pioneering necessary and how can it be successfully accomplished?

Since we can never do more than speculate about the behavior of our extinct primate ancestors, comparative research on the behavior of contemporary primates provides the only possible avenue toward understanding how human behavior evolved.

During the past twenty years, research on primate behavior has been one of the more active fields of biological science and has yielded spectacular discoveries. Some of them, such as findings about the social life and capacity for speech of chimpanzees and gorillas, have been well publicized. Four aspects of behavior have rightly received major attention—reproductive strategy, aggressiveness, social structure, and means of communication.

Has primate reproductive strategy, both nonhuman and human, evolved along lines dictated by a strict interpretation of the principle of Darwinian natural selection? Is a genetically determined drive to produce the largest possible number of vigorous and fecund offspring the chief force driving this evolution? Is human aggressiveness, particularly organized warfare, an innate, genetically determined trait of our species? How distinctive is human language? What is its genetic basis, and to what extent is the ability to use abstract symbolic language responsible for human thought, artistry, science, and philosophy?

There is enormous diversity among primates with respect to the four main aspects of behavior. The least developed social organization is found in tree shrews, lorises, tarsiers, and some lemurs, which are mainly solitary; males join females only during the mating season. In other species of lemurs and New World monkeys, as well as in gibbons and siamangs, males and females form pairs that may endure for a lifetime. Each pair defends its territory against all intruders, just as song birds and other animals do.

Only a small minority of primates form lifetime pair bonds. More common are reproductive strategies dominated by competition between males for possession of females. Some of the losers may win females in later competition, but many males never have access to females. Competition may be relatively mild, such that the loser gives up and accepts a lower position in a dominance hierarchy before being seriously injured by the winner. In other instances, battles between males become so serious that injury or death results. As a result of these contests, natural selection of successful males—sexual selection—has evolved species among baboons and gorillas in which the males may weigh twice as much as the females and are correspondingly much more powerful.

Among the various patterns of reproductive strategy that are dominated by competition between males, three modal conditions stand out. One of these is the troop made up of a single male and his harem, in which the dominant male guards the females that he controls, aggressively chasing away male intruders and biting any female that attempts to stray. Harems are found among the patas monkey of Africa and in various species of baboons. A second mode is the multimale troop. Here dominant males tolerate the presence of other males and allow them to mate with females who have not reached the peak of their estrus cycle; secondary males, often the sons of a dominant male, may aid in defense of the troop, warding off the attacks of predators. The gelada baboon of Ethiopia has this kind of reproductive social structure.

Many species of monkeys, such as the gray langur, the macaque, and the rhesus, have societies organized on a more changeable pattern. Fights between males and dominance hierarchies exist, but these are temporary. Males can and do move from one troop to another, and dominant males are often displaced by younger, more aggressive members of the troop. Males are not much larger than females and lack distinctive markings peculiar to their sex. This third mode of reproductive pattern gains added significance because it is characteristic of chimpanzees— apes that, in anatomical features, body chemistry, DNA sequences, and individual behavior, including the rudiments of language, are more like humans than are any other primates.

In flexible multimale bands such as those of langur monkeys and chimpanzees, raising the new generation becomes a communal project. In langurs, stable groups of females persist even when males are moving from troop to troop. Primatologists Richard Curtin and Phyllis Dolhinow report that when a baby is newly born into a troop, it becomes a center of attraction for all the females. The new arrival may be handed from one waiting female to another more than fifty times in four hours during the first day of life.

Jane Goodall describes similar examples of interest in newborn young among chimpanzees, where immature females mother their brothers or sisters much as young human females play with dolls. Similar behavior has been observed in gorillas.

Primatologist Harry Harlow found that captive rhesus females that had been deprived of maternal care from birth developed abnormally. Aggressive toward humans and other monkeys, they themselves became negligent mothers. Apparently, in many species of primates, including those that are the most closely related to humans, intensive care of the young is necessary for the success of a family or troop.

Among nonhuman primates, nearly all aggression between members of the same species is caused by competition for mates. Group aggression, where many members of the same band or troop fight in a united way, is most often directed against other species, such as leopards, hyenas, or dogs. Anything that resembles organized warfare between different troops or bands belonging to the same species is absent from most species of nonhuman primates and uncommon in others. Savanna-living baboons of Africa defend troop territory to a limited extent; usually, however, troops that occupy neighboring territories either avoid each other or engage in relatively mild ritual combat rather than serious fighting.

The aggressiveness of individual humans, particularly males, is firmly rooted in the behavior patterns of primates in general. Aggressive pursuit of females, including competition of all kinds between male rivals for the most desirable members of the opposite sex, must have some degree of genetic foundation in humans as in most other primates and mammals in general. The assumption that intertribal and international warfare is a necessary outcome of this kind of aggression is, however, not only highly questionable but dangerous as a basis for social theorizing. This subject is discussed further in Chapter Thirteen.

Social organization among primates is governed almost entirely by two basic factors. One is sexual behavior, and the other is adaptation to the environment. A social structure consisting of bonded male–female pairs that defend their own territory is most adaptive, though not always present, in forest-dwelling species that are always able to find plenty of food, are relatively well protected from predators, and face chiefly the danger of overpopulation. If the number of reproducing pairs is controlled by the number of territories that are available to them, population size automatically retains a stable balance with the environment. All the pair-bonding species, including lemurs and their relatives as well as gibbons and siamangs, live in tropical rain forests.

In regions where food is scarce, particularly during unfavorable dry seasons, natural selection is likely to favor a social structure that consists of the smallest number of individuals that can maintain the reproductive potential of the population. Since a male can inseminate several females, single-male troops are the most successful under these harsh conditions. A dominant male can control not only his harem of females but also the water holes and the places where food is abundant. Males who have lost the battle for dominance are expendable. They can become socially sterilized through lack of access to females or can starve to death without loss to the reproductive efficiency of the species as a whole.

In intermediate or mosaic habitats that present a variety of ecological opportunities, various kinds of social structures may be equally successful. In a highly seasonal climate, a good adaptive strategy consists of changing the social structure according to season. The gelada baboon inhabits the highlands of Ethiopia, where lush rainy seasons alternate with dry ones. During the rainy season, several dominant males and their harems may share the abundant grass on the same mountain ledge, while subordinate males gather around the edge of the troop, looking for opportunities to displace one of the patriarchs. If enemies such as leopards or dogs approach the troop, the males band together to fight off the intruders. As the rainy season wanes and food becomes scarce, the social structure changes. Each dominant male herds his females to a particular feeding ground, and the subordinate males separate from each other until the next rainy season.

The ecological bases of social structure in the gorilla and chimpanzee are unclear. Both species inhabit lush tropical forests, where individual pair-bonding and defense of the territory are believed to be the most adaptive kind of social organization. Their Asian relatives, the gibbon, siamang, and orangutan, which live in similar habitats, have this kind of organization, yet both the chimpanzee and gorilla exhibit more complex social structures. In the gorilla, male dominance is pronounced. Males may be twice as large as females, as in savanna-inhabiting baboons. In the chimpanzee, harems are smaller and less permanent, and a female normally mates with more than one male during her lifetime or even during a single week of sexual receptivity.

The chimpanzee organization resembles that of Old World monkeys living in the richer savanna or mosaic environments. Its adaptiveness may stem from the fact that chimpanzees subsist to a large extent on fruit, which is not always easily available. In large stretches of forest, fruits may be completely absent, but when a jungle tree does produce fruit, it yields large amounts. Relatively large social groups can more easily join in the benefits of this occasional bounty than could individual pairs confined to separate restricted territories.

The social structure of gorillas is hard to explain on this or any other ecological basis. Gorillas are strict vegetarians, feeding chiefly on bamboo shoots, which are almost always abundant in their forest habitat. Of what advantage to the gorilla is the presence of huge males with exclusive harems? Why should they have a social organization resembling that of the savanna-inhabiting baboon? Could it be that their ancestors lived in savannas, evolved the dominant males, and transmitted this biological trait to the modern species in the absence of strong selective pressure against it? Or did gorillas evolve their social structure because of a chance combination of characteristics that arose at some time in the past and set in motion a chain reaction of male competition, resulting in strong sexual selection for large dominant males? One might speculate that large size in males is adaptive in defending the troop against predators—but if so, why didn't it evolve among chimpanzees? Clearly, the orgins of different social structures in primates are by no means well understood.

With respect to language, the fourth aspect of primate behavior that has been intensively studied, scientific opinion has undergone a revolutionary change during the past decade. The old belief that symbolic language is a peculiarly human trait no longer commands credibility. Psychologists B. and R. Gardner, a husband and wife team, raised a female chimpanzee, Washoe, in their home and taught her to "speak" in deaf-and-dumb sign language. Her vocabulary included words like *sweet*, which express general properties and so are symbols rather than names of particular objects or actions. A. and D. Premack taught a simple language to another chimpanzee, Lana, using plastic symbols. These experiments have greatly altered the viewpoint of anthropologists about the differences in behavior between apes and humans.

Although the partial mastery by nonhuman primates of sign language and language based on plastic symbols has shaken the foundations of anthropology and evolutionary science, the significance of such language experiments is still unclear. With respect to language capacity (as well as anatomy and biochemistry), the African apes resemble humans much more than they do Old World monkeys and other primates. Moreover, extinct human ancestors with brains as large or larger than those of gorillas and chimpanzees must have possessed at least a rudimentary capacity for expressing themselves in a symbolic fashion. Probably the shift from the strictly animal method of communication by means of cries and gestures about objects and phenomena close at hand to human methods dominated by symbols with abstract meanings took place over a very long period of time, either gradually or in a series of quantum bursts.

What bearing does the diversity of primate behavior have on the origin of human culture? A satisfactory answer to this question seems impossible, because it is unlikely that we will ever learn what social structures and behavior patterns existed among the extinct ancestors of our species. Even among existing hunter–gatherer tribes of humans, many different kinds of social organization exist, and there is no way of telling which, if any of them, is the most original or most basic to our species.

As an example, let us examine one characteristic of human societies that has also been intensively studied among primates—reproductive strategies. We have already seen that different species of primates evolved different reproductive strategies, probably in response to differing environmental challenges. Among humans, native American Indian chiefs had more than one squaw, although "harems" were rarely large. Among Australian aborigines, most young couples expect to remain married to each other for life, and monogamy is the rule. In these tribes, rules for the selection of mates are highly formalized and rigid. On the other hand, social life among most inhabitants of Tahiti, Samoa, and other islands of Polynesia is much more permissive. Young people are encouraged to form experimental mating bonds and have much more flexibility in the choice of mates.

The ideal morality of most modern societies may dictate that couples remain married for life and assume mutual responsibility for the care and upbringing of their children, but there is no indication that this ideal has ever been followed regularly, either in modern society or in historical times. Societies that sanction polygamy (the control by a dominant male of several wives or concubines) have existed in some parts of the world throughout recorded history, and the reverse condition of polyandry (the control over several males by a dominant female) has existed in some Polynesian societies. Perhaps this diversity of human reproductive strategies mirrors the diversity of primate strategies, and perhaps we should look to different environmental conditions for the causes of different human reproductive strategies. Since behavior patterns are not preserved in the fossil record, we will never be able to trace the origin of such complex behavior patterns with any certainty.

We can speculate, however, that patterns of social behavior may be altered in positive ways by pioneering individuals who respond to new environmental conditions. A fascinating example of adaptive modification is recorded for a troop of macaque monkeys living in Japan. This troop had to be removed from their original home, which was being destroyed by human interference, to an island where they were supported in part by artificial feeding. Their social structure was intensively studied by Japanese scientists. One day, a remarkably innovative female was discovered. She was observed to take one of the potatoes that had been set out as food, carry it to a nearby stream, and wash sand and dirt off its surface before she ate it. After she had done this several times, young monkeys tried to copy her, but they were restrained by their elders, to the extent that the potato-washing female became almost an outcast. Gradually, however, other monkeys began washing potatoes also, and finally the custom spread to the entire band.

Disruption of social behavior patterns due to environmental stress has been found in some bands of the gray langur, a monkey common throughout the Indian subcontinent. The first three primatologists who did research on the behavior of this species reported a most distressing condition. Over a period of four years, 83 percent of the infants died, and many of them were seen to be killed by adult males that invaded and raided neighboring troops. Observing that invading males immediately took possession of the mothers of the infants that they had killed, and believing this to be a normal practice among langur monkeys, one investigator (S. B. Hrdy) concluded that natural selection had favored the infanticidal, aggressive males because they were spreading their genes at the expense of those derived from the fathers of the dead infants. Other investigators asked themselves: If this is true, how has the langur monkey been able to persist as one of the commonest animals on the Indian subcontinent?

The problem was resolved by further research. Richard Curtin and Phyllis Dolhinow studied langurs in parts of India far from the locations of the original observations. They observed monkeys in jungles that had been little altered by human interference, where natural predators were still abundant. During several months at three different localities, they often watched fighting between males, but it was rarely lethal, and vanquished males were often able to return to the troop after a period of isolation. Males were never seen to kill infants.

Similar observations were made by another observer in Sri Lanka, the southern limit of the langur's range, as well as in Nepal, its northern limit. The conclusion reached by these later observers was that langur monkeys are ideally adapted to undisturbed jungles, where numbers are kept down by predation and the populations are in equilibrium with their habitat. There, competition between males is relatively mild. Although males often vanquish rivals and take over their females, the vanquished male usually is allowed to leave without serious harm, and the infants that he has sired are not disturbed. By contrast, the three localities at which excessive violence was observed are all in or near densely populated human communities, where predators have been killed and forests destroyed, and where the monkeys have been crowded into small areas where people regularly feed them. Under such conditions, frequent contacts with strange males and their families arouse aggressiveness, battles between males become violent and lethal, and infants are often harmed or killed during the struggle. There is no evidence to indicate that the males have intentionally become baby-killers. The formerly peaceful monkeys have become violent as a result of unnatural conditions, particularly excessive crowding.

One could conclude that neither violence nor infanticide are genetically conditioned traits among langurs but that longstanding patterns of conflict have been altered by changes in the environment. Similarly, we may conclude that it is fruitless to search for an original or basic primate behavior pattern; rather, the diversity of primate behavior lies in the diverse responses to differing environmental challenges.

Part Three

HUMAN EVOLUTION: BIOLOGICAL AND CULTURAL

A reconstruction of Peking man. (The American Museum of Natural History.)

Chapter Eleven

THE BRIDGE BETWEEN
APES AND HUMANS

THE FOSSIL RECORD OF
HUMAN ANCESTRY

The extinct Dryopithecines, a variable collection of apelike animals that lived in Africa, Asia, and Europe during the Miocene epoch, between 6 million and 20 million years ago, gave rise to several radiating lines. The oldest fossils that show distinctive resemblances to humans consist of a large number of teeth, a few fragments of jaws, and still fewer fragments of limb and body skeletons. Since skulls and brain cases are completely lacking, one can only guess at the actual size, brain capacity, and intelligence of these creatures. They are about 9 million to 10 million years old and are called *Ramapithecus*. This name, derived from that of the Hindu god Rama and the Greek word *Pithecus*, meaning ape, reflects the fact that the first fossil teeth were found in the Punjab.

Ramapithecus has teeth that resemble those of other humans in composition and structure, with canines smaller than those of chimpanzees and gorillas but larger relative to the other teeth. Fragments of skeletons suggest that *Ramapithecus* was smaller than a chimpanzee and may have walked on all fours. The remains are associated with those of other animals, including species best adapted to live in open savannas, indicating that *Ramapithecus* lived at least part of the time on the ground and away from dense forests. Fossils contemporary with the later Dryopithecines have been found in Pakistan, Turkey, and East Africa.

Associated with *Ramapithecus* in the Sivalik deposits of Pakistan and India are teeth and other skeletal fragments of two larger forms that may be related to it. Those called *Sivapithecus* apparently belonged to an animal about the size of a chimpanzee. The name *Gigantopithecus* is given to an even larger form, the overall structure of which is unknown. A reasonable speculation is that all three of these species belonged to a complex from which humans, chimpanzees, and gorillas evolved.

Between *Ramapithecus* and the next oldest human ancestor lies a gap of 4 million to 6 million years unrepresented by fossils. Then came the Australopithecines, also called southern apes because their first remains were discovered in South Africa. The dramatic story of these discoveries, beginning with the skull of a child found at Taungs by Raymond Dart, is told in an exciting fashion by Richard Leakey and R. Lewin in their book *Origins*. The principal actors in this drama have been the late Louis Leakey, whose discoveries in East Africa convinced scientists that Australopithecines are in fact "missing links" between apes and humans, and his son Richard, who has shown that they were a diverse assemblage of creatures, probably representing several different and independent lines that were evolving in the direction of humanity. A few years ago, Australopithecines at first seemed to consist of a relatively simple succession of forms. One of them, called *Australopithecus africanus*, was relatively slender and agile, with a brain slightly larger than that of a chimpanzee, although with a smaller body size and a tooth structure that suggests a diet including meat as well as fruits and seeds. This ancestor is believed to have lived between 5 million and 2 million years ago. Older fossils that were first called *A. africanus* have recently been recognized as a distinct species, *A. afarensis*.

Contemporaneously with *A. africanus* in both East Africa and South Africa lived a larger, robust form with heavier bones, a skull with conspicuous brow ridges, and a slightly larger brain—*Aus-*

tralopithecus robustus—for which Louis Leakey coined the name *Zinjanthropus*. Also included among the Australopithecines is a series of fossils called both *Australopithecus habilis* and *Homo habilis*, because at present these fossils can be regarded either as the latest known prehumans or the earliest humans. *Homo habilis* had a brain considerably larger than that of *A. africanus* and equal to some individuals of a later species, *Homo erectus*, which is generally regarded as a true although primitive human. Nevertheless, the brain sizes of all these forms differ little from each other. If we assume that individual brain size within populations varied as much as it does in modern humans and chimpanzees, we can infer that the brain size of all these forms overlapped (Figure 11-1).

The remains of *Homo habilis* are associated with chipped stone tools of several different kinds. Because tools can be made much more easily from wooden branches and other perishable materials than from stone, it is reasonable to assume that perishable tools were being used even before stone tools were invented.

Australopithecines and *Homo habilis* had some features in common. Their teeth differed little from those of modern humans but differed a great deal from those of gorillas and chimpanzees. The shape of their skeletons—particularly the junctions between the head and vertebral column and the bones of the pelvis or hip—was much more like that of humans than that of apes, showing that they walked erect on their hind legs in some manner. They could manipulate objects between their thumb and forefinger more easily than a gorilla can, although not as easily as modern humans can.

Australopithecines lived throughout eastern and southern Africa, and evidence in the form of tools similar to those associated with East African fossil bones suggests that *Homo habilis* ranged as far as northwestern Africa, western Europe, China, and Southeast Asia.

Contemporaneous with and following the latest Australopithecines, the species *Homo erectus* spread through Eurasia and Africa. First discovered by the Dutch physician Eugene Dubois on the island of Java in 1898, the remains were first called *Pithecanthropus erectus*—the Java ape man. A somewhat later and more extensive

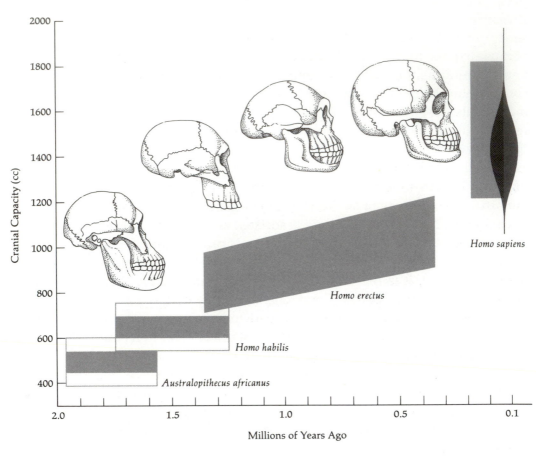

Figure 11-1.

Chart showing the increase in cranial capacity or brain size during the past 2 million years, as represented by *Australopithecus africanus*, *Homo habilis*, *H. erectus*, and *H. sapiens*. The shaded areas represent the time during which each species lived and its range in brain size. For *A. africanus* and *H. habilis*, the shaded rectangles are flanked by open rectangles that represent the probable ranges of brain size that would be revealed if a large number of individuals were known, on the assumption that this range is comparable to that known to exist in populations of modern humans and chimpanzees. The rectangle for *H. sapiens*, which encompasses estimates for Neanderthal and more recent fossils, is flanked by a shaded spindle that represents the range and relative frequency of cranial capacities in modern humans. (Data from P. V. Tobias, B. Campbell, and others.)

series of skulls and skeletons was discovered by anthropologist Davidson Black in a cave near Peking, China, and called *Sinanthropus pekinensis*—the Peking man—a name that persisted until careful analyses showed that Java and Peking man differed from each other no more than do different races of modern humans. A fossil jaw found near Heidelberg, Germany, could also be placed in the same species, and further remains of *H. erectus* were found by the Leakeys in East Africa.

Homo erectus had heavier bones than most modern humans, but their skeletons had similar proportions. Members of the species walked and ran on well-developed legs and feet and could manipulate tools with the same dexterity as humans. Average brain size was about three-fourths that of modern humans, but the largest-brained individuals had brains as large as those of the smallest-brained normal living humans. Brow ridges were heavier than in modern humans; jaws protruded somewhat, and chins were poorly developed. Although *H. erectus* had a somewhat brutish appearance, it was not apelike.

The most remarkable feature of *H. erectus* fossils is their constant association with stone tools of a certain style. Finds of these tools, whether associated with *H. erectus* fossil bones or not, have allowed scientists to trace the wanderings of the species from France to China, southward to Java in the east, and to South Africa and Morocco in the west. The first wave of expansion coincided with the first advance of the glacial ice in northern Eurasia at the beginning of the Pleistocene epoch, when climates in lower latitudes became much colder and wetter, so that even parts of the Sahara Desert were habitable by humans. Stone tools made by *H. erectus* are found in many spots that are now barren, lifeless wastes.

Although far less refined than were tools made by later humans, the Acheulian tools of *H. erectus* were by no means crude (Figure 11-2). Learning how to make them could not have been easy. The artisan had to select a smooth stone of the proper size and shape, then another stone to serve as a chipper. He then had to flake chips off both sides of the tool stone in a symmetrical (bifacial)

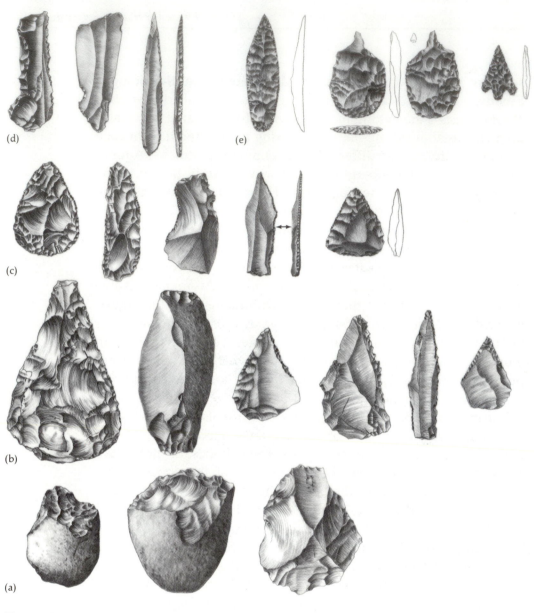

Figure 11-2.

Stages in the evolution of stone tools. (a) Two choppers and a hand axe made by *Homo habilis* a million or more years ago. (b) Chipped hand axe, scrapers, blades, and a point made by *Homo erectus* between 1 million and 400,000 years ago, belonging to the Acheulian style. (c) Scraper, point, and thin-edged tools made by flaking rather than chipping, the handiwork of Neanderthal men between 50,000 and 100,000 years ago, Mousterian style. (d) Finely chipped and thinly flaked tools made by contemporaries of Cro-Magnon man, between 22,000 and 15,000 years ago (Perigordian Age). (e) More recent tools (Solutrean Age). (From F. Bordes, *The Old Stone Age*, World University Library.)

fashion, making the edges of the cuts sharp enough to inflict a serious wound on the animal prey. Anthropologist Sherwood Washburn of the University of California tried an experiment to see how rapidly intelligent modern humans could learn how to fashion tools according to the Acheulian style. He presented students enrolled in an advanced course in anthropology with a model of an Acheulian tool, with stones that could be fashioned, and with "chipping" stones with which to do the work. All the students required several weeks to complete the project. Youths of *Homo erectus*, having only three-fourths the brain capacity of modern humans, may have taken even longer to achieve the same degree of proficiency.

Acheulian bifacial tools did not appear suddenly. Their evolution from the previous style used by *H. habilis* can be followed in successive fossil deposits. Those laid down in the Olduvai Gorge, Tanzania, where Louis and Mary Leakey made their landmark discoveries, were carefully analyzed and dated. The lowest strata in the gorge are barren of tools. Then come layers in which are embedded tools associated in several localities with *H. habilis*, which are crudely chipped on one side only (Figure 11-2). There is no definite style; each pebble was chipped in a slightly different way from the others. Higher layers contained successively tools that are more completely and regularly fashioned, approaching the Acheulian style. Finally, the uppermost layer contained only typical Acheulian tools. The tools found in any one layer equalled or surpassed the workmanship of the crudest tools found in the layer below it. The four layers of strata were laid down over a period of about 100,000 years.

Fossil bones associated with the remains of *H. habilis* and *H. erectus* are clues to their lifestyles. Those at Olduvai were identified by paleontologists as the remains of small animals. Moreover, they were broken in such a way as to suggest that their carcasses had been eaten. By contrast, deposits of *H. erectus* bones or of Acheulian tools are regularly associated with bones of large animals—rhinoceroses, horses, wild cattle, or buffalo. Anthropologists therefore infer that *H. habilis* was able to hunt and kill only small animals, although he may well have been able to steal the kills from larger predators by chasing them away. On the other hand, *Homo erectus*, hunting in bands with Acheulian tools, must have been able to kill large game and thus provide the band with ample food as the result of a single hunt.

The fossil deposits in China where *H. erectus* was found also contained charred pieces of wood, suggesting that campfires had been built. These remains are about 500,000 years old, from the middle of the period when *H. erectus* flourished. Were such fires used only to ward off predators, or was cooking practiced at this early date? We do not yet know the answer to this and many other questions about the everyday life of hominid populations.

In finds from the middle and later Pleistocene Ice Age, Acheulian tools and fossils are rare. The principal fossils from this period are parts of a skull dug up at Swanscombe on the Thames River, near London, another found at Fontéchevade in southern France, and similar fragments from Steinheim, Germany. These suggest that humans living about 200,000 years ago had brain cases nearly as large as those of modern humans, and so they might be classified as *Homo sapiens*.

This transitional period, about 400,000 to 300,000 years ago, is the date of the earliest recognizable human dwellings. To be sure, cleared sites that may have been the foundations of primitive shelters are found even in the Olduvai region of Africa, dated as 2 million years old. Tribes of *H. erectus* that lived outside of caves may often have protected themselves from wind and rain by simple temporary structures. Nevertheless, those found near Nice, France, and at Terra Amata, Spain, and dating from the beginning of the transition from *H. erectus* to *H. sapiens*, are the first shelters elaborate enough to qualify as genuine dwellings. Averaging about 35 feet long and 15 feet wide, these shelters must have been made of branches that formed a sort of dome, with a hole at the top to let out smoke from cooking fires (Figure 11-3).

Skeletons of humans belonging to the subspecies called Neanderthal have been unearthed in many parts of Eurasia and Africa— France, Germany, Yugoslavia, the Middle East, Central Asia (Uzbekistan), South China, Southeast Asia, and South Africa. Neanderthal humans existed for about 60,000 years—from 100,000

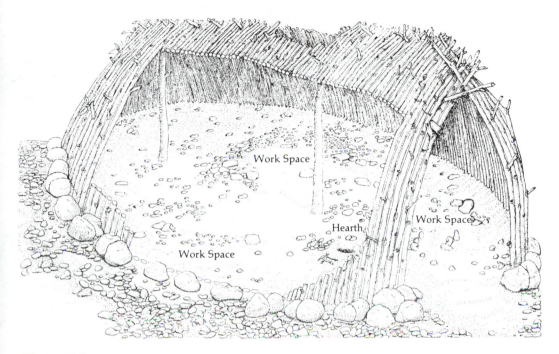

Figure 11-3.

Oval huts, ranging from 26 feet to 49 feet in length and from 13 feet to 20 feet in width, were built at Terra Amata, Spain, by visiting hunters. This reconstruction shows that the walls were made of stakes about 3 inches in diameter, set as a palisade in the sand and braced on the outside by a ring of stones. Some larger posts were set up along the huts' long axes, but how these posts were jointed to the walls is unknown; the form shown is conjectural. The hearth was protected from drafts by a windscreen made of pebbles. (From H. de Lumley, "A Paleolithic Camp at Nice." Copyright © 1969 by Scientific American, Inc. All rights reserved.)

years ago to 40,000 years ago. Although the skeletons from Europe have many primitive features, they resemble those of modern humans more than they do those of *Homo erectus*. Skulls from the Middle East, particularly Israel, are even more modern in appearance.

This impression of modernity is strongly reinforced by the cultural remains associated with Neanderthal skeletons. Their stone tools, made according to a style called Mousterian, were no longer produced by striking chips from the margins of large stones and retaining the core as a weapon, but rather by striking the flattened surface of the central core so as to knock off a thin flake with a point much sharper than the core itself (Figure 11-2).

Careful analysis by anthropologists Lewis and Sally Binford suggests that Neanderthals used Mousterian-type tools for a variety of purposes—hunting game, scraping and boring holes in hides, preparing food from plant materials, and suspending meat over an open fire. The occurrence of particular groups of tools of certain kinds in restricted areas suggests that women, who spent most of their time near campsites, used tools different from those that the men carried on hunting expeditions.

Some of the remains associated with Neanderthal-type bones show that these people had acquired some of the most distinctively human qualities—reverence and spirituality. They apparently buried their dead with ritual and ceremony. Neanderthal graves have been found in regions as widely separated as France, Iraq, and Uzbekistan. In all these burial sites, the skeleton or skeletons were found in the same characteristic position—knees drawn up in front and an arm curved back under the head in a sleeping position. With the skeletons were laid stone tools, and in the cave at Le Moustier, France, the skeleton is surrounded by charred animal bones, possibly the remains of a burial feast. Dust from around such a grave near Shanidar, Iraq, was found to contain pollen from the kinds of wildflowers that still bloom near the cave—hollyhock, bachelor's button, wild hyacinth, and groundsel. It is unlikely that pollen of this kind could have been blown into the cave, and so we may assume that flowers were placed near the bodies before burial. Since these flowers are odorless, they would not have been used to alleviate the smell of the rotting corpses. Like most modern humans, Neanderthals must have felt that one last way of paying tribute to their dead was to surround the remains with beauty.

There is also some evidence that Neanderthals cared for the old people of their tribes. One of the skeletons unearthed in Shanidar Cave was that of a very old man, plagued by arthritis, who had evidently been crippled early in life. His right arm was deformed and he was blind in the left eye. His worn teeth suggest that he may have spent his days near the campsite, using his teeth in place of his missing arm to help with the processing of hides and perhaps with other domestic tasks. He was buried with a heap of stones over his grave, accompanied by remains of animal food.

Evidence of religious rites has been found in several caves containing Neanderthal-type remains. One such site in the Swiss Alps contained a stone enclosure about a yard square, covered with a stone slab, on which had been placed seven bear skulls, facing the entrance to the cave (Figure 11-4). In Lebanon, deer meat had been

Figure 11-4.

Reconstruction of a Neanderthal bear ceremony. (From F. Poirier, *In Search of Ourselves*, Second Ed. Minneapolis: Burgess, 1977.)

laid out on a stone and sprinkled with reddish earth. Symbols of this kind are still used today in some primitive societies.

Cannibalism may have been practiced by the Neanderthals, perhaps as a part of intertribal wars, perhaps in symbolic rites intended to confer on the eater the wisdom or courage of the dead person.

The question of whether Neanderthals had a language is still open. Ashley Montagu insists that some kind of symbolic communication must have evolved even as early as *H. habilis*, because teaching the young to make a variety of tools would have been difficult or impossible without it. Philip Lieberman notes that some Neanderthal skulls are so constructed that they could not have supported the muscles we use to make open vowel sounds like *ah* or *oh*, but that other skulls of the same age do have the right construction for this purpose. He concludes that, although earlier hominids may have communicated in some way, spoken languages comparable to modern ones evolved with *Homo sapiens*. This opinion is shared by Kenneth Oakley, who believes that the burst of cultural inventions between the Mousterian (Neanderthal) and Magdalenian (Cro-Magnon) periods could well have been triggered by the development of grammatical speech.

Communication among early hominids may have been a combination of meaningful facial expressions, gestures with the hands, and short, expressive vocal exclamations. The voice may have been more useful during collective hunting expeditions, while the face and hands were used for teaching and learning at the home base. This theory implies that language evolved during hundreds of thousands of years, chiefly when *H. erectus* was the dominant hominid.

About 30,000 years ago, Neanderthals were replaced in Europe and southwestern Asia by people who in every detail of their skeletons were indistinguishable from ourselves. Their best-known remains consist of several such complete skeletons found in central France. They bear the name *Cro-Magnon*, a locality of that country. The nature of the transition from the Neanderthal to the Cro-Magnon race of *Homo sapiens* is somewhat in doubt. A common theory is that Cro-Magnon invaders from some unknown part of Eurasia displaced the less efficient Neanderthals, causing them to become extinct, presumably by conquest and slaughter. Others postulate that one race was gradually transformed into the other by natural selection of new gene complexes. Skeletons were found in a cave on Mount Carmel in Israel that are intermediate between Neanderthal and Cro-Magnon races. They have been variously interpreted as transitional forms that support the genetic replacement hypothesis, hybrids resulting from contact between two distinct races, and a self-perpetuating race of hybrid origin containing a mixture of Neanderthal and Cro-Magnon characteristics. The exact sequence of events that gave rise to modern human races may never be known, but the superficial nature of differences between modern human races has been biochemically established. The presence of intermediate individuals at Mount Carmel suggests that the genetic differences between Cro-Magnon and Neanderthal man may have been nearly as superficial as those between modern races of humans.

The cultural differences between Neanderthal and Cro-Magnon man were far greater than the anatomical differences. The Aurignacian and Magdalenian cultures, which date from 30,000 to 15,000 years ago, are as rich and diversified as the cultures of many modern hunting and gathering peoples. The cavemen of this period fashioned thin, bladelike spear points of careful design, much like the arrowheads of recent American Indians (Figure 11-2). They were also expert carvers of bone (Figure 11-5). During the latter part of this period, fish hooks were made, the first evidence that fishing was used to supplement hunting as a way of getting food.

Among the most famous examples of these cultures are the cave paintings preserved in Altamira (Spain), Lascaux (France) and several other sites. Some are so lifelike that they give the impression of motion; others include abstract symbols as well as figures (Figure 11-6). Another drawing at Lascaux shows an animal that no zoologist can identify (Figure 11-7). Perhaps it represents a legendary or mythical animal, like the unicorns and the dragons imagined more recently.

The Aurignacian and Magdalenian sites have also yielded elab-

Figure 11-5.

Various art forms from Upper Stone Age humans of Europe. (Photo courtesy of American Museum of Natural History.)

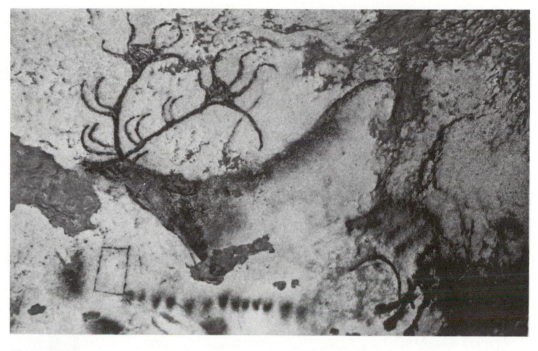

Figure 11-6.

This painting of a red stag appears on a cave wall at Lascaux. Both the stag and the two abstract signs below it, a rectangle and a row of dots, were painted on the rock surface with manganese pigment. (Photograph by Jean Vertut.)

orate bone carvings. Some of these may represent mythical sub-jects—for example, a bird-headed man associated with bison and rhinoceros forms, and female figures much like those represented as mother goddesses by modern Siberian tribes. Other bones are ornamented with a regular series of checkmarks. Alfred Marshack has analyzed these marks and compared the configurations found on several different bones. He concludes that each mark may rep-resent a day of a lunar cycle. The first calendars appear to have been carved on pieces of bone.

This evidence suggests that all the genes necessary for civiliza-tion probably existed in human populations 25,000 years ago (see Table 11-1). Since that time, humans have responded to environ-mental challenges only to a limited extent by changing their gene

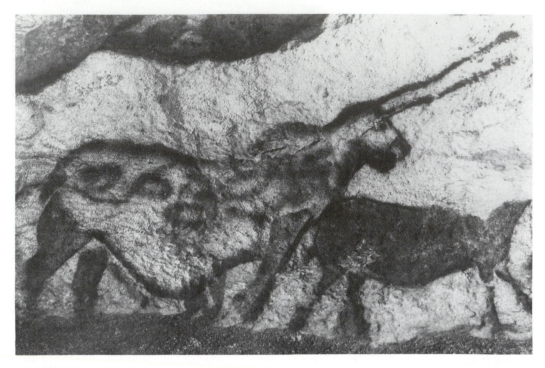

Figure 11-7.

The mysterious mythical animal painted on the wall of the main hall of Lascaux Cave. (Photograph courtesy of Department of Anthropology, University of California, Davis.)

pools. Rather, the chief response has been to exploit in new ways the genetic potentialities they already possessed. Biological evolution has not ceased, but it has been pushed into the background by humanity's capacity for cultural change. Newly acquired biological adaptations have largely been invisible, consisting of such characteristics as resistance to disease, greater tolerance of noise, and flexibility, or the capacity to be trained to perform a great variety of tasks.

The major crisis and evolutionary challenge that humans faced came soon after the end of the last Ice Age, 20,000 to 11,000 years ago. When the last glacial advance was at its height, conditions in regions south of the ice sheet were ideal for hunting game and

Table 11-1. The fossil record of human ancestry.

Species or Genus	Time of Appearance (years ago)	Time of Extinction (years ago)	Average Cranial Capacity (cc)	Other Characteristics
Homo sapiens				
Modern Races	30,000	Still living	1330	Magdalenian, Aurignacian
Cro-Magnon				and modern cultures.
Neanderthal	100,000	30,000	1470	Mousterian culture. Earliest rites and ceremonies.
Archaic	300,000	200,000	1300	Late Acheulian culture.
Homo erectus	1,300,000	400,000	950	Acheulian culture. First use of fire.
Homo habilis	2,000,000	1,600,000	700	Oldovan culture. First use of chipped stone tools.
Australopithecus robustus	3,000,000	2,000,000	550	Walked erect, no known tools. Vegetarian(?), not ancestral.
Australopithecus africanus	3,000,000	2,000,000	500	Walked erect, no known tools. Omnivorous(?), possibly ancestral.
Australopithecus afarensis	3,500,000	2,500,000	425?	Walked erect, probably ancestral.
Ramapithecus	14,000,000	10,000,000	?	Walked on all fours. Teeth approach hominid type.
Dryopithecines	22,000,000	12,000,000	?	Knuckle walker? Possible common ancestor of apes and humans.

gathering edible plants. In the forests of southern Europe and Asia Minor, game, fruits, nuts, berries, and roots were plentiful. Farther south, the glacial climate brought rain to parts of the Sahara and Middle Eastern deserts, which became savannas that could support nomadic life. Human populations may have increased during this period.

The retreat of the last ice sheet was followed by a long period of warm, dry weather—the postglacial xerothermic period. Botanist R. L. Whyte, an authority on the origin of grasslands and crop plants, depicts this period as one of relative famine, when edible plants became scarcer, game perished, and humans starved. The few tribes that survived must have been those that were prepared for the emergency by their pre-existing culture, so that they could exploit their environment in new ways. Tribes that lived near large lakes, such as Lake Urmia in Iran or Lake Van in Turkey, would have been the most likely to develop fishing as an alternative to hunting. Being sedentary, they would have become particularly well acquainted with valleys that were rich in edible seeds, such as wild wheat, barley, rye, lentils, and peas. At the same time, they would have been favorably placed for domesticating the orphaned young of dead sheep, goats, and other animals. Agriculture may have begun in these regions under climatic stresses, between 10,000 and 15,000 years ago.

The margins of the great valley of the Tigris and Euphrates Rivers are particularly favorable to small-scale agriculture. They form a "fertile crescent" that contains sites of ancient villages, where archeologists have unearthed the oldest known grains of wheat and barley as well as forms that are intermediate between cultivated varieties and their wild ancestors.

These "famine and drought" explanations of the origins of agriculture are not accepted by all authorities in the field. Anthropologist M. N. Cohen has pointed out that evidence of the earliest cultivation of crops has been found in many different localities in Asia. He believes that agriculture evolved independently in widely separated regions with diverse climates. In his opinion, the common underlying cause of the shift from hunting and gathering to cultivating and herding was a steady and widespread increase in the size of human populations. By their more efficient use of fire and weapons, humans had eliminated the climatic hardships and the attacks of predators that had previously kept population size down. As each tribe increased in number, it needed more space, but this was thwarted by the presence of neighboring tribes. Success in warfare, followed by takeovers of territory claimed by neighboring tribes, was one option. Another strategy was to obtain a larger amount of food from a smaller area. The shifts to agriculture and animal domestication were the principal ways by which tribes adjusted themselves to the new constraints. He believes that humans settle down as farmers through necessity, not choice.

Soon after humans had begun to till the soil and live in permanent villages, they started to organize states and build cities. In the great valley of Mesopotamia, the kingdoms of Sumeria became the founts of urban culture and civilization. The period between 10,000 and 7000 years ago marks a great turning point in the evolution of humanity. Here evolution and early prehistory, which are studied by biological evolutionists and anthropologists, gave way to late prehistory and history, which are studied by archeologists and historians. I discuss history from the point of view of evolution in Chapter Fourteen.

INTERPRETATION OF THE HOMINID FOSSIL RECORD

The previous sections of this chapter contain only a factual review of human ancestry. In reviewing this record, no attempt was made to answer the central question: Why did humans diverge from apes and develop our most distinctive trait—dependence on an elaborate culture? This question is very difficult to answer because much essential information is lacking and cannot be obtained from the fossil record. Nevertheless, we can speculate about the origin of the distinctively human way of life.

With respect to human origins, the discoveries made during the past fifteen years present a complex picture. The facts do not support the hypothesis of a simple progression *Ramapithecus* → *Australopithecus* → *Homo habilis* → *H. erectus* → *H. sapiens*. Instead, they are best interpreted as reflecting a series of radiations. Most of the radiant lines became extinct; only a few led to more advanced forms. Among apes and humans, each successive radiation produced less genetic difference between forms than in the preceding radiation. The radiations that took place during the Oligocene and early Miocene epochs—between 35 million and 25 million years ago—produced the Old World monkeys and apes. Radiation during the late Miocene and Pliocene—between 15 million and 5 million years ago—led to gorillas, chimpanzees, and humans. Both Dryopithecines and *Ramapithecus* and its relatives were involved in this radiation, but their relationships to each other and to modern apes and humans are not yet clear. Radiation between 5 million and 2 million years ago produced two or more species of Australopithecines, one of which was the ancestor of *Homo erectus*. The final radiations that took place between 300,000 and 30,000 years ago culminated in modern humans as well as Neanderthals.

The nature of the transitions between *H. erectus*, Neanderthals, and modern humans (as exemplified by Cro-Magnon man) is still a matter of debate. Some anthropologists cling to the hypothesis that they are a single evolutionary line, so that later forms were the product of gradual transformation from earlier ones. Steven Stanley argues strongly against this hypothesis, maintaining that both Neanderthal and modern humans originated from earlier forms by a relatively rapid process of speciation that took place in small populations. The earlier forms are supposed to have existed side by side with the later ones for a short while, after which they became eliminated by unfavorable competition or by being defeated in battle. Until more fossil evidence becomes available, there is no firm basis for accepting either hypothesis.

What selective pressures caused certain apelike animals, about 5 million years ago, to evolve in the direction of tool-making, culture-dependent humans, while contemporaneous related animals that apparently lived in very similar habitats evolved into forest-loving apes, highly socialized but without a tool-based culture? A commonly repeated "explanation" that appeared to be valid only fifteen years ago—that the ancestors of apes lacked the genetic potentiality for evolving into humans—has been exploded. Chimpanzees use tools to catch termites, crack nuts, and obtain bananas that are out of their reach, while both chimpanzees and gorillas can be taught to use sign language that involves some abstract symbols. Given strong enough selection pressures, these animals might be able to evolve into tool-users capable of more elaborate symbolic communication. Why haven't they done so during the millions of years that they and their ancestors have existed?

The separation may have begun when ancestors of apes and humans adopted different evolutionary strategies for surviving in forest margins or savannas. This speculation is based on the fact that both *Ramapithecus* and the Australopithecines had similar tooth enamel. Anthropologist Clifford Jolly suggests that this thick enamel was acquired as a result of selective pressure to cope with a diet of seeds and nuts. Jolly emphasizes grass seeds rather than nuts as the crucial factor, but I would emphasize nuts, for several reasons. First, nuts grow on trees and undoubtedly would have been eaten by forest-dwelling primates along with fruits, which are the principal food of modern chimpanzees. Second, nuts have to be cracked. Fashioned tools are not necessary for this purpose, but dexterity in handling unfashioned pebbles or stones would have great adaptive value for a nut-eater. In addition, nuts can be stored. Perhaps the ancestors of *Ramapithecus* or the Australopithecines acquired the habit of hiding nuts in caches and saving them for dry seasons when food was scarce. Such a habit would place a premium on ingenuity in finding good hiding places and on memory to recall them. Finally, some nuts are sweet, others are bitter; some nuts are good to eat, others are poisonous. The ability to distinguish between good and bad nuts would have been a matter of life and death.

Records of plant fossils indicate that, in India and Pakistan during the critical late Miocene epoch, nut-bearing trees belonging to the pea or legume family were increasing rapidly at the expense of fruit-bearing trees belonging to other families, and it may be that nuts became a more important part of the primate diet than was previously thought.

Once a race of apes had become dependent on using rocks to crack nuts and grind up grass seeds, bulbs, and tubers, they would be ready to abandon the practice (still characteristic of chimpanzees and gorillas) of crossing open country only to get from one tree or forest to another. Instead, they could have spent most or all their time in open country, going from nut trees to rock piles and caves where caches of food were kept. Their main problem would then be predators. Being adept at handling rocks for preparing their food, and being capable of running for at least short distances on their hind legs, they could have used rocks as missiles to ward off or kill predators. As their aim improved, more access to animal meat could have brought about a change in diet. A greater dependence on meat could have raised the adaptive value of fashioning rocks with a more lethal impact. Meanwhile, sticks may have been used as tools for digging roots, extracting termites from their nests, and various other purposes. This would have increased the adaptive value of more efficient and erect walking and running, which would have freed the hands for holding weapons and manipulating tools.

In short, it appears that hominids increased in intelligence while apes did not because ancestors of humans relied on hard foods and on the tools that made these foods easier to prepare and eat. In this way they became better adapted to life in open savannas, while ape ancestors lived in areas that provided soft fruits and plant shoots, for which no tools were needed.

Judging from the fossil record, the Oldovan culture, represented by partly and irregularly chipped pebbles, lasted about half a million years. During a period of about 200,000 years, it evolved into the Acheulian culture, which in turn lasted for more than a million years. Its replacement by the Mousterian (Neanderthal) culture required about 100,000 years, after which a period of 70,000 years saw little change in human culture. Then came a short period of change that lasted not over 15,000 years, during which developed the much more elaborate cultures that are associated with Cro-Magnon man, who closely resembled modern humans. Their type of culture persisted until modern times in Australia, South Africa, and parts of the Americas, but in western Eurasia and North Africa it was replaced after about 25,000 years by cultures based on agriculture and livestock raising. This replacement required about 2,000 years. This record shows clearly that cultural changes took place by a series of quantum jumps.

What caused the transition from each of these cultures to another? All of them were dominated and controlled by one overarching series of events—the advances and retreats of the glacial ice sheets. Humans living in the north could survive only if they learned how to withstand bitter cold and to obtain food under harsh conditions. Farther south, the same climatic changes that produced the northern glaciers gave rise to long rainy spells that changed completely the savanna landscape, including the game animals that lived in it.

Anthropologists know so little about the transition from the Oldovan to the Acheulian culture, and therefore from *Homo habilis* to *H. erectus*, that they cannot determine how much climatic change influenced this stage of evolution. Perhaps the cooler climate that prevailed even in the tropics made it easier for people to store meat without decay, and so favored those tribes that could benefit from the supply of food provided by killing large animals.

With respect to the origin of Neanderthal-type humans from *Homo erectus*, the basis of theory is somewhat better. The transition apparently took place in temperate regions at the time when the last ice sheet was advancing. The consequent march of the seasons—from warm summers to bitterly cold winters—placed a great premium on such cultural advances as stone tools that could be used to kill game more quickly during the relatively short season when hunting was possible, using and making fire, making clothing from skins, and driving dangerous animals from the caves that afforded the only reliable shelter from winter storms.

As the last ice sheet was retreating, forests and meadows were replacing arctic tundra, game animals and edible plants increased in numbers, and a time of plenty was at hand. Bands or tribes of humans were well organized and better fed. Those tribes that could capitalize most effectively on the environment by diversifying their cultures and evolving higher levels of organization were those likely to prosper.

Even harder to understand than the origin of tools is the origin of language. If apes can be taught to communicate in sign language, and even to invent new ways of expressing themselves that include the use of abstract symbols, why don't they talk in sign language to each other?

Indirect evidence from observations of human infants favors the hypothesis of an ancient origin of language. One of the most distinctive behavior patterns of a human baby is its constant babbling.

What parents haven't listened with excitement and wonder at these babbling sounds, waiting in anxious curiosity for the first of them that resembles a word? In marked contrast, infant chimpanzees and gorillas don't babble at all. They remain completely silent until they can imitate the cries and calls that their elders make. This distinctive trait of human infants must be, to a large extent, innate and genetically determined.

Obviously, the evidence suggesting that the ability to speak is of ancient origin and is encoded in our genes does not explain how these genes could have become established in human populations. Selection theory would favor the idea that genes of this kind conferred an adaptive advantage to hominids living in open savannas but were adaptively neutral or disadvantageous to populations of apes living in forests. Perhaps their value to hominids was associated with cooperative hunting, particularly with making plans for a hunt. However, evidence for or against such a hypothesis cannot be obtained.

A COMPARISON BETWEEN HUMAN EVOLUTION AND THAT OF OTHER ANIMALS

We are now in a position to suggest answers to the questions that were posed at the beginning of Chapter Ten. First, we can be reasonably sure that human evolution has been speeded up by interactions between humans and other organisms, both plants and animals, as well as interactions between different groups of humans. The hominid shift to savanna life was triggered by a change in climate, but the first steps toward humanity were taken not so much because of this shift but because of the way in which it was carried out. The hominid strategy based on tools, cooperation, and an increasing development of foresight and planning stands in sharp contrast to the strategy adopted by other primates, particularly baboons, to make the same shift. The change in tooth structure that is evident from the teeth of *Ramapithecus* and the Australopithecines indicates new food preferences and therefore new interactions between other animals and plants. Dependence on tools and related cultural beginnings were responses to the value of hunting large animals and the need for protection against predators, which gave importance to interactions between individuals and tribes.

We can also be reasonably sure that earlier preadaptive trends were largely responsible for the major trend of human evolution—increasing reliance on learned cultural traits as a means of survival. Whether manipulation of unfashioned tools began as a way of obtaining new kinds of plant food, or whether dexterity increased as a result of the need for killing game and warding off predators, the earliest stages of manipulation were possible only because forelimbs were adapted for this use by their anatomical structure. The common ancestor of chimpanzees and humans may have already been as adept at using tools and as interested in them as are modern chimpanzees. This whole argument rests on the belief that, with respect to the use and fashioning of tools, behavior was the pacesetter of evolution. Early hominids started to use tools when they were no more intelligent than modern gorillas, who use them very little. The beginning of humankind's spectacular increase in intelligence followed adoption of and dependence on tool-using as the normal way of life.

With respect to the most important change in hominid characteristics—the increase in brain size—quantum jumps followed periods of relative stability. The pattern of punctuated equilibria prevailed, but, as Darwin postulated, the jumps themselves were slow enough in terms of human generations that they can easily be explained on the basis of relatively mild selection pressures acting on a gene pool consisting of many small individual differences.

The common ancestor of chimpanzees, gorillas, and humans probably had a cranial capacity of about 400 cubic centimeters, whereas the average capacity in modern humans is about 1400 cubic centimeters. Since this common ancestor lived at least 4 million years ago and perhaps much earlier, a total increase of 1000 cubic centimeters took place at a rate of not more than 0.00025 cubic centimeters per year, assuming that the increase was continuous throughout this period. If we further assume that the average length of generations in the ancestral line of hominids was 20 years, this increase amounts to 0.005 cubic centimeters per generation. Experiments that exposed the fly *Drosophila* to a new environment that differed from the old one only with respect to a somewhat lower temperature (16°C instead of 20°C) showed that mild natural selection can alter body size and wing length at a rate that, if extrapolated for human brain size, would amount to 0.5 cubic centimeters per generation. If we assume that brain size increased at this rate, then the recorded increase could have taken place by means of a series of bursts, each one lasting for only a few thousand years. Ten quantum bursts, each one lasting 4000 years, would have produced the total increase in cranial capacity.

The fossil record tells us that, about 3 million years ago, the cranial capacity of our ancestors was about 500 cubic centimeters; about 2 million years ago, it was 700 cubic centimeters; and about 500,000 years ago, it was 900 cubic centimeters. These data fit the speculation that one quantum burst lasting 4000 years took place between 5 million and 3 million years ago, two such bursts took place between 3 million and 2 million years ago, and two more between 2 million and 500,000 years ago. Each of these bursts would have been separated from each other by periods of stability that lasted for hundreds of thousands of years. The later increase was much more rapid. Nevertheless, it could be explained by assuming five quantum bursts, each one lasting 4000 years. These could have been separated from each other by periods of stability lasting from 80,000 to 100,000 years.

The more rapid change that took place in cranial capacity during the last 500,000 years is best explained by the environmental challenges that glacial climates presented to some human populations. The expected response of humans to these challenges and the advantage of predicting changes in the seasons would have been increased intelligence based on larger, more efficient brains.

With respect to all changes in anatomy, evolution from apelike animals to humans was definitely of a mosaic character. Australopithecines were a mosaic of human and apelike characteristics. Their skeletons, particularly their hip bones, vertebral column, and hands, were more like modern humans than apes, but the earlier Australopithecines had brains not much larger than that of a chimpanzee. The bodies, legs, and arms of *Homo erectus* resembled those of modern humans, but their skulls and brains had not evolved much more than half the distance from Australopithecines to *Homo sapiens*.

Modern fossil evidence has completely fulfilled the prediction of Charles Darwin, that forms intermediate between apes and humans would be found. The most striking example of an intermediate form is the small group of fossils called by some *Australopithecus habilis* and by others *Homo habilis*; but all the fossil remains of hominids that have been dated between 3 million and 500,000 years old are intermediate between apes and humans in one way or another.

One feature in which human evolution has been strikingly different from that of all other animals, including nonhuman primates, is the relatively small amount of adaptive radiation. Before humankind evolved, each successive group of animals that achieved success on a worldwide basis capitalized on that success by generating a great variety of anatomically different forms adapted to different ways of life. Apes and early hominds, however, radiated and diversified most actively when they were uncommon compared to other mammals. Increasing success of humans has brought about a continual decrease rather than increase in the amount of adaptive radiation. Humans did not approach their present pinnacle of success until the final stages of the Old Stone Age, immediately before the beginnings of agriculture and civilization. By that time, the most likely hypothesis is that humans were already as much diversified anatomically as they are now. Counterparts of all modern races already existed or could have been generated from the existing gene pool.

Why should the evolutionary history of humans have been so much different from that of animals in this particular respect, whereas with respect to other features human evolution was much like that of animals in general? The answer to this question lies in the different methods by which humans interact with their environment. The difference was stated accurately more than a century ago by naturalist Alfred Russell Wallace, a contemporary of Darwin and codiscoverer of natural selection as an important mechanism of evolution. He pointed out that animals become adapted to new environments or to new strategies for exploiting existing environments by means of hereditary modifications of their bodily structure. Humans adapt by means of tools and other manufactured artifacts that either extend the power of their bodies to cope with an unaltered environment or enable them to change the environment to make it more congenial.

In accordance with the nut–seed hypothesis, the first unfashioned tools may have been adaptive counterparts of the powerful gnawing teeth of squirrels and other rodents and the sharp canines of gorillas that are used to tear apart tough parts of plants. Tools with artificially sharpened edges, such as Oldovan pebbles and Acheulian hand axes, had the same functions for their makers as do sharpened, retractile claws and tearing teeth evolved by lions and tigers. As cultural artifacts became more diverse during the late Stone Age, humans could use tools to obtain food of all kinds by a variety of methods. Each of these was as effective for a particular function as is the sole method of food-getting used by various animals that are the end products of adaptive radiation. Both ancient and modern hunter–gatherers possess a versatility that makes them equal not to just one kind of wild animal but to a whole array of different animals. Even before the advent of agriculture and animal domestication, humans had substituted cultural versatility for evolutionary anatomical changes, and in so doing had achieved ecological success greater than that possessed by a whole series of adaptive radiants in other orders of mammals, birds, and reptiles.

The advent of agriculture, animal domestication, and life in settlements ushered in an entirely new way of dealing with the external environment—the wholesale alteration of the environment to suit the ecological needs of humanity. Prior to the evolution of *Homo sapiens*, this ability appeared to only a limited degree. Nest-building is widespread, not only in birds but also in a few fishes and a number of mammals. Insects and spiders build rudimentary houses in the form of cocoons, and a few spiders can soar for scores of miles by means of parachutes that they make from their own silk. All the social insects build nests or hives within which temperature and moisture are carefully regulated by various devices. Ants domesticate slaves that belong to other ant species, and they subsist on the nutritious secretions of aphids, which they move to locations near their nests. Some ants even cultivate fungus gardens within their nests.

All these activities are particular attributes of species that have geographical and ecological ranges no broader than those of related species that do not modify their environment in these ways. Nests, termitaries, and beehives function for only a single generation. Each successive reproductive individual or pair of birds or social insects must survive the vicissitudes of a harsh external environment before it can provide its own artifacts that permit modifications of its immediate environment. Consequently, evolution of these animals has followed patterns of adaptive radiation that are much like those found in animals that do not modify their environment at all.

Humans, however, have combined their much greater ability to modify their environment with intensive care of their young up to a relatively advanced age—a trait they inherited from their non-human primate ancestors. In stable agricultural regions, successive generations of farm families care for and watch over their homes, fields, flocks, and herds so carefully that the lifetime of an individual farm spans several generations of farmers. Most urban dwellers, from ancient Sumeria to modern times, have lived their entire lives in cultural environments that their parents, elders, and ancestors provided for them. This continuity of culture has permitted humans to live successfully in temperate climates, deserts, arctic wastes, and high mountain plateaus without altering the anatomy or physiology of their tropically adapted bodies.

Humans have evolved an unprecedented richness of ecological opportunity by means of intelligence and social organization, in spite of the otherwise limited capacity possessed by a relatively feeble body. Once this capacity had appeared, evolution through adaptive radiation with respect to anatomical structures and physiological functions became a far less efficient strategy for dealing with environmental challenges than cultural amplification through inventions and technological progress.

Has this revolutionary way of interacting with the environment brought to the earth a new and permanent way of life? Will future evolutionary change depend chiefly on cultural adaptation rather than biological adaptation? Or are 10,000 years of human cultural achievement nothing more than a sort of "biological supernova," destined to become extinct in a few hundred or few thousand more years, leaving the earth's biota to continue its evolution in the old way? These questions are discussed further in the following chapters but are not resolved until the final chapter of this book.

HOW DISTINCTIVE IS HUMANITY?

The facts reviewed in this chapter tell us that, with respect to any single characteristic that one might choose to analyze in depth, humans differ only quantitatively from apes. Chimpanzees and gorillas have a degree of intelligence that is nearer to our own than it is to that of lemurs or other early primates. They can be taught the rudiments of symbolic language. Chimpanzees use a variety of tools and fashion some of them for more efficient use. These apes form strong social bonds with each other. Their societies are highly structured. As in human societies, rank and position are of great importance. Some of these societies are rigidly structured, providing little opportunity for individuals to rise from the ranks to positions of dominance. Others, particularly those of chimpanzees, are relatively open, so that aggressive young males can rise from the ranks to positions of leadership. Ape mothers show great affection for their young, and they usually take good care of them from birth almost to maturity. Many male chimpanzees and gorillas are almost equally interested in and solicitous of the younger generation.

Anthropologist W. Baumgartel, after having spent several months among gorilla bands in the mountains of Africa, says this about them: "Gorillas are like us in so many ways. They live and die, copulate and reproduce like us. They get sick from the same diseases as those we suffer from. They belch, cough, hiccup, sneeze, pick their noses and break wind just as humans do. They love, protect, care for and discipline their children. They like and love one another. Mother love, in particular, is very pronounced, as it is with all mammals. Love is an essential emotion in the lives of these animals."

Given this great degree of similarity, must we conclude that humans are only sophisticated apes endowed with enormous technical prowess but otherwise only quantitavely different from animals? I think not. My opinion is based upon the conception of novelty that is explained in Chapter Five of this book. All the qualities that we regard as novel in any species emerge from large quantitative changes with respect to several different characteristics. The essential meaning of Dobzhansky's concept of transcendence is that novelty must be judged on the basis of predictability rather than by analysis of component differences between novel conditions and their ancestral states.

If novelty is defined chiefly on the basis of how well we can predict the properties of a structure or system of organization on the basis of a thorough knowledge of its predecessors, then human society must be regarded as containing many novel traits. The family life of humans, including our attitudes toward relatives and friends, could be reasonably well predicted on the basis of a thorough knowledge of chimpanzee and gorilla societies. Nevertheless, these apes have not developed any form of organization that resembles even remotely the human division of labor, disciplined armies, religions, nations, and international organizations. Modern human societies are qualitatively as well as quantitatively different from all existing animal societies.

I recognize three distinctly novel human characteristics—artisanship, conscious time binding, and imaginal thinking.

Artisanship is the transcendent outcome of the evolution of toolmaking, already present in rudimentary form in chimpanzees. One could not predict the Parthenon, the pyramids, or the Empire State Building from watching pre-Oldovan hominids "fishing" for termites with sticks.

The concept of conscious time binding, developed by the Polish philosopher Korzybski, expresses the extraordinary ability of humans to plan for the future while profiting from the memory of past experiences. Its forerunners in apes and other animals are at best rudimentary. To be sure, many animals appear to plan for future emergencies. Ants and bees store up food in their nests, birds migrate southward to escape the cold of winter, and bears and other mammals go into hibernation before winter begins. Careful analyses of these behavior patterns have shown that they are automatic, stereotyped, and relatively inflexible. These animals have little ability to modify their behavior when unusual and unexpected changes in circumstances appear, as do intelligent humans when carrying out a pattern of behavior that they themselves have planned.

Behavioral patterns that anticipate the future are in animals largely controlled by inborn gene-based systems of reflex reactions, with relatively little learning involved. Conscious time binding in humans, on the other hand, is based to a much greater degree on memory and learning. Its beginnings were probably based on cooperative hunting and taming of fire by *Homo erectus*. The first definite evidence of conscious time binding comes from the carved animal bones on which early *Homo sapiens* recorded the first lunar calendars.

Imaginal thinking, a concept developed by J. Galanter, is based on the fact that the mind of a human being is rarely blank. To this simple fact must be added an additional feature of the human intellect—our ability for imagining situations and events that do not exist, have never existed, and may never come to pass. Throughout history, this activity has enriched human life. The ancient Greeks listened in wonder to the tales recorded in Homer's *Odyssey* of strange beasts, enchantresses having magic powers, and heroes who could shoot arrows into the air with such speed that they caught fire. Renaissance Italians were captivated by the horrors of Dante's *Inferno;* English people of the seventeenth and eighteenth centuries were equally absorbed in Dean Swift's tales of Lilliputians, giant Brobdingnagians, and horses who behaved like people. In contemporary America, tales of Star Trek voyages to outer space are the smash hits of television. As was mentioned earlier in this chapter, the appearance of unidentifiable animals painted on the walls of the cavern at Lascaux suggests that tales of imagination and myth were already being told 15,000 years ago. Most probably, all elaborate ceremonies for burial of the dead spring from imaginal thinking. If so, the rites practiced by the flower people of Shanidar suggest that imaginal thinking is as old as our own species, *Homo sapiens.*

We can never be sure that a chimpanzee or gorilla sitting in the forest with nothing to do has a mind that is at rest and not thinking of anything, or whether, like our own minds, it is always active. We can, however, be reasonably sure that whatever activity may be going on in their minds does not resemble in any way the imaginal thinking of humans. Except for actions that are based on memories recalled by obvious cues, the behavior of these apes consists entirely of reactions to their immediate surroundings. Most probably, both conscious time binding and imaginal thinking are possible in humans only because our species has a rich and highly developed symbolic language. Imaginal thinking is not just an amusing pastime—it is even more necessary for cultural progress than is conscious time binding because it is the essential precondition for planned achievement.

The quality we call humanness consists primarily of these three qualities—artisanship, conscious time binding, and imaginal thinking. From the biological viewpoint, these characteristics are based on nothing but a quantitative increase in the number of cells that form the surface layers of a single organ, the neocortex of the brain. This genetically determined change, interacting with the rudiments of a culture that was already present when it began, triggered a direction of cultural evolution that brought about novel and transcendent social behavior.

These three qualities are the foundations of the three principal realms of human knowledge. Natural science and engineering are the outgrowth of prescientific artisanship. History, political science, and other branches of the social sciences are fundamentally ways of directing social behavior to avoid disaster and to improve the material state of mankind. The humanities—literature, the arts, and philosophy—are extensions of imaginal thinking. The fact that artisanship, conscious time binding, and imaginal thinking evolved gradually by means of quantitative physical changes that affected a complex of behavioral characteristics does not detract in any way from the novel or transcendent quality of the human way of life.

The distinctive feature of humanity—culture based on artisanship, time binding, and imaginal thinking—is illustrated by this sculpture of *King and Queen* by Henry Moore, placed against the backdrop of two of the world's most outstanding monuments of architecture, the cathedral and Giotto's tower in Florence, Italy. (Courtesy of Henry Moore; photograph by Errol Jackson.)

Chapter Twelve

THE GENETIC AND CULTURAL HERITAGE OF HUMANITY

In his book *Beast or Angel?* biologist–philosopher René Dubos wrote: "In the final analysis, one is human to the extent that, while remaining an animal, one transcends those aspects of behavior which are deterministically governed by animality." In the preceding chapter, it was suggested that this transcendence was achieved by the development of three distinctive qualities—artisanship, conscious time binding, and imaginal thinking. How did these qualities evolve? Was their evolution an epigenetic succession of responses to environmental challenges through natural selection? To what extent was their evolution based on changes in genes as a result of biological evolution, and to what extent are they cultural extensions of already existing biological capacities?

HUMANITY'S TWOFOLD EVOLUTIONARY HERITAGE

Humankind has evolved by means of two entirely different interacting processes. One process, based on changes in the genes, is responsible for our bodily structure and our capacity for acquiring knowledge and culture. A succession of genetic changes, established in populations by the action of natural selection modified by chance events, has provided us with a wide range of physical and mental capacities. The second process, which determines how we use these capacities, depends on transfer of information from one individual to another. Patterns of imitation, teaching, and learning have evolved—sometimes gradually and at other times so rapidly as to constitute social revolutions. They translate genetic capacities into actual performance and thus are largely responsible for what we think and do. These patterns differ greatly from one society to another.

Genetic and cultural evolution have been going on for millions of years. The example of potato-washing in macaque monkeys mentioned in Chapter Ten suggests that imitation had already become an important vehicle for transmitting cultural information in the common ancestor of apes and monkeys. It may have played a vital role in the beginnings of cultural development throughout the evolution of the higher primates. Moreover, biological and cultural evolution have always been bound together by a continuous web of interaction. As humans relied more and more on artisanship for survival, genes that increased the capacity for making better tools, shelters, and clothing and for cooperation through sharing and division of labor spread more and more widely because of their increased adaptive value. At the same time, societies that used their capacities more efficiently could acquire more food and defend themselves better against predators, and thus improved their chances for survival and reproduction.

In this way, two kinds of selection began to acquire almost equal importance for human evolution. Superior individuals became leaders of their groups and through access to more females spread their genes at the expense of their less fortunate competitors. At the same time, more efficient groups crowded out less efficient ones, either by taking over their sources of food or by defeating them in battle. Their greater efficiency was perpetuated by cultural transmission. Genetic inheritance was complemented by patterns of imitation, teaching, and learning.

Genetic and cultural change probably interacted to increase the size and complexity of hominid societies. Collective defense against animal and human enemies may be a cultural heritage handed down to us from Australopithecine or other early hominid ancestors. Building shelters for a family or larger community is a cultural trait that may have begun with *Homo habilis.*

The earliest evidence of crude shelters was discovered by Mary Leakey in the Olduvai Gorge site of East Africa. It consists of stones arranged in such a way that they appear to be foundations of some kind of shelter. Whatever kind of shelter Australopithe- cines or *Homo habilis* erected on these foundations, their very exis- tence shows that the builders had a home base. While the men were away hunting, the women may have been closer to the shel- ter, digging roots and tubers, gathering wild nuts and berries, and caring for children. A home base is valuable as a place where food is shared by different members of a family or other cooperating group. Gorillas, chimpanzees, and other primates do not share their food in this way. Along with shelter, carrying and sharing are the principal bases of the more elaborate social organization on which cultural processes of evolution are based.

As humans evolved, they became better artisans, not through improved genes for mental capacity but through transmission of increasingly complex cultural practices and techniques. The shift from chipped to flaked tools appears to have followed the increase in brain size from *Homo erectus* to *H. sapiens*, and it may have been made possible by the evolution of the brain. However, the elegance and diversity of Cro-Magnon artifacts arose without an increase in brain size over that possessed by Neanderthals. Thus, cultural diversity may well have arisen through improvements in imitation, teaching, and learning.

The evolution of agriculture and civilization, by which human- kind acquired a greater capacity to dominate and control the environment, was triggered not by a quantitative improvement in artisanship but by the evolution of conscious time binding. Early hominids began to plan for the future by organizing and planning collective hunting expeditions that could bring down large animals too swift and powerful for a single individual to capture and kill. With the use of fire, acquired about 400,000 to 500,000 years ago, planning for the future became essential for survival. The complex task of handling and perpetuating fire would have been learned and taught only by people who could recognize its future value. Without fire, tropically adapted bodies could never have withstood cold winters. Fires and coals had to be kept going year after year to ensure communal survival.

Tribes of humans could live continually in seasonal climates only by learning to predict in autumn that a cold winter was on the way and that spring would eventually return. Looking toward the future by recalling memories of the past became a necessity of life. Once this cultural trait had become deeply ingrained, societies were prepared to advance in two new and revolutionary directions—the practice of agriculture and the building of planned cities. Once these revolutions had transformed society, conscious time binding became even more important. The interaction between planning and artisanship gave rise eventually to architecture, engineering, and the sciences.

For complex societies living in regions having seasonal climates, the evolution of artisanship and conscious time binding can easily be explained according to principles that apply to previous stages of plant and animal evolution. Genes that made people more intelligent (in the sense of being able to master more complex technology) would have spread by natural selection soon after they appeared in a population. Inventions that were made as a result of increased mental capacity would then have been perpetuated by patterns of imitation, teaching, and learning, and would have spread by cultural selection.

THE EVOLUTION OF MENTAL ABILITY

The evolution of the third and most distinctive human trait—the ability to think continuously and abstractly—is by far the most difficult to understand in terms of mutation and natural selection, even supplemented by imitation, teaching, and learning. One difficulty is that mind and consciousness are almost impossible to define or to quantify. They cannot be analyzed or measured by scientific procedures.

Each of us knows that we are thinking throughout waking hours, but we can only infer that others are doing the same. The question of whether animals think is much more difficult to answer. A bird looking for a worm is displaying highly adaptive alertness, and the most alert robins undoubtedly spread their genes at the expense of those less alert, but conceptualization and reflection are quite different from alertness or attention.

The problem is made more difficult because it is almost impossible to imagine what might be the advantage of conceptualization at a time when an animal is well fed and safe from predators. The sum of these difficulties is such that some scientists believe that the evolution of consciousness is impossible and that consciousness must be a given characteristic of all animals, including the one-celled amoeba.

There is as yet no way to be sure that nonhuman animals lack imagination. We assume that some animals dream, because we observe the dog, for example, exhibiting sleep behavior that suggests dreaming. But dreaming is probably based on memories rather than on new images—that is, mental images of things as yet unseen or unexperienced.

We cannot analyze the minds of extinct humans. Consequently, we can never trace the evolution of imagination itself, but we can follow the evolution of the visual expressions of imaginal thinking as embodied in ritual and art. We first see it in the flower-bedecked gravesites of Shanidar. It reappears in the form of the strange, two-horned beast that the Cro-Magnons painted on the walls of their caves at Lascaux. Throughout the world, the earliest cities were adorned with representations of gods and various creatures. Moreover, every modern tribe has a rich repertoire of myths and symbols. This evidence leads us to believe that the capacity for imaginal thinking evolved among the earlier races of *Homo sapiens* and may have existed in a rudimentary form even in ancestral *H. erectus*. Finally, imaginal thinking is widespread throughout the species, regardless of cultural advancement in terms of artisanship and time binding. This universality itself is evidence that imaginal thinking must in some way have played an important adaptive role in human existence.

The concept of mind is a slippery one, notoriously hard to define. Many psychologists reject it altogether. According to philospher A. J. P. Kenny, "To have a mind is to have the capacity to acquire the ability to operate with symbols, in such a way that it is one's own activity that makes them symbols and confers meaning upon them." This definition almost equates mind with the capacity to acquire imaginal thinking—a capacity determined to a great extent by our genes, particularly those that code for the large size of the human brain and the development of the cerebral cortex. On the other hand, mind is certainly developed by learning, particularly during the first six years of life. The mind is analogous to a computer. Its expression in terms of thinking is possible only when the computer has been programmed by conditioning and learning. So mind has a genetically determined physiological basis in the brain, but the social processes of imitation, teaching, and learning are essential to its development.

The parts of the brain that underlie the subjective phenomenon of mind have two properties of basic importance. First, there is the capacity to reinforce the connections between certain groups of nerve cells in such a way that some combinations or patterns may be activated more easily than others. This capacity is the basis of memory. It exists in a rudimentary form in all animals that have brains and is well developed in several species of mammals, such as great apes, elephants, and dolphins. Second, the human brain is capable of separating parts of learned patterns and generating from these elements new patterns that constitute original thoughts or ideas. This succession of processes is the essence of creativity. Nonhuman animals cannot carry out such a succession or can do so only in a rudimentary fashion.

The untrained mind (the unprogrammed computer) is the product of a person's genes and of the ways by which these genes interact during development with each other and with the external environment. In these respects, mind is like an organ of the body. One might expect, therefore, that the capacities recognized as mind evolved in much the same way as did the stomach, kidney, liver, or respiratory system. Natural selection does not operate on gene-determined capacities, however, but only on the phenotype—that is, on the physical expression of these capacities in an individual of reproductive age. As a consequence, the effectiveness of selection depends greatly on the kinds of interactions between genes and environment that take place during an individual's development. In humans—and probably also in species of hominids ancestral to *Homo sapiens*—interactions between the developing brain and its environment are totally different from those that affect the development of other bodily organs. The developing mind of the infant and young child can function normally only through learning experiences. Moreover, the way it operates is greatly influenced by parents and other caretakers and by the environment. These factors are expressions of the culture in which a child is raised. Thus the operative mind—the organ vital to survival and reproduction, and the only mental organ on which natural selection can operate—is a product of the interaction between genes and specific culture-based environmental influences.

The ability of parents to teach and of offspring to learn is by no means a distinctive human quality. Nevertheless, human teaching and learning is unique in one respect—the use of symbolic language. Although apes appear able to handle a few abstract symbols, there is no evidence to suggest that these animals use such symbols to teach their young. The perfection of symbolic language lifted teaching and learning to an entirely new level of complexity. Without it, neither conscious time binding nor imaginal thinking could have evolved beyond a rudimentary level.

At the genetic level, the evolution of the human brain can be described as an increase in the frequency of genes that code for larger numbers of neurons in certain parts of the brain, particularly those areas known to be most important for making associations between different sense impressions and learned ideas. At the level of natural selection of phenotypes, however, at least the later stages of this evolution depended on the superior adaptive value of both new combinations of genes and the kinds of cultural influences that interacted with the brain to produce functioning minds. So, the basic question we would like to answer is: What kinds of cultural influences would have favored selection of genes for larger brains that had the capacity for being trained to generate original ideas?

This question cannot be definitely answered on the basis of present knowledge. Nevertheless, speculations about it are possible and fruitful for orienting future research. They must, however, be made in the context of the stages of human evolution during which the postulated increase in brain size presumably took place. As has already been suggested, these stages encompass the entire course of hominid evolution from the earliest Australopithecines to races of *Homo sapiens* that existed about 30,000 years ago. Four different kinds of adaptive advantages may have promoted the spread of imaginal thinking in human societies.

First, groups of hunters could have planned the next cooperative hunt by recalling and discussing past experiences. Mental manipulation is a highly efficient way of sifting out different plans and deciding on the one that is most likely to succeed.

The second adaptation may have been the development of continuous alertness as the most effective defense against predators. If tribes lived in caves, they would have had to maintain a fire at the cave's entrance to ward off bears or other predators. During the long night hours, someone would have to tend the fire and chase away predators that came too close. The survival of the tribe would depend on the alertness of such sentinels. What would they do to keep themselves alert during the long night hours? They might invent simple games, they might look at the stars and imagine the form of familiar animals among the constellations, or they might try to imagine what animals were responsible for the moving shadows they saw from time to time. Under such conditions, the passage from the real to the unreal and imaginative would have been particularly easy. Furthermore, if the next day the sentinel told the tribe about the imagined danger from which he had rescued them, he might win approval from his superiors and advance his position in the tribe's hierarchy. In this way, he would gain access to a larger number of females and spread his genes more widely. The activity of guarding and protecting could aid the spread of genes that promoted imaginative capacity because of its advantage both to individuals and the tribe as a whole. Moreover, particular flights of the imagination could be told and retold as myths and thus become part of the culture of the tribe.

A third adaptive advantage may have grown up in connection with play and training. Many of the animals painted on the walls of the caverns at Altamira and Lascaux are represented as having spears or arrows stuck in their sides. Most anthropologists believe that such drawings symbolize the desire for success in future hunts. Anthropologists have suggested that youths were trained to become better marksmen by being asked to throw spears or other projectiles at targets of some kind. There may also have been spontaneous games in which the youngsters conducted imaginary hunts. Thus imagination aided education for survival.

Another important adaptive value of imaginal thinking may have been its association with tribal ritual. People do not live happily together because of a reasoned "social contract," as Jean Jacques Rousseau and his eighteenth-century followers believed. They are held together by ties of emotion, affection, respect, and, when necessary, coercion. This is just as much true of a tribe or nation as a family. As long as hunting and gathering were relatively simple, units no larger than a nuclear family or an extended family of fifteen or twenty people could be held together by ties of affection and still function relatively efficiently and harmoniously. But big game hunts and the maintenance of fire demand cooperation from groups of larger size. This new environmental challenge could be met only by adopting various devices for increasing the size of the tribe and for keeping it together. One strategy may have been raiding other family units, capturing their younger members, and raising them as foster children. Another strategy may have been the amalgamation of two previously separate extended families into a single larger unit. In either case, unity could have been achieved by establishing a single object or person as a rallying point. In modern tribes of hunter–gatherers, who live in a fashion much like that of early *Homo sapiens*, living or dead chiefs often become godlike symbols. This suggests that one of the earliest functions of religion was to unify newly formed tribes. The word *religion* itself denotes a tying back, a restraint on unbridled individualism. The restraining process would have been greatly aided by ascribing imaginary powers to the god–chief, a practice that was often observed. In modern hunter–gatherer societies, the young are taught hunting skills and the art of warfare during religious ceremonies that make use of ritual, dance, and elaborate, imaginative costuming.

The alliance between religion and the creative arts, a form of cultural coevolution that has stimulated both of them, may be one of the oldest adaptive cultural traits of the human species. Equally old and adaptive has been the alliance between religion and government.

Can we believe that the biological basis of the complex attributes we see in humankind today derives from nothing more than quantitative changes in the size and structure of our brains? For instance, does the creative ability of a modern genius, such as that of Charles Darwin to assemble a whole host of facts and concepts to generate a new theory of evolution, differ qualitatively or only quantitatively from that of chimpanzees Washoe and Lana to put together concepts such as "sweet" and "fruit" and apply the symbols "sweet fruit" to a watermelon? My answer, that the difference in ability is only quantitative, is based on the concept of transcendent novelty discussed in Chapter Five. These two extreme examples are connected to each other by a whole continuum of intermediate examples of varying degrees of complexity. Darwin's theory, like all modern theories about the nature of the world, was new and original only with respect to the combination of facts and pre-existing concepts that he put together to form it. All so-called original human creations are made up of pre-existing elements put together in new and original ways.

SOCIOBIOLOGY

The discussion in this chapter has emphasized the fact that the most distinctive characteristics of humanity—artisanship, conscious time binding, and imaginal thinking—have evolved via a combination of genetic and cultural changes. Their evolution can be understood only by means of a synthesis of knowledge based on genetics and Darwinian natural selection that is reinforced by a full understanding of human culture and how it can be acquired through imitation, teaching, and learning. The avowed aim of the new discipline of sociobiology is to provide such a synthesis. The term *sociobiology* was coined by zoologist E. O. Wilson in a monumental volume of the same name. Wilson's masterly review of behavior patterns in a great variety of animal societies has justly won the admiration of most of his colleagues. Nevertheless, during the past seven years, the term *sociobiology*, as well as the activities of sociobiologists, has sparked one of the most violent controversies in the biological community that it has seen during the present century. Why should this storm have arisen?

This question is easily answered. As presented in his book published in 1975, particularly the final chapter that deals with human societies and their basis in genetics and evolution, Wilson's synthesis is not an impartial gathering and synthesis of facts, followed by an equally impartial series of theoretical conclusions. From the start, it is colored by the assumption that human evolution is little different from that of higher animals, so that conclusions reached by studies of animal behavior can be applied with little modification to interpret the nature and direction of evolution in human societies.

Thus, from its beginning, sociobiology has had a dual nature. One aspect is exemplified by Wilson's general definition, "the systematic study of the biological basis of all forms of social behavior, including sexual and parental behavior, in all kinds of organisms, including humans." The other aspect of sociobiology arises from the fact that Wilson and his school believe firmly in one particular principle that they regard as the most important connecting link between animal and human behavior, and therefore as the keystone of sociobiology. This is the principle called by W. D. Hamilton "inclusive fitness." Those who hold to this principle maintain that, in both animals and humans, new patterns of behavior become established only if they enable certain individuals or families to produce a larger number of vigorous offspring. The controversy, therefore, is not over the desirability of producing a synthesis that fits Wilson's definition as stated above, but over whether or not inclusive fitness should be the foundation of the synthesis.

In evaluating this aspect of sociobiology, few reviewers are neutral. Opinions range all the way from the highest praise to outright rejection and ridicule. Animal behaviorist David Barash begins his critique with the sentence "A revolution is under way in the study of behavior." Anthropologist Marshall Sahlins, in a highly critical book, labeled sociobiology "the abuse of biology." Another anthropologist, Sherwood Washburn, supports his contention that "sociobiology creates confusion" by quoting a passage from a leading sociobiologist, W. D. Hamilton, and labeling it an "absurdity." Although other reviewers are more moderate, opinions are strongly polarized.

Most of the admirers of Wilson and other leading sociobiologists, such as W. D. Hamilton and R. E. Trivers, are themselves research scientists in animal behavior. They agree almost unanimously that Darwinian natural selection and inclusive fitness are basic guiding principles that make sense out of a great mass of observational and experimental data on all kinds of animals from amoebas and jellyfishes to insects, fishes, amphibians, reptiles, birds, and mammals. Factual evidence supports their contention that behavioral patterns can be analyzed in the same way as anatomical structures and physiological reactions at the cellular level and are subject to the same principles of biological evolution. Their research leads inevitably to the question: Is *Homo sapiens* just another species of animal that has evolved in the same way as have the various species of monkeys and apes?

The answer to this question is a qualified "no." As was stated in Chapter Eleven, modern human behavior and culture could never have been predicted, analyzed, or understood solely on the basis of analyses of animal populations and the behavior of animal species. Cultural evolution, acquired by nongenetic transmission of complex patterns of behavior, has risen in importance during human evolution to the extent that it now overshadows genetic change, at least with respect to interactions between human populations and their environment. This fact is as important to human evolution as the fact that the biological characteristics of our species are the results of genetic changes guided by natural selection.

Critics of sociobiology maintain that its proponents have made too many unwarranted assumptions derived from analogies between animal and human behavior. By itself, this criticism is not valid. As philosopher David Hull has pointed out, Darwin's theory as proposed in *The Origin of Species* had little or no factual basis but was based principally on analogies. Natural selection could not be tested experimentally until Mendelian genetics had reached a level of sophistication that made possible the construction of experimental models. At that time, the role of natural selection in evolution was amply verified.

The tenets of sociobiological theory may be verified as well when better methods are developed for analyzing the genetic and developmental bases of human behavior. The ability of sociobiologists to adjust their theories to fit new facts will be the final test of the value of sociobiology as a discipline.

Some critics further say that sociobiology is a revival of the concept of genetic determinism, a theory used at the turn of the century to justify appalling social inequities. In replying to this charge, Wilson disavows any attempt to foster genetic determinism or the inevitability of the status quo. Moreover, he criticizes strongly those who believe that human aggression is an internally determined trait that will break out into open conflict unless it is artifically restrained. Wilson is not committed to the theory that lethal war is inevitable because of the biological basis of human nature. Nevertheless, his pronouncements remind many of his readers of a warning that scientists are beginning to heed. Speculation about purely theoretical matters, such as the biological basis of bird songs or the evolution of sexual dimorphism in baboons or gorillas, are relatively harmless and may be valuable as stimuli to further research. However, when such speculations are extended to patterns of behavior that are fundamentally human, they can be harmful if they are based on few facts and unsound reasoning and are later shown to be wrong.

In particular, several critics have objected to the tendency of sociobiologists to apply such concepts as slavery, altruism, coyness, machismo, negligence, devotion, and spite to animals. Socio-biologists reply that ascribing human behavioral traits to animals is as old as Aesop's fables and that such words are used only figuratively, as a kind of shorthand. Nevertheless, such phrases as "genes for altruism" and "genes for spite" give a misleading impression to readers who are not fully acquainted with modern genetics. This point is further discussed in the next chapter, where we examine in more detail the contribution of sociobiology to the search for answers to basic questions about the interaction of biology and culture in determining human behavior.

This photograph of women construction workers operating a cement mixer in New Delhi, India, illustrates a common feature of many modern cultures—a striking mixture of the old and the new (© Kent Reno/Jeroboam, Inc.)

Chapter Thirteen

SOCIOBIOLOGY AND HUMAN EVOLUTION

The value of sociobiology to the study of human evolution depends on the answers it can provide to some basic questions. The first of these is: How much and what kind of variation exist in natural populations with respect to genes that code for behavioral traits? Unfortunately, this question cannot be answered directly, because geneticists cannot experiment with people as they do with mice, flies, and bacteria. Genetic variation with respect to human blood groups, enzymes, and other biochemical differences can be recorded and quantified, and visible traits can be traced through genealogies, but philosopher Michael Ruse is essentially correct in stating that "we are not much beyond the starting point with respect to direct evidence for the genetic control of behavior."

GENE POOLS FOR BEHAVIOR

Geneticist John L. Fuller, based on a lifetime of research on animal behavior, confirms by hard scientific evidence what every animal lover intuitively feels to be true—every individual dog, cat, or horse has its own unique personality, despite similar upbringing and training. There is no reason to believe that humans are different. Every human individual has a unique combination of genes. Moreover, the range of genetic variation is probably as large with respect to behavior as it is for height, weight, body type, hair color, and other visible characteristics. We thus assume that in human populations, the amount of variation is similar at all loci, whether for biochemistry, appearance, or behavior.

Human races could probably be selected and bred for "desirable" characteristics in much the same way that other animals are bred, on a purely genetic basis. These "desirable characteristics" could include genetically controlled behavior patterns, like those found in many breeds of dogs. Such selection would, however, take many generations to perfect any desired pattern, and it could be implemented only by long-established totalitarian controls.

In evaluating the potential for natural selection in human populations, we must consider not only the amount but also the kind of genetic variation that is present. It is probably similar to the kind of variation that exists for visible characteristics. A relatively small number of genes code for distinctive physical characteristics, such as albino color, blue eyes, hemophilia (absence of blood clotting substance), nervous disorders such as Huntington's chorea, and perhaps some forms of schizophrenia. At a much larger number of gene loci, alleles probably exist that differ only slightly from each other in their effects on appearance or behavior. These have been called *multiple factors, polygenes,* and *modifying factors*. In dairy cattle, for example, two alleles of a single gene determine whether or not the animal will have black markings on a white background; but the size of the markings is determined by a large number of modifying genes, the effect of any one of which is so slight that it is equal to or less than the amount of variation likely to be produced by differences in the environment of the fetus or young animal. Individual genes exerting such small effects cannot be identified, even when they affect easily visible or measurable traits, and genes that affect normal behavior patterns are still more difficult to recognize. With respect to this problem, Fuller writes: "Striking genetic differences have been found in the brain structure of normal laboratory mice, but the psychological significance of these differences is still undetermined."

For many quantitative and particularly behavioral characteristics, the primary product or protein for which the gene codes directly is separated from the adult characteristic on which natural selection acts by a long and complex sequence of developmental processes. The longer and more intricate this sequence is, the more likely it is to be modified by the effects of the environment.

Phenotypic characteristics differ greatly from each other with respect to their constancy under different environmental conditions. At one end of the continuum are biochemical differences, such as blood groups, enzymes, and immune reactions, produced by differences between proteins that are the direct product of translation by the genetic code. For them, the pathway from gene to character is so short and direct that it can be deflected or modified by the environment only under extreme, usually abnormal or pathological, conditions. At the other end of the continuum are such characteristics as obesity, for which the phenotype is somewhat influenced by genes—some people can eat indiscriminately without getting fat, while others have to watch their weight carefully—but nevertheless the environment, in this case the diet, is of paramount importance. The existence of these latter differences is the main reason why many geneticists object strongly to such phrases as "genes for altruism" or "genes for spite," which imply a direct connection between the gene and the behavioral trait. Both in their writings and in their mathematical models, sociobiological theorists fail to distinguish between classical Mendelian genes that have direct and easily visible effects and those that have relatively slight modifying effects on one or more behavioral characteristics.

The pathway from gene to character with respect to behavioral traits has another distinctive property that must be carefully considered. For most anatomical or physiological traits, either the entire pathway is completed before birth or environmental influences that affect it after birth are reversible. Brain development is distinctive in that the functioning of the mature brain depends to a large extent on its "wiring"—the synaptic connections between nerve cells, or neurons. Trillions of these connections are required to permit the full development of thought, mind, and memory. A large proportion of them are formed after birth, particularly during the first six years of infancy and childhood. Experiments with young rats have shown that raising them in a more stimulating environment causes them to develop more complex "wiring" than that of rats raised in a relatively deprived environment. The "computer" that exists in the mammalian brain differs from manmade computers in an important respect—it begins to be programmed before it is fully wired, and the programming process exerts a feedback influence, causing the brain to be wired in either a more complex or a simpler fashion. No one knows how specific these feedback instructions are, but their existence must be considered when one attempts to compare genetic and environmental influences on human behavior.

GENETIC DIFFERENCES BETWEEN GROUPS

The second question to which sociobiology might help to provide an answer is: What kinds of differences exist between populations with respect to genes that affect mental capacity and behavior? This question is important because evolution consists of divergence between populations with respect to gene frequencies. The concept of population is used here to designate small groups of people, such as tribes of hunter–gatherers and village communities, as well as regional groups, nations, and races. At all population levels, evidence for differences in genes that affect behavior is indirect and circumstantial, because specific genes that affect behavior cannot be identified and their frequency in populations cannot be estimated. Quantitative comparisons both within groups and between groups have been possible for biochemical differences, and several analyses of this kind of variation have been made. All show that genetic differences between groups are small as compared to those within groups. B. D. H. Latter's study of national and racial populations showed that variation between populations within a geographic region is about one twenty-fifth of that within a population, while differences between major racial groups—Caucasian, Mongoloid, African, Amerindian, and Australian—are only one-eightieth of those within populations. There is no reason to believe that comparisons based on genes that affect behavior would give significantly different results.

One way of obtaining indirect evidence about behavioral predisposition is to compare the behavior of newborn infants of different races. Behaviorist Daniel Freedman compared the behavior of infants of South Chinese origin with North European infants born in the same San Francisco hospital. He found that North European babies cry more easily and appear more responsive to external stimuli. Chinese babies are less likely to cry, are more easily consoled, and will lie more easily in different positions. A similar study of Navajo Indian newborns indicated that they resemble Chinese infants in certain behavioral traits, but are even more adaptable. In another study, psychologist N. Warren found that African, Afro-American, and native Australian (aborigine) babies acquire better muscular coordination during the first few days of their lives than do Caucasians or Orientals.

Although the evolution of different races of humans appears to have been accompanied by divergence with respect to behavioral traits as well as outward appearance, we do not know how great the behavioral differences are between races as compared to differences between individuals belonging to the same race. Nor have we identified the adaptive value that would have caused such differences to be promoted by natural selection.

These two questions cannot be answered until many more data are available. Comparative studies of the behavior of newborn infants may give sociobiologists new insights into the diversity of the human personality, but as yet there are no proven answers.

Gene-based physiological differences do not entirely explain differences in behavioral patterns, not only between races but between members of the same race who are raised under different sociocultural conditions. Genetic and cultural differentiation have progressed hand in hand. It may be that gene-based differences were relatively important during the early differentiation of human races and that cultural differentiation has gained the upper hand more recently. Whatever the actual situation may be, throughout the course of evolution humans have benefited from our unique gene-based capacity for behavioral flexibility and opportunism.

THE ADAPTIVENESS OF DIFFERENCES AMONG HUMAN POPULATIONS

The third basic question is: To what extent are the differences among human populations based on either adaptation to specific environments or on the history of the population, race, or nation? Unless differences are adaptive, they cannot have evolved by natural selection. However, differential adaptiveness is not necessarily the result of biological evolution alone. If a learned trait can be transmitted faithfully by cultural means from one generation to the next, it is also subject to selection and can spread because of its adaptive value. Furthermore, a culturally transmitted trait can spread and alter the way of life of a population much more rapidly than one based on genes. The spread of genes is greatly retarded by the length of human generations. If subject to selection pressures similar to those in insect populations, a human gene would require hundreds of years to become fully established in a population of moderate size. By contrast, cultural traits have spread through large populations in less than a decade. Faced with new and drastic environmental challenges, human populations can become adjusted far more rapidly by cultural means than by biological evolution. Sociobiological theory has paid too little attention to this basic fact.

CRITERIA FOR ESTIMATING THE INFLUENCES OF HEREDITY

Many sociobiologists believe that if a behavioral trait is found in many diverse cultures, it must be firmly based in our genes. The fallacy of this reasoning has been pointed out by behaviorist R. A.

Hinde. Complex adaptive traits can be handed down over many generations by imitation, teaching, and learning as faithfully as by genetic templates. If such traits are valuable in a great variety of societies, they are likely to be discovered many times and will spread from one society to another by *cultural diffusion*.

Two good examples are toilet training and the handling of fire. For tens of thousands of years, humans have lived in dwellings or caves. Soiling the nest has been not only disagreeable but harmful as a potential source of disease. Thousands of successive generations of infants and children have been trained to delay excretion, to urinate and defecate only in designated places. Yet there is no evidence that modern children are easier to train than were their remote ancestors. One sociobiologist, David Barash, comments that children are more difficult to toilet train than are dogs and cats. However, this difficulty and the parent–offspring antagonisms that it generates may be peculiar to western culture.

Before fire was tamed by *Homo erectus* 500,000 years ago, humans must have had a deadly fear of it, as have all mammals. It could be used for cooking and defense only by those who had become thoroughly familiar with ways of handling fire without getting burned. By the time *Homo sapiens* had evolved, its use must have been universal and essential. If widespread use and adaptive necessity inevitably lead to genetically controlled adaptation, then genes that would render the intelligent use of fire nearly or quite instinctive should surely have evolved with our species thousands of years ago. Yet, as every mother knows to her distress, children learn how to use and respect fire only after the most rigorous training. This example shows how universally adaptive traits can arise and be perpetuated by cultural transmission as well as by genetic means.

The criterion that comes nearest to distinguishing between traits that are strongly based in the genes and those that must be relearned by every generation is that of *changeability*. If a trait has persisted for many generations, its constancy may be due either to unchanging genetic heredity or to equally constant and faithful cultural transmission. If, however, behavior patterns have been known to have changed during one or a few human generations—

let us say, a hundred years or less—then they cannot be strongly influenced by genes, because genes cannot spread through even a small population in two or three generations unless subjected to very strong selective pressures that require high rates of mortality. On the other hand, strong indoctrination can change the culture-based attitudes and behavior patterns on a large scale during the course of a single generation.

Another possible guideline, which holds true for many examples, is that generalized behavior patterns operate more under genetic control than do specialized forms of behavior, which are more likely to be conditioned chiefly by the environment. An example is language. The child's ability to learn a language may well be coded to a considerable extent by the genes, but the particular language that a child speaks—that is, hears and imitates—is controlled almost entirely by the environment.

Another example is the tendency of both humans and other primates to show curiosity about and to explore their surroundings. Most children show curiosity at a very early age and continue to be curious and exploratory unless blocked and inhibited by anxious parents. This tendency probably has a strong genetic basis and its adaptiveness for human evolution is obvious. If not completely inhibited, the tendency to explore and look for answers can be expressed in various ways, all of which depend greatly on culture and much less on genes.

With respect to adaptiveness or neutrality—selection versus chance—basic behavioral patterns have evolved within the species *Homo sapiens* in much the same way that structural and physiological differences have developed among other animals. Humans are unique, however, in the high proportion of differences—both adaptive and neutral—based chiefly on tradition, imitation, and learning.

GENETIC AND CULTURAL BASES
OF ADAPTATION

The fourth basic question is: To what extent are adaptive differences between modern human societies based on differences in genetic capacity and to what extent on differences in acquired culture? This is a most controversial question. Sociobiology holds that human behavior is dominated by a genetically based urge to produce the maximum number of successful offspring, either directly or by favoring one's next of kin—brothers, sisters, and children. This is called *the urge toward maximum inclusive fitness.* Thus sociobiologists regard nepotism as unavoidable. The behavior of scores of animal species fits perfectly with this concept. If behavioral patterns in harmony with it did not exist, Darwinian natural selection could not operate. Is the human species an exception to this rule?

The answer to this question is unclear. There are good reasons for believing that human behavior has been strongly influenced for millions of years by a tendency to favor one's kin. Nevertheless, this tendency has evolved in the same dual fashion as have other human traits. In many societies, families are held together by both genetic and cultural ties. In most modern societies, children are taught to respect not only their parents but all their relatives. Moreover, tribal customs often cause family ties to deviate strongly from those that would be expected if they were determined strictly on the basis of the proportion of genes that relatives have in common. Marshall Sahlins reports that, in Polynesia, unrelated children are often adopted into a family, mothers may practice infanticide on their own children, and a father may welcome as his own a child conceived by his wife during his absence. The adaptive value of these cultural deviations from the biological principle of inclusive fitness is shown by the fact that they have persisted in well-adapted, harmonious societies that flourished for hundreds of years until disrupted from without by political intervention.

THE IMPORTANCE OF CULTURAL TRANSMISSION

Human intelligence and memory—aided over the past few millennia by writing, printing, and computer systems—have evolved a new method of transmitting cultural information from one generation to the next that can be as constant and faithful as the transmission of biological capacities by genes. The importance of this thread of cultural transmission has been recognized by both biologists and social scientists. Cell physiologist P. B. Medawar has called it exosomatic transmission, and geneticist Richard Dawkins has coined the word *memes* for the units of transmission.

The relationship between genetic and cultural transmission can be understood by analogy with templates. A template is a replicable pattern that specifies and serves as a model for some complex object. Typical *cultural templates* used in everyday life are blueprints for houses and machines, patterns for dresses, and other devices based on copying. The process used with these templates in making a photographic positive from a negative employs the same principle used in copying genetic DNA by complementary base pairing. Thus we can regard DNA as a *genetic template* for transmitting hereditary biological capacities. With both kinds of templates, the replication process can be extremely faithful. Occasional errors made in copying from genetic templates—mutations—are eliminated from biological heredity by natural selection unless they happen to be useful to the organism, just as errors in the replication of cultural templates are eliminated by conscious human selection.

Note that in both biological heredity and cultural transmission, only the template is copied. An adult characteristic is produced from a genetic template by a long sequence of developmental processes. In the case of some characteristics, such as eye color and blood type, this sequence is influenced relatively little by the environment, and the adult character is transmitted to the offspring with great faithfulness to the genetic template. In other cases, the pathway from gene to character is long and indirect and can be greatly influenced by the environment—this is particularly true of behavioral traits. A safe generalization is that genetic templates transmit *potentialities* or *capacities* rather than adult behavioral traits.

In the modern world, cultural transmission of behavioral traits can be even more faithful than their genetic transmission. This principle can best be understood by expanding somewhat the concept of a cultural template. Many stereotyped behavior patterns are specified with great precision by written or printed instructions. Examples are religious ceremony, military drill, and choreography. The directions for executing these performances—both the copying of the instructional template and its translation into action—can be transmitted from one generation to the next with great fidelity. With respect to human patterns of behavior, we cannot assume that genetic transmission of adult behavior is of necessity more accurate than cultural transmission. This is the basis of the statement by sociobiologists R. D. Alexander and G. Borgia that culture "can be simultaneously more heritable than genes and more abruptly changeable."

Modern peoples who do not use written communication, such as Australian aborigines and African bushmen, preserve and pass on their often elaborate rituals and rules by means of oral tradition. For groups of humans governed by a relatively small number of cultural artifacts or behavior patterns, transmission of these patterns by highly trained memory forms a system of cultural templates. This is in fact the basis of education.

How far back in human evolution can we trace the results of accurate cultural transmission through memory, imitation, teaching, and learning? Our best clue is the existence of stone tools and other artifacts made according to a fixed design that persisted for hundreds of thousands of years. This tells us that complex cultural patterns have been transmitted in this fashion ever since the appearance of *Homo erectus*. The cultures that flourished between 30,000 and 15,000 years ago reveal a diversity of artifacts that were faithfully reproduced generation after generation. The later cultures begin to show geographic diversity. Tribes in different parts of western Eurasia used templates that differed slightly but significantly from those used by neighboring tribes. This kind of diversity could form the basis for the evolution of cultural templates by group selection as one tribe gained ascendancy over another.

Homogeneous cultures did not begin with *Homo sapiens*. As we mentioned in the last chapter, the chipped hand axes of the Acheulian culture that persisted for hundreds of thousands of years before our species evolved were remarkably constant throughout Eurasia and Africa. Transmission of cultural templates, which forms the basis of cultural evolution, developed hand in hand with biological evolution throughout this long period. Feedback interactions between them must have been occurring constantly. Better memory-based templates meant better tools, more food, and better protection from enemies. These templates could evolve only in larger brains having more ample storage capacity for memories. Natural selection then made sure that genes coding for these better brains would increase their frequency in human populations.

Biologists recognize an important difference between biological and cultural templates. Biological templates are all of a single kind—information-carrying sequences of nucleotides of DNA. Diversity of function is acquired first by the primary products of gene action, the proteins, and later by specific organs. The earliest cultural templates, based chiefly on memory, were essentially similar to biological templates.

When humans began to live in cities and adopted both writing and drawing on parchment, the cultural templates themselves became diversified. Anthropologists Cornelius Osgood and David Bidney recognize three aspects of culture that are transmitted separately—material culture, based on artisanship; social culture, which consists of rules of behavior; and mental culture, including inventiveness and all kinds of creative activity. In modern culture, printed instructions serve as partial templates for all three of these activities, but additional templates exist for all of them. Examples of templates for material culture are drawings, blueprints, and plans. Examples of templates for social culture are such symbols as the cross, the crown, and judicial robes. Examples of templates for mental culture include musical scores and instruments, carefully matched pigments used by artists, and computers. The very recent evolution of this rich diversity of cultural templates, based on gene-controlled mental capacities that must already have existed 15,000 years ago, is perhaps the most dramatic evidence of the transcendent novelty that the human brain has achieved.

Some forms of cultural transmission may have existed earlier than the direct ancestors of *Homo sapiens*. They may have begun in the first nonhuman primates that developed a complex social organization, perhaps preceding the common ancestor of humans, chimpanzees, and gorillas. Evidence for this belief comes from the research of Harry Harlow on rhesus monkeys. He has shown that female infants who are raised apart from their own mothers, even when well fed and cared for, develop abnormal personalities as they grow up, and they become very poor, negligent mothers. Even in these monkeys the genes do not by themselves ensure the development of normal maternal behavior. Their action must be supplemented by the effects of a separate thread of cultural transmission that each successive mother learns from her mother and passes on to her daughters. This environmental influence is essential for the development of maternal care, one of the most important behavior patterns that the animal has. This kind of influence, which may have existed in our ancestors as long as 15 million to 20 million years ago, suggests strongly that biological and cultural phases have interacted with each other throughout the entire course of evolution from nonhuman social primates to modern humans. Cultural transmission was subordinate to biological heredity up to the appearance of *Homo sapiens* and became equal to it during the evolution of humans as hunters and gatherers. Finally, as humans adopted agriculture, animal domestication, and the urban way of life, invention and spread of cultural templates became the predominant way of responding to new environmental challenges.

A COMPARISON BETWEEN BIOLOGICAL AND CULTURAL EVOLUTION

Biological and cultural evolution have in common several important features. Both of them consist of an orderly sequence of responses to environmental challenges, epigenetic in nature, including many examples of autocatalysis and coevolution. For both of them, successful responses can be made only if the population possesses a store of variability of the right kind that preadapts it to evolving toward a new adaptive peak. In both of them, the actual response consists of the selection of appropriate traits and their recombination into new adaptive complexes or syndromes.

The differences between them are nevertheless great—in some ways even greater than their resemblances. Coevolution via cultural templates occurs not between humans and wild species living in undisturbed habitats but between cultivators of crop plants and the food and fiber plants that they have gradually improved by both conscious and unconscious selection, as well as between animal husbandmen and the domestic animals that they have bred. Modern humans are better fed and better clothed than their ancestors largely because through thousands of years of history such crop plants as wheat, barley, rice, corn, cotton, and flax have evolved greater productivity through human efforts, and similar improvements have been selected in cattle, sheep, swine, chickens, and other animals.

The variability necessary for biological evolution is the gene pool. Cultural evolution depends on a different kind of variation. In the early centuries of agriculture and urbanization, cultural evolution progressed rapidly not only because of the great environmental challenges that were present but also because, during the immediately preceding millennia, humans had built up a vast store of diverse artifacts, probably accompanied by an equally extensive store of rules and principles to guide social behavior. This store of cultural variation can appropriately be called a *cultural pool*.

Selection brings about cultural evolution as well as biological evolution. This can happen in two ways. Within a band or tribe, certain individuals may learn more easily than others how to hunt with skill, to cultivate the land efficiently, to locate and gather wild plants or to harvest crops, to care for and train children, or to defend their family or tribe. Such individuals are more likely to survive and become the parents of successful offspring than are laggards and weaklings. They will transmit their genes through biological heredity, and they will transmit their learned skills by training and teaching their children.

But in contrast to natural selection in biological evolution, cultural selection need not be based entirely on the success of certain more reproductive individuals. From ancient times up to the present, cultural templates of all kinds—instruments of war, designs for buildings, the templates of language and of religious ritual—have been spread by conquering armies. Selection on the basis of competing groups, which biologists have shown to be ineffective in biological evolution, nevertheless plays an important role in cultural evolution.

Another distinctive way by which both recombination and selection contribute to cultural evolution is the process anthropologists call *cultural diffusion*. In the normal process of trade and commerce between tribes and nations, newly invented artifacts may be transported for long distances and either accepted or rejected by the group that receives them. Sometimes they are accepted because some member of the group recognizes that the use of the new artifact is to his advantage. At other times, group acceptance is ensured by edicts from a king, the advice of priests, or in modern times by advertising campaigns. Cultural diffusion has its counterpart in biological evolution in the form of a much slower and less efficient process, migration and genetic hybridization.

The most important difference between biological natural selection and cultural selection is that the latter does not require either differential mortality or differential reproduction. Genes, being part of our bodies, live and die with us. They can be spread only through our offspring. Cultural templates are separate from our bodies. They can be multiplied, spread, and substituted for older, less efficient templates during the lives of single individuals. Cultural evolution is, therefore, far more flexible than biological evolution. Given a stable, unchanging environment and a society that is well adapted to it, the society can either retain an ancient culture for hundreds of thousands of years or undergo a revolutionary pace of cultural evolution.

Another important difference between genes and cultural templates is that new genes that result from mutations must replace old ones. However, new templates can be added to the cultural pool while old ones still persist. Examples of the coexistence of old and new cultural templates abound in the modern world; only isolated cultures retain old cultural templates to the exclusion of new ones.

In biological evolution, mixing or hybridization between radically different gene pools belonging to different species produces organisms that are either weak or, if vigorous, are sterile (like the mule) and unable to pass their mixture of genes to the next generation. The templates of cultural evolution can be mixed with each other much more freely without damaging results.

In Chapter Three, I reviewed the theory of Sewall Wright, who says that biological evolution proceeds most rapidly in subpopulations of a species that are isolated from each other long enough so that they can acquire gene pools having distinctive adaptations and then mix to produce new combinations of genes that enable them to meet the challenge of changing environments. In another chapter, I suggested that Wright's theory, which is now supported by facts derived from many different species of animals and plants, might explain the rapid evolution of *Homo sapiens* from *Homo erectus* during the early to middle part of the Ice Age. Wright's theory applies even more to cultural evolution than to biological evolution, since the results of mixing drastically different cultures are less harmful than the results of mixing genes that belong to different animal species.

CULTURAL FITNESS AND THE DESIRE FOR APPROVAL

Earlier in this chapter, I discussed the concept of inclusive fitness. Genes are spread following the reproductive success of certain well-adapted individuals. This success is based on both a higher reproductive rate and, particularly in nonhuman primates, better care of the young. Individuals or families who are superior in this respect are said to have superior inclusive fitness. Their genes are better represented in following generations than are the genes of those having inferior inclusive fitness.

If societies or families respond adaptively to environmental challenges not by natural selection of new combinations of genes but by adopting new cultural templates that they can develop and use on the basis of existing genetically controlled capacities, then the key to survival and evolutionary change is a different kind of fitness, called by behaviorists Richerson and Boyd *cultural fitness.*

Cultural fitness depends on qualities that are very different from those that promote inclusive fitness. Most sociobiologists have assumed that, because of the adaptive value of inclusive fitness, human populations are dominated by genes that cause people, particularly men, to desire more children. Women, who traditionally bear the main responsibility of caring for offspring, may succeed better if they are more interested in quality rather than quantity. Arguing from the standpoint of biological evolution through natural selection, one might suppose that genes for increasing the desire to have more children would become associated with genes for maleness, while genes favoring maternal behavior that includes greater care and consideration for quality would become associated with genes for femaleness. If, however, we look at society from the standpoint of improvement through cultural change, our concepts of the kinds of qualities that promote evolution become much broadened. Improved artisanship increases the resources of a tribe or nation and enables it to support a larger number of children, thus enlarging the gene pool of the group. In wars between tribes or nations, both genes and cultural templates that promote better artisanship were spread even more widely as conquering armies took over, often killing enemy males and mating with their women. Adaptiveness involved not only greater ability to fashion improved tools but also the accompanying motivation to spend the long hours of learning and practice that are essential for developing a successful artisan. Two kinds of motivation were necessary—a desire for power and a desire for approval. As cultural evolution became more important than biological evolution, the genetic capacities that made these motivational qualities easy to develop must have had greater adaptive value than the genetic capacity for reproductive fitness. The desire for power would have increased greatly in importance as tribes became large enough to include distantly related or unrelated individuals or families. The desire for power may well have increased through interaction with the desire for reproductive fitness. Better hunters and warriors would be more likely to become chiefs and thus would be able to choose the most desirable women for their mates. Their contribution to the group as a whole would increase the group's ability to compete with and dominate other tribes. The transition from family patriarch to chief or king must have taken place repeatedly in this way.

A gene-based capacity for developing the desire for approval probably already existed in the common ancestors of apes and humans. A hierarchy of rank exists in many societies of apes and monkeys, and it is adaptive because it reduces aggression among members of the group who have different positions in the society. When individuals of lower rank approach their superiors, they give signs of submission and may perform small favors, such as grooming or offering food. Behavior of this kind may have been practiced throughout the early stages of human evolution. As societies became larger and increased division of labor became more adaptive, young people having this kind of genetic capacity would have become more easily trained to become leaders than their peers. With the advent of cities and trade, cleverness and the capacity to win approval, acquired either by favorable genes, training, or a combination of the two, would have promoted individual success. These qualities could have been developed first by natural selection and later by interaction between natural and cultural selection throughout the course of human evolution, both before and after the appearance of our own species.

We can now attempt to answer the fourth basic question asked earlier in this chapter: To what extent are adaptive differences between modern human societies based on differences in genetic capacity and to what extent on differences in acquired culture? Cultural transmission has become increasingly faithful, and cultural traits are more easily adopted or rejected than are genetic differences. This means that cultural evolution can be far more rapid than biological evolution. Consequently, human societies respond to environmental challenges far more often by cultural change, based on cultural selection, than by changes in their gene pools brought about by Darwinian natural selection. In humans, cultural fitness, based to a large extent on a desire for approval by

society, has both a genetic and a cultural basis. It often overshadows inclusive fitness, which depends on a gene-based desire to
produce a maximum number of fit offspring and to favor the offspring of one's relatives.

HUMAN AGGRESSION AND
SOCIAL HARMONY

A fifth basic question facing sociobiologists is: Are gene-based
trends toward individual and family selfishness so strong that
evolution toward a harmonious social order is a hopeless dream?
This question is asked chiefly because of the obvious fact that
evolution toward social harmony by changing mankind's total
gene pool is far too slow.

I consider humanity's future in the final chapter of this book.
At this point, the relevant questions are: Can human aggression
be curbed and kept within reasonable bounds by cultural means?
Do humans indulge in fights, murders, rape, and organized warfare because they can't help it, or are they incited toward these
acts of aggression by social stimuli that can be reduced or sup-
pressed?

The opinions of scientists on this issue are sharply divided. The
idea that a genetically determined instinct for aggression is so
strong that it will inevitably lead to violence unless inhibited or
given an outlet by activities such as violent games and sports has
been held by psychologists Sigmund Freud and Erich Fromm,
ethologist Konrad Lorenz, and many of their followers. Their conclusions are based largely on the almost universal existence of
violence and fighting in animals and humans and the enthusiasm
with which violent and even bloody sports have been supported
by civilized people.

Anthropologists and sociobiologists have cited much evidence
against this point of view. If Freud, Fromm, and Lorenz are correct,
then substitute aggression, in the form of violent sports and sham
battles, should be found most commonly in countries that have
minimal crime rates and are the least likely to use war for political

purposes. These substitutes should have reduced the amount of actual violence. If, on the other hand, violent sports are a part of a general aggressive tendency on the part of a society, whatever may be its basis in genes or culture, then they should be most strongly developed in societies that are the most prone to military adventures and that have the highest crime rates. Anthropologist Richard Sipes has pointed out that the latter is the case. Ancient Rome, which witnessed the bloodiest sports events that history records, was constantly fighting enemies around the borders of her vast empire. In the modern world, nations like Sweden, Switzerland, and Sri Lanka are peaceful both politically and with respect to the kinds of sports and games that their people prefer.

While recognizing that the capacity for aggression is encoded in the genes, sociobiologists admit that its realization depends on environmental factors. E. O. Wilson believes that human beings are innately aggressive, but he defines this innateness as "the measurable probability that a trait (aggressiveness) will develop in a specified set of environments, not the certainty that the trait will develop in all environments." If this principle is accepted, then the possibility of controlling aggression and warfare depends on finding out how widespread and necessary the environmental conditions are that promote the development of aggression and restricting these environments as much as possible. If specified sets of environments are of primary importance, violent aggression can be thwarted more easily by environmental control than by trying to alter the gene pool. Sociobiologist W. Durham maintains that primitive warfare, recognized as a "cultural tradition" rather than a genetic necessity, evolved because the people engaged in it believed, either consciously or subconsciously, that it would increase their inclusive fitness—their ability to produce succcessful offspring would be improved by winning battles against enemy tribes. In this view, modern warfare is seen as a legacy of this primitive conviction.

Most psychologists and anthropologists support more strongly the belief that overt aggression is partly a learned trait that becomes expressed in certain environments. Psychologist L. Berkowitz bolsters his case by an experiment showing that even animals learn normal aggressive tendencies. The experimenter tested learning by observation and imitation in kittens. He divided the kittens into three groups. Kittens in the first group were raised in an environment where they never saw a rat. Those in the second group watched their mothers kill rats, and the third group was raised in the company of rats and never saw one of their rat playmates being killed. As adults, 45 percent of the "mother watchers" killed rats as adults; none of the kittens that had rat playmates turned into rat killers. This shows that a highly adaptive trait, the potentiality for which must exist in cat genes, can either be heightened or inhibited altogether by relatively mild manipulation of the environment.

Behaviorist R. N. Johnson recognizes the fact that environmental stimulation is needed to convert gene-coded tendencies for aggression into aggressive actions, but he provides evidence to show that other stimuli than learned behavior can be responsible. Experiments by various workers have shown that animals as different as rats, cats, and monkeys can be made aggressive by electrical stimulation of the lower part of the brain, particularly the hypothalamus and the basal portion of the cerebrum. The location of these centers is highly significant because they are also the seat of emotional responses.

In another series of experiments, monkeys were subjected to a minor and painless operation under anaesthesia in which the experimenter placed on the animal's head a special instrument designed to permit him to locate particular regions of the brain. He then made a tiny hole in just the right place, lowered a fine insulated wire or electrode into the brain, and firmly anchored the electrode to the skull. A few days after recovery from anaesthesia, the animal was ready to test. By applying a few millionths of an ampere of electric current to the electrode, the investigator could change dramatically the animal's behavior. The kind of change— whether eating, drinking, gnawing, or copulating—depended on the spot in the brain into which the electrode was inserted. If the electrode tip is situated in parts of the hypothalamus, a "sham rage" can be induced. The "rage" includes attacking or even killing other monkeys. It differs from real rage in that no external stimulus was received by the sense organs, and the experimenter could turn it on or off at will.

Would humans react in the same way if they were the objects of such experiments? We are enough like monkeys that such an inference would be logically sound. Moreover, indirect evidence supports this view. In 1966, people throughout the United States were shocked and horrified by the mass murders committed by Charles Whitman at the University of Texas. Notes of confession, written by Whitman before he climbed to the top of the university tower and shot at every person in sight (killing 14 and wounding 24), showed that he was highly intelligent and aware of his impulse to kill but could do nothing about it. A postmortem examination of his brain revealed a tumor as large as a walnut situated next to the hypothalamus.

These experiments and the Whitman case have general applications. Brain centers are normally stimulated by electrical impulses that come from other parts of the body. Most impulses are generated by sense impressions from the eyes, ears, nose, and sensitive regions of touch. Others are generated or amplified within the body, particularly by such hormones as epinephrin, which is secreted by the adrenal glands. Adrenal secretions are closely associated with such emotional responses as fear, frustration, and the desire to dominate others. The human body undeniably contains a well-regulated mechanism by which gene-determined capacities for aggression can be quickly converted into overt acts by emotional stimuli of almost any kind.

Genes can influence not only the capacity for acts of violence but also the mechanism by which it is expressed. The environmental influence on violence, however, is equally well demonstrated by the experiments on monkeys. The experimenter could turn off aggression as easily as he could turn it on. Charles Whitman, between his fits of uncontrollable rage, bore no hatred against the people he killed, including his mother and his wife. Persistent anger and continued violent aggression require either persistent stimulation or strong conditioning. With respect to humanity as a whole, the following statement by R. N. Johnson is particularly appropriate: "As a species, man must confess to unspeakable violence and brutality, yet he is distinguished by altruistic behavior and a remarkable capacity to learn and to adapt."

Faced with this situation, one naturally asks the question: How often and under what circumstances is the human capacity for aggression converted into aggressive acts of violence? This question cannot receive a single answer. Three kinds of aggression must be considered separately—aggression against other species, individual or organized aggression against other members of the same community, and group aggression or warfare against another tribe or nation.

Human aggression against other species of animals requires little comment. Before the origin of animal domestication thousands of years ago, killing animals was highly adaptive for obtaining food and protecting human communities against predators. Now it persists as a form of sport that is becoming more and more carefully regulated. It persists in urban cultures chiefly as a way of escaping from a dull environment and an opportunity to get trophies about which a man can boast to his associates.

Individual aggression in the form of violent crime has plagued urban society throughout history and persists in most parts of the world. In spite of a steady flow of sensational reports in the newspapers (which are published because crime pays in the form of increased newspaper sales), statistics show that the proportion of U. S. citizens who actually commit crimes of violence is remarkably small, and in most other countries it is still smaller. In Boston during the year 1969, 137 cases of murder were reported among a population of 3.5 million people. At least one-fourth of these were the result of family quarrels in the home. Most of the crimes were committed in slums and ghettos by young men between 15 and 24 years of age in the lower income brackets that include the unemployed and unskilled laborers.

From these and many similar statistics, one can conclude that, in modern prosperous societies, the conditions that make for overt, aggressive violence in the form of murder, rape, or serious injury only occasionally constitute a serious social problem. For people who are fortunate enough to belong to such societies, the

risks posed by violent aggression are far less than the risks associated with driving automobiles. The reduction of violent aggression in most societies therefore seems clearly linked to the age-old
problem of reducing or eliminating poverty, discrimination, and
inequality.

Is group aggression, particularly organized warfare, an inevitable practice of human society, or can it be reduced or even eliminated by providing a favorable environment for world peace? A
stock answer to this question is that warfare has been practiced by
all tribes and nations since *Homo sapiens* evolved, in spite of urgent
and continued pleas for peace. Anybody who thinks it can be
stopped, even though its continuation might mean the extinction
of the human species, is illogical and utopian.

Is this answer valid? To be sure, the earliest written records
describe repeated wars between the city–states of Sumerian Mesopotamia, and sculpture on the oldest Egyptian tombs shows a
Pharaoh standing proudly among the decapitated corpses of his
vanquished enemies. Hunter–gatherer and primitive agricultural
tribes as diverse as Polynesian Maoris, Australian aborigines, Central African Masai, and North American and Amazonian Indians
have engaged in warfare as a way of life until forcibly pacified by
European colonists. Contemporary tribes and nations that are
basically peaceful form an insignificant minority of the world's
population. The case against a peaceful future for humanity
appears to be unassailably strong.

Yet rays of hope exist. The art of preagricultural humans, particularly the cave paintings of Europe and similar murals made by
African tribes, depicts hunts of animals but little or no aggression
against other humans. Can it be true that, for 65,000 years after
Homo sapiens evolved imaginal thinking as exemplified by Shanidar
Cave, warfare between tribes was so uncommon that it was not of
major concern? If the animals were painted on cave walls in order
to assure success in the forthcoming hunt (either by magical
means or by training the young), wouldn't similar paintings have
been executed of enemy warriors if these had been a serious threat
to the tribe or if conquering them would have been of advantage
to the tribe's own warriors? To be sure, some Neanderthal hearths

in eastern Europe contain remains that suggest cannibalism, a practice that among many primitive tribes is associated with violent warfare. In some instances, however, modern cannibalism is associated with peaceful religious rites.

Anthropologist Irving Devore has suggested that serious warfare began with agriculture, when tribes first led a sedentary life and acquired property, food stores, and domestic animals that had to be defended. If, as M. N. Cohen has suggested, agriculture evolved at least in part as a result of population pressure and crowding, this hypothesis becomes highly plausible.

If warfare arose in connection with population growth, crowding, and defense of property, then the few peaceful tribes that exist today (or existed before European invasions) should have in common one or both of two attributes—sufficient isolation so that threats from other tribes or nations were rare or nonexistent, and little property to defend. To what extent is this true?

In the New World, the most peaceful Indian tribes were those of California, who were separated from their neighbors by high mountains and desert wastes, and those of Fuegia, the southern tip of South America, separated from the rest of that continent by dense forests on its west coast and the barren steppes of Patagonia on the east side. In the Pacific area, the now-extinct Tasmanians were separated from their Australian cousins by stretches of ocean that were nearly impassable for people who knew nothing of seafaring. Completely peaceful tribes in the center of the Philippine Island of Mindanao are separated from coastal tribes and raiders by forests so dense that very few potential enemies even knew of their existence. The island of Sri Lanka, one of the most peaceful countries of the world, is close to the Indian Peninsula, but until Europeans arrived with firearms, its people could easily escape marauders by retreating into the dense forests and mountains. In Africa, the peaceful pygmies of the Ituri Forest and the bushmen of southernmost Africa were similarly protected.

These few societies, even though constituting a tiny minority of the human species, nevertheless teach us a lesson. Many and perhaps all humans will remain peaceful if they believe that doing so is to their advantage. As W. Durham has pointed out, people engage in warfare either because they have concluded that it is to their own advantage in terms of inclusive fitness to do so, because they have been stimulated or forced to do so by others whom they respect and admire, or because they are forced to do so by powerful leaders whom they must obey. Not only is warfare nearly universal among humans, but propaganda and other forms of stimulation toward group violence are equally common and widespread. The Yanomamo Indians of South America are often cited by scientists as primitive warlike people who fight over the possession of women so that the victors' genes are spread far more widely than those of the vanquished. Nevertheless, Yanomamo boys at the age of six or seven become violently aggressive only after a prolonged, intense initiation ceremony, described in dramatic fashion by anthropologist Napoleon Chagnon. These ceremonies, and many like them throughout the world, most probably stimulate the aggression centers in the limbic–hypothalamic region of the brain, just as did the artificial electric current applied to monkeys and the tumor with which Charles Whitman was afflicted. Similar stimulation has been the necessary prelude to warlike activity throughout history. Well-known examples are the detailed description of gory battles in Homer's *Iliad*, the victory celebrations and display of captive chieftains practiced by Roman generals, the sanctification of violence against "unbelievers" by Muslim priests who incited the conquests of the eighth and ninth centuries, the zeal for crusades on the part of medieval Christian bishops and kings, the glorification of conquest and empire by British nineteenth-century writers like Tennyson and Kipling, the pomp and ceremony of mass rallies that were the essence of the Fascist and Nazi creeds, and the incitement to violence based on "kill, kill, kill" and other chants used by drill sergeants in the United States.

The capacity for violence in humans most certainly has a genetic basis. Violent behavior, however, develops only when this capacity is reinforced by external stimuli from society. Human nature is flexible enough that complete control of lethal and destructive violence is at least possible from the genetic and biological point of view. The question remains whether or not human societies can control the social forces that throughout history have converted the capacity for aggressive violence into its expression.

THE FUTURE OF SOCIOBIOLOGY

I believe that the discipline of sociobiology is here to stay. For the good of humanity, the questions we have asked in this chapter must be explored and answered as thoroughly as possible. They can be answered only by drawing from both biological research and social research. The perspectives of biology, anthropology, sociology, and psychology must become united in this quest. The complementary perspectives of reductionism and holism can be combined to produce a final synthesis. The necessary attitudes have been well expressed by John T. Bonner: "It seems to me fairly obvious that it is important for the biologist, and more particularly the sociobiologist, to realize that his flashes of insight that have come, for instance, through the aegis of kin selection, will not solve all the problems of the social sciences, but may shed some bright light on aspects of human social behavior. The social scientist, on the other hand, must face the possibility of some biological information being extraordinarily useful to him, and certainly it should not be rejected for doctrinaire reasons." In other words, sociobiology will flourish and increase its usefulness to the extent that it becomes an impartial quest for knowledge and a synthesis through careful, logical deductions based on all relevant facts rather than an attempt to bolster and justify certain preconceived principles or ideas.

The ideas discussed in the last two chapters, including my own judgment about the nature and future impact of sociobiology, can be summarized as follows. The behavior of modern men and women depends on both the potentialities that exist in their genes and the way these potentialities have been developed by conditioning and learning, especially during the years of infancy and childhood. Both genetic and cultural sources of behavior are equally valid; the idea that there was once an original unspoiled human nature that existed in hunter–gatherer tribes or any other humans is an unfounded myth. At present, no universal theory or formula can be devised that will enable scientists to determine how much of any characteristic or behavior pattern is contributed by genetic inheritance and how much by cultural inheritance. Such a formula may never be devised; perhaps separate evaluations will always be necessary for any particular characteristic.

The most distinctive feature of human genotypes is their enormous phenotypic flexibility. Behavior patterns of humans can be modified by conditioning, teaching, and learning to a greater degree than can those of animals. This is because (1) the brain of the newborn human is less mature than that of any animal, and (2) the human capacity for memory storage makes possible a great amount of cumulative learning, the direction of which can be modified at any stage of development.

As human cultural evolution increased in importance, the adaptive value of societies larger than an extended family increased correspondingly. This adaptiveness is related both to inclusive fitness (production of a maximum number of vigorous offspring) and to the power of the group, particularly its control over vital resources. As civilizations evolved, power and control (and the pattern of group organization that promoted power and control) acquired greater adaptive value than inclusive fitness. This shift in adaptive values required new responses based on phenotypic rather than genotypic adjustment, since the whole time span of civilization is too short to permit major alterations of human gene pools that (indirectly) control behavior patterns.

The major problem of ethical human behavior involves resolution of conflict between the primeval urges toward maximum inclusive fitness and maximum power, which have been highly adaptive in competition between groups of all sizes, and the need for harmonious coexistence in an age when major conflicts can result in the extinction of the entire human species. No single scientist, least of all myself, is in a position to suggest how this conflict might be resolved. In the last two chapters I shall consider some aspects of this conflict and discuss them in relation to the principles of biological evolution that have been developed in earlier chapters.

The great empires of ancient Egypt and the Middle East mark the dawn of recorded history and the end of human biological evolution as the main strategy for responding to environmental challenges faced by our species. This portion of a wall painting from the tomb of Rekhmire, vizier to Pharaohs Thutmose III and Amenhotep II (XVIIIth Dynasty, 1470–1440 BC), illustrates the high level of artistry and craftsmanship that existed during this period. (Courtesy of Kent Weeks.)

Chapter Fourteen

AN EVOLUTIONIST
LOOKS AT HISTORY

From the perspective of human evolution, recorded history is that phase of cultural evolution for which written records exist. The time elapsed is about 6000 years, representing the most recent five percent of the biological and cultural evolution of *Homo sapiens*. When historical records began to be kept, humans had already completed two-thirds of the evolutionary time span that has elapsed since the Cro-Magnon type of culture appeared.

Can this last phase of human evolution be analyzed according to the same principles that governed the prehistoric evolution of humanity and the animals that preceded mankind? Is world history a viable topic for such research? Can an application of biological principles aid scholars in their interpretation of history? Or are the particular sequences of events that constitute history so far removed from evolutionary principles that attempts to apply these principles will be more of a hindrance than a help to our understanding?

During the twentieth century, historians have held sharply contrasting opinions about the answers to these questions. Books and essays on world history have become increasingly numerous. During the first half of this century, Oswald Spengler's *Decline of the West*, H. G. Wells's *Outline of History*, and Arnold Toynbee's *A*

Study of History were enthusiastically received by the intellectual public. Their authors were acclaimed by many as apostles of a new history, just as Albert Einstein, Ernest Rutherford, Niels Bohr, Claude Bernard, Louis Pasteur, Emil Fischer, Jacques Loeb, Hugo De Vries, and T. H. Morgan were recognized as leaders of new trends in science. Historians, however, quickly detected flaws in these great works, with respect to both factual detail and fundamental conception. They have already lost their luster and in our time have little influence over either scholarly research or teaching of history.

Meanwhile, new studies of world history have appeared. Some of them, such as *The Rise of the West* by J. H. MacNeill, are primarily historical chronicles of political events and the fates of nations. Others, such as *Oriental Despotism* by K. A. Wittfogel, *Origins of the State and Civilization* by E. R. Service, and *Humanity and Society* by K. N. Cameron, look at history from a sociological viewpoint. Anthropologists have contributed their interpretations, exemplified by C. R. Coon's *The Story of Man*, the books of Marvin Harris, and R. Bigelow's *The Dawn Warriors*. C. D. Darlington's *The Evolution of Man and Society* contributes the perspective of a geneticist and biological evolutionist. All these writers regard history as a chronicle of whole societies rather than of power struggles between their leaders. Their purpose is not only to tell a story but to discover why events have occurred.

These world histories contain material that the evolutionist can use to build bridges between the disciplines of human prehistory and cultural anthropology. Nevertheless, none of these interpretations can be accepted uncritically. All of them must be examined in order to detect and eliminate poorly founded opinions that are based on personal bias. The evolutionist looks particularly for false or weak analogies to biological phenomena. For instance, both Spengler and Toynbee compare societies to living organisms and attempt to record their "birth," "growth," "maturity," "senescence," and "death." The modern biologist looks askance at this analogy. The life spans of individual humans as well as animals are limited by the gene-determined nature of their cells, organs, and organ systems. Environmental influences can lengthen or shorten life only within narrow limits. Human communities and populations of animal species are not subject to such restrictions. Examples of plant and animal species that are essentially immortal have been presented in earlier chapters. In contrast to the death of an animal, the extinction of a species depends much more on disharmonies of interaction between populations and their environments than on disharmonies that are inherent in the populations themselves. The historical events that, according to Spengler and Toynbee, are examples of the death of societies, such as the extinction of ancient Greek culture and the fall of the Roman Empire, were brought about by flaws of organization that became critical only when the societies were faced with new and unforeseen challenges. Moreover, when a human being or an animal organism dies, all of its cells, organs, and organ systems die with it. When a society "dies," however, its component individuals continue to transmit to their offspring both the genes and at least some of the cultural traits on which the "dying" society was based. Both the genes of the noble Romans and individual elements of their society, such as the Latin language and Roman law, persist to this day. Looking into the future, we can confidently predict that, even if all modern societies should "die" in the sense of Toynbee or Spengler, many of the social inventions that they have generated, particularly those having adaptive value under new conditions, will live on. The question of the survival of the human species is discussed in the next chapter.

Evolutionary biologist C. D. Darlington's attempt to interpret history as a continuation of human evolution suffers from a different bias—an extreme belief in genetic determinism. According to him, both the nature of society and the great events of history are determined by the genetic nature of the people who brought them about. The course of history of medieval Europe, he believes, was governed by the mating habits and marriage contracts of the noble families who ruled it. Hybridization between Viking leaders and "the newly established governing class of Capetian France . . . produced the race and phenomenon we know as Normans. The plainest effect of this hybridization was that the Normans, without altogether losing their knowledge of the sea, had acquired an understanding of the horse with harness and stirrups . . . ." In the paragraphs that form the context from which this quotation was extracted, Darlington makes clear his belief that these capacities for specific knowledge and skills were transmitted by genes rather than by cultural means.

Many other world histories, such as those of historians K. A. Wittfogel, E. R. Service, and K. N. Cameron, as well as those of anthropologists such as R. Bigelow and Marvin Harris, look upon the course of history as the result of interactions between human populations and their environments. Their broad interpretations of the concept of environment include not only physical factors such as climate, available resources such as water and minerals, and degree of isolation from other societies because of physical barriers such as mountains and deserts, but also biological and cultural factors. Among the biological factors are danger from predators, the presence of plants that could be cultivated and animals that could be domesticated, and freedom from or susceptibility to disease. Cultural factors include previous structures of society, degree of advancement with respect to tools and other artifacts, and availability of a labor force of humans or domestic animals. Described and analyzed in this fashion, history is an account of relatively recent human cultural evolution.

HISTORY AS POPULATION– ENVIRONMENT INTERACTIONS

As a biological evolutionist looking at history, I recognize that the course of history has been much like the course of biological evolution as modern evolutionists interpret it. Repeatedly, human bands and tribes have radiated outward by migrating from centers that had previously achieved new advances in artisanship or time binding. In doing so, they became diversified by adopting new crops for cultivation, new animals for domestication, new methods of warfare, new styles of architecture and dress, new power structures, new religions or gods, and new ideas or ideologies. Societies have become diversified largely as a result of responses to environmental challenges. The particular kinds of responses made by any band, tribe, or nation were motivated by the desire for power, wealth, and production of successful offspring. The adage "Be fruitful and multiply" expresses the principle that has guided biological evolution for more than 3 billion years. Moreover, among the alternative responses that could have been made at any time in history, tribes or nations usually adopted and followed the one that was the easiest to follow, causing minimum disruption of social organization. Like natural populations of animal species, most human tribes, nations, and societies are basically conservative. Faced with a challenge, they meet it by altering their ways as little as they can and still continue to survive.

The results of responses to environmental challenges have varied among human societies in much the same way as among species and many evolutionary lines of animals. As Toynbee pointed out repeatedly, societies have become extinct, in the sense that their way of life has had little effect on later generations. Other societies, such as hunter–gatherers and primitive agricultural societies, have persisted in only a few remote corners of the earth. Relatively few cultures have formed the mainstream of progress toward more complex and sophisticated social structures. Individual steps toward more complex social structures have been taken in widely different parts of the earth and at widely separated intervals of time. Agriculture based on irrigation arose independently in the Middle East and in Mesoamerica. Independent city–states arose following the switch to agriculture in Mesopotamia, India, and China, but not in Egypt, where imperial dynasties ruled by divine authority over villages of serfs and peasants. The arch became a basic structural principle of architecture in the Roman Empire and the cultures that followed it, but not in other urban civilizations. The course of history, like biological evolution, has been dominated by responses to environmental challenges, based partly on chance, but based even more on genetic or cultural preadaptation and conservative modification along the lines of least resistance. Neither biological evolution nor cultural evolution is inevitable, although they sometimes appear to be so because of widespread, inexorable changes in the environment. Genetic determinism, evolutionary determinism, and historical determinism are all illusions that result from a human tendency to prefer simplistic interpretations of phenomena to those that consider complex interactions. The simplistic dichotomy of determinism versus chance is as false and misleading with respect to cultural evolution as it is when used to interpret biological evolution. Chance operates at the lowest levels of social adjustment, but at higher levels change is strongly influenced by either natural selection or cultural selection and by the effects of migration, including cultural diffusion. Evolutionary opportunism includes a whole spectrum of situations intermediate between blind chance and rigid determinism. The pathways followed both by evolutionary lines of animals and by sequences of cultures have been so numerous and diverse that, at least with our present state of knowledge, they cannot be interpreted by any all-inclusive general laws of evolution.

HISTORY AND HUMAN PURPOSE

Many historians might wish to separate historical trends from those of prehistory and biological evolution because they believe that the course of history was greatly influenced by human purpose. Such historians believe that the will and foresight of leaders of certain tribes and nations may have determined the course of events much more than did opportunistic interactions between populations and their environments. How valid is this difference? With respect to modern history, it is very much open to question. Even before the end of this century, recent events have dashed completely the hopes and desires of early twentieth-century leaders, regardless of their political position and ideology. As designers of a future world order, Kaiser Wilhelm, Georges Clemenceau, Woodrow Wilson, Lenin, Mussolini, and Hitler have all been complete failures. The leaders of the West during the nineteenth century—Napoleon I and Louis Napoleon, Metternich, and Bismarck, as well as the British imperialists from Clive and Hastings to Lord Salisbury—all had little effect on the course of events during the twentieth century after World War I.

Has our age been unusual in this respect? Did great leaders have more influence on ancient history than on the events of modern times? Classical historians, from Xenophon and Thucydides to Macaulay and Breasted, have all focused on generals and statesmen, as if their influence had determined the course of events. Sargon of Sumeria, Sennacherib of Assyria, Darius of Persia, Alexander of Macedon, Maurya of India, Hannibal, Scipio Africanus, Caesar, Attila the Hun, Mohammed and later leaders of Islamic armies as well as Charlemagne and Ghengis Khan are all placed in the historical limelight as if they were not only rulers but also guiders and planners of human events. Modern historians, particularly those mentioned at the beginning of this chapter, have shown by their research that these leaders were more often the products of the cultural climate in which they lived than the arbiters of future events.

SIMILARITIES BETWEEN HISTORICAL AND EVOLUTIONARY TRENDS

Modern world historians are seeking answers to entirely different questions. Their questions and answers are similar to those discussed by biological evolutionists and by anthropologists who study prehistory.

For example, to what extent was the separate origin and evolution of urban civilization in different parts of the world determined by people and their genes in a more or less purposeful manner, and to what extent was it a product of population–environment interactions, acting principally by selecting cultural differences? This question can be answered by comparing the different ways that progress toward civilization has taken place in different parts of the world. If biological or personal factors have been in control, then the initial genetic differences between populations that were isolated from each other should have caused them to react differently to similar environmental factors. The early stages of civilization in various parts of the world should have differed from each other, even where environmental factors were similar. During later stages, better communication between different cultures should have produced cultural diffusion of inventions, ideas, and behavior patterns, and hence should have promoted the convergence of different cultures. On the other hand, if gene-determined potentialities and personalities were similar in most or all human populations that moved toward civilization, or if genes had little effect on the course of change, then differences in artisanship and particularly social structure should have reflected chiefly the different environments with which different populations were interacting. Wherever environments were similar, cultural evolution should have followed a similar course. Later divergence might be expected as societies developed different traditions and themselves generated particular environmental challenges to which they responded in different ways.

If we apply this criterion only to modern times, some support can be obtained for the genetic hypothesis. During the past two centuries, cultural diffusion has caused all civilizations to resemble each other in many respects. More crucial, however, is the question: How much alike was the course of cultural evolution in the early stages of civilizations that developed under similar conditions? To what extent can differences between them be ascribed to environmental factors?

Historians Wittfogel, Service, and Cameron have presented interpretations that support strongly the environmentalist hypothesis for the origin of urban civilization. Based on the accounts of specialists on the history of various nations and cultures, they show that the social and economic structure of all societies that first acquired urban civilizations can be reduced to a few characteristic patterns, plus intermediate conditions. Each pattern is clearly associated with a particular pattern of environmental circumstances that was repeated in different parts of the earth. The model patterns are (1) feudalism under the direct control of the state and church, (2) feudal domains controlled by large private landholders, and (3) commercial civilizations in which city merchants held power and acquired wealth by commerce and trade.

Direct control by state, church, or imperial monarchs developed in regions having broad, flat valleys, such as the Nile (Egypt), the Tigris and Euphrates (Sumeria, Babylon, Assyria), and the Indus

(India), where large-scale irrigation was essential, as well as in China, where irrigation and the building of canals were essential for continued production. In those regions, the system of agriculture usually broke down under the control of private landholders, who were constantly feuding with each other. Only the kind of despotism that Wittfogel calls "the hydraulic society" could survive.

In regions having smaller fertile valleys, such as Asia Minor (home of the Hittites) and Persia, state control, although nominally present, dwindled until private landholders gained almost unlimited power. Feudal Europe has a similar geography and had a similar history.

The third type of socioeconomic structure—commercial civilizations based on maritime empires—developed in regions having long coastlines, several or many islands, and relatively few and small fertile valleys that could be cultivated. Seafaring traders dominated the empires of the Phoenicians, the Greeks, the Arab city–states, and, much later, those of Portugal, the Netherlands, and Britain. The Roman Empire and the Arab Empire of the seventh to fourteenth centuries both began as commercial civilizations but became increasingly feudal as they conquered other countries that possessed large areas of fertile land. The same was true of the Portuguese and Spanish colonial empires in the New World.

A second important question is: To what extent have the people of a society determined their own fate, and to what extent was the direction of culture determined by a self-perpetuating elite or ruling class?

Reading the accounts of such historians as Cameron, Harris, Wittfogel, and some of the specialists whose works these authors cite, I became strongly impressed with three facts. First, the majority of the people in urban societies until very recently have lived harsher, more brutal, and more degrading lives than did their ancestors before civilization came into being. The leaders of society lived in luxury or at least comfortably by virtue of the constant toil of others in ancient Egypt, Mesopotamia, India, China, Greece, Rome, the medieval Arab states, Europe, Olmec and Aztec Mexico, and Inca Peru. These exploited masses were feudal serfs—small farmers who groaned under usurious debts and taxation and could always be conscripted for forced labor. Many of those who toiled in fields and mines or on construction projects were chained together day and night and were not allowed to marry and have a family life. Surely, no one would choose to live as these people did. In ancient times, the degrading life of the majority was imposed on them by a ruling elite.

Second, the literature, arts, architecture, engineering, science, and other mental activities that lifted human civilizations above the level of ancestral illiterate cultures could not have developed without the labor of the toiling masses, since only in this way could the intellectuals have enough leisure for their pursuits. Greek philosophers were well aware of this inexorable fact, and they mentioned it many times in their writings. Organized urban society requires much more work for its maintenance than does the looser organization of hunter–gatherers or farmers. Domestic animals can contribute some of this labor, but not enough to maintain a high state of culture. The minority can enjoy the amenities of civilization only if the necessary hard work is performed either by other humans or by power-driven machines. The intellectual progress that was essential to forming and maintaining complex, highly organized societies was by necessity retained and enjoyed by a ruling elite.

The third fact is that, once urban civilizations had become organized, they acquired a remarkable capacity for survival with little change. Ancient Egypt lasted, with a few interruptions, for 2500 years. The Mesopotamian civilizations changed rulers repeatedly, but the socioeconomic structure, including the power of landlords and the debased condition of small farmers, serfs, and slaves, remained unchanged for 4000 years, after which the area suffered a general decline under Turkish occupation. India also changed rulers by successive conquests, but its brand of feudalism remained much the same from 2500 BC until the British conquest in the nineteenth century AD. Even the intellectual pursuits of the ruling leisure class changed very slowly, as did the progress of invention and material knowledge. Coal was burned in China during the time of Christ, but it did not provide power that could replace human labor until eighteenth- and nineteenth-century England. Gunpowder was known in China for centuries before its use became diffused to the medieval West, where it became an agent of destruction. The Greeks invented the steam engine, but they used it as a toy to amuse children. More than two millennia passed before steam power became a substitute for human labor. Mathematics became highly sophisticated in the Indian Empire during the fifth century BC. A hundred years later, Euclid developed his geometry, after which the development of mathematical knowledge stood still until Leibniz invented the calculus in 1676. Art and literature have changed through the centuries and have acquired distinctive forms in different countries, but from the cave art of 15,000 years ago through the arts of ancient Egypt, Greece, and the European Renaissance up to the present, one can detect little that could be called progressive change or improvement. Music likewise changed little until the European Renaissance brought the innovations of organized choirs and orchestras and its advances in technology made possible the modern organ, pianoforte, and the various instruments of a symphony orchestra.

This very brief resumé highlights two basic characteristics of historical change that resemble the evolutionary changes discussed in earlier chapters of this book. One is adaptive radiation in response to changed population–environment interactions. The other is the occurrence of a few rapid, almost sudden changes in the form of quantum jumps, following the pattern of punctuated equilibria rather than the pattern of continuous progressive change.

CULTURAL ENVIRONMENTS AND THE GROWTH OF POWER

If the course of historical change was governed by the same principles that governed evolution of plants and animals, we are justified in asking: Were the quantum jumps of historical change triggered as responses to environmental challenges, as were the quantum jumps of evolution?

The first quantum jump to be considered is the origin of urban civilization. It appears to have occurred between 8000 and 6000 years ago. By comparing early civilizations with modern agricultural peoples, Marvin Harris has argued in favor of response to environmental challenge as a major cause. As is true of nearly all challenges that *Homo sapiens* has faced and is facing, this challenge was generated at least in part by humanity itself. Beginning before the advent of agriculture and continuing up to the present, the persistent, inexorable growth of human populations beyond the supportive capacity of land available to them has posed a succession of challenges. The same kind of challenge appeared independently among sedentary farmers living in fertile valleys throughout the world. In each of these areas, the challenge was met in a similar way—the growth in power and prestige of one or more leaders, who became hereditary chieftains, kings, priests, or demigods. A model of the way this may have happened is provided by anthropologist Douglas Oliver's study of "bigmanship" among a tribe of Polynesians, the Siuai inhabiting Bougainville, one of the Solomon Islands. A young man can acquire prestige as a "big man" or *mumi* by working harder than everyone else and by saving some of his food rather than eating it. He eventually has enough coconuts, meat, and other delicacies so that he can hold a big feast, enlist the aid of his admiring family, and entertain a circle of neighbors, who by enjoying the feast become obligated to him. Further work and saving enable him to build a clubhouse in which he can entertain an even wider circle of friends. He becomes admired generally as a good provider. Less ambitious associates are happy to be entertained and do not worry about future obligations.

The success of a *mumi* brings him into direct confrontation with other *mumis* who have arisen in a similar fashion, and fights ensue. Before Europeans arrived, the most successful *mumis* enlisted small private armies from the associates who were indebted to them. The conversion of a powerful *mumi* into a petty chief would have been a small step, one that probably occurred many times. Given sufficient power, a chief who had followed the *mumi* path could have forced his subjects into accepting his son as his successor, particularly if the young man, because of either superior physique or better training, could have led the *mumi*'s army and defended their subjects against other chiefs. Hereditary power would then have replaced power acquired by hard work.

In Polynesia, the upper limit of power was the rank of petty chief. This limitation arose because the people could abandon their chief and make their living in other ways if they became dissatisfied. Pacific islands have a favorable climate for fishing and growing yams throughout the year, so that nobody needs to rely on a store of food that a chief has hoarded. However, in the great agricultural valleys where urban civilizations first appeared, dissatisfied people found themselves trapped once they had submitted to the provident generosity of a chief. They could not escape and farm elsewhere, since all available farmland was already owned by a neighboring tribe or chief. If they had not hoarded their own stores, or if these had been taken by the chief, they would starve but for the chief's bounty. In a way, they were already subjugated.

The hereditary chief was also trapped in his role. He was constantly threatened by neighboring chiefs. If he conquered a neighbor, he could kill his rival, kill or enslave the rival's men, and take over their land, thus gaining both temporary safety and better resources of food and labor. If he lost a battle, he himself would be killed, his wives would be taken over by the victor, and his chiefdom would disappear. Obviously, the most successful chiefs would be those whose chiefdoms were the best organized, those whose men and women worked the hardest, and those who were in a position to acquire the largest number of implements for both war and agricultural production through trade with friendly chiefs and, later, traveling merchants. The end result of this inflationary spiral of labor and organization would have been first a city–state and later an empire.

The "big man" theory fails to explain one important fact about the earliest stages of urban civilization in Sumeria. The first rulers of these city–states were priests, who claimed to be interpreters of the city's god. The lands that they controlled were designated as belonging to the god and his church. Forced labor on these lands was demanded as a service to the god, and women who were designated as servants of the god actually became concubines of the priests or prostitutes under their control. Later, priestly control gave way to that exercised by secular monarchs.

Given these facts, one can hardly escape the interpretation that religious beliefs played a very important role in establishing the power of the first urban rulers. As is mentioned in Chapter Eleven, religion and politics have played interacting roles in governing most human societies ever since societies became larger than a single family.

If this account is even approximately faithful to the events that actually preceded the orgin of urban civilization, then we must discard forever the idealistic social contract theory of eighteenth-century philosophers such as Jean Jacques Rousseau. We must substitute an epigenetic sequence carried out by people who tried to look into the future but who had not the slightest conception of where their course of action would finally lead. The parallel with the epigenetic principle of evolution is very close.

Once city–states and empires had concentrated power in the hands of a ruling class, further change was initiated principally by these rulers, until they faced enemies or environmental challenges they could not cope with. To be sure, peasant revolts broke out occasionally, often aided by dissident minor chiefs. Massive peasant wars took place in ancient Iran, and occasional rebellions by peasants and minor vassals are recorded by Indian historians before the time of Christ. The Chinese peasants, who most of the time could be beaten or killed by their landlords at will, rose up in AD 18 to stage the "Revolt of the Red Eyebrows." Men and women together marched on the capital city of Peking, defeated the armies of the king and their landlords, and carried on full-scale warfare against them for six years. In ancient Rome, the gladiator Spartacus led a revolt that in modern times has been widely publicized in books and movies. In England, peasants revolted against their masters in 1381, while several prolonged peasant revolts broke out in Germany between 1430 and 1525. Peasants and serfs staged short revolts against landholders during the French Revolution. All these attempts at overturning the established order were put down after a few years, usually with extremely barbarous cruelty. They all failed for essentially the same reason. A large, complex state can be run only by those who have talent and training for maintaining an organization. Leaders of revolutions are usually ill fitted to build and maintain organized power.

The ruling classes of every nation and empire have often been deposed and replaced by other rulers having similar aims and philosophies. Most historians have been concerned with recording the details of these transfers of power, but the outcome of these struggles only occasionally had a profound effect on the majority of their subjects. Theoretically, leaders of nations and empires have been free to look ahead and plan purposefully for their future and the future of their subjects. Actually, they did little such planning, or if they did, their plans were usually thwarted by unforeseen events. Nations and empires acquired their initial social and political structure largely through an adjustment between prevailing ecological conditions and the cultural inventions by which their leaders exploited the opportunities that these conditions offered. Societies changed their structure when the ruling class found that they could lead a better life by adopting new ideas, customs, and rules of conduct. Adaptive improvements of this kind led to advances in artisanship, conscious time binding, and imaginal thinking. The three model social structures—state feudalism, control by private landlords, and commercial civilizations—promoted the innovations that aided their particular way of life.

Since state-controlled feudalism was aided by precise predictions of seasons and weather, the preoccupation of feudal rulers with calendars was to be expected. Irrigation projects require careful surveying and engineering techniques. These sciences became highly developed in ancient Egypt and China. The social structure of state-controlled feudalism could be kept stable only by maximal attention to law and order. Armies and police systems, both overt and secret or spy-based, were highly developed in Egypt, Mesopotamia, and the Mauryan Empire of India about 300 BC. The elaborate law code of Hammurabi, ruler of Babylon in Mesopotamia, is justly famous as an innovation that aided state-controlled feudalism. On the other hand, the abundance of peasants, slaves, and other forced labor reduced the adaptiveness of labor-saving machinery, and none was developed. The most elaborate mechanical inventions of feudal societies were engines of destruction, which were used against walled cities.

Private landholders and the states dominated by them had less need of elaborate social and political organization, and their relatively small-scale agriculture was well served by methods that had been used by small farmers before chiefdoms and serfdom arose. In these countries, religion was still a powerful force for keeping the community together and was further developed. In state-controlled feudalism, reverence for deified rulers, such as the Pharaoah and his high priests, was the most acceptable form of religion. In societies based on private landholders, the ideal God was above politics and temporal power. His "just laws" were interpreted by priests and prophets, who could succeed only if they were friendly to the dominant owners of land, flocks, and herds. The image of one supreme God, who is above and rules the petty affairs of humans, was first given the name *Ormazd* or *Ahura Mazda*. He is the same God who is personified as the Jewish Yahveh or Jehovah and the Mohammedan Allah. Both Judaism and Islam developed in countries that were dominated by large-scale herdsmen or by private landholders. In these societies, the word of God was revealed not by a temporal ruler but by a deified prophet (Moses, Christ, Mohammed).

In Asia, religion is tied to social structure in much the same way. India, a nation of state-controlled feudalism, adopted Hinduism as a state religion, although it tolerated Buddhism as an escape mechanism for ascetics and their followers. In China, Buddhism was also tolerated as an escape mechanism, but the other Chinese religions, Confucianism and Taoism, were fostered by private landholding intellectuals, who were opposed to domination by the state. These religions emphasize an ethical way of life and are not concerned with a divine ruler of human affairs.

In addition to spiritual, often mystic religions that appeal primarily to human emotions, societies dominated by a private, independent, noncommercial ruling class fostered other intellectual activities with strong emotional appeal. Examples in literature include the poetry of the *Rubaiyat* of Omar Khayyam, Chinese poetry, and the Song of Solomon. In the decorative arts, the quiet, imaginative landscapes and ferocious wild beasts that appear in Chinese paintings and vases stand in contrast to the formalized, factual, social representations of the Egyptians and Greeks. In architecture, the soaring cathedrals of medieval Europe have greater emotional appeal than any other great monuments that humans have built. Under the patronage of wealthy private landholders, the emotional creativity of artistic expression has flourished as it can never flourish under a state-controlled bureaucracy, whether controlled by a king, an emperor, a dictator, or a politburo.

The natural conditions that favored commercial civilizations were the combination of relatively small arable valleys, deposits of minerals, particularly ores of metals, and above all safe harbors that could be the sites of seaports close enough to other civilized nations so that seaborne commerce was possible. These conditions favored urbanization, particularly of seaport cities, and industry based on skilled artisanship, particularly metallurgy. The pioneer in this direction was the Minoan civilization of Crete, which lasted from about 2500 BC to 1400 BC but which is very poorly known since Cretan writing has not yet been deciphered. Most probably, however, the Greek city–states were the cultural descendants of Crete. The cities of Ionia, Miletus, and others on the west coast of Asia Minor, on islands such as Samos and Chios, as well as Athens, Sparta, and colonies established in southern Italy and Sicily, formed a commercial civilization that flourished from about 900 BC until 359 BC, when it was swallowed first by the Macedonian Empire and later the Roman Empire. That this civilization was the intellectual source of modern western culture, including the arts, philosophy, and science, is common knowledge. After Greece lost its independence, civilization was dominated by imperialism for sixteen centuries, until in AD 1250 the city–states of northern Italy began their 250 years of freedom from imperial rule. Their civilization had much in common with that of ancient Greece, including a flourishing of the arts, literature, and philosophy as well as inter-necine wars between individual states, which contributed to their downfall and loss of independence.

In the late thirteenth and particularly the fourteenth century, seaport cities of western and northern Europe and the states that they supported, particularly Spain, Portugal, Britain, the Neth-erlands, various German states, and those of Scandinavia, founded the commercial civilization that would eventually lead to global exploration, colonialism, and our modern worldwide civilization based chiefly on commerce. In this civilization, the arts, literature, and philosophy also flourished, but they were overshadowed by science and technology, much of it directly useful to mining, metallurgy, and ocean-borne commerce. The rise to dominance of this civilization was no accident. It was the inevitable result of conditions favorable to a commercial civilization—long coastlines, small areas favorable for agriculture, and abundant mineral resources plus the legacy of culture that it received from ancient Greece and the Italian Renaissance.

THE AGE OF HARNESSED ENERGY

In the review of history we have presented to this point, we have largely avoided reference to events of the past two centuries. This is because recent history has been dominated by the Revolution of Harnessed Energy. I believe that this phrase expresses the change that has taken place more accurately than the more familiar expression, *Industrial Revolution*. Industry based on animal and human labor is as old as civilization. Its latest phase, based on harnessing human laborers to a treadmill, was much like the steam-based industry that began contemporaneously. The real revolution of the late eighteenth and early nineteenth century in Europe was the replacement of human labor in industry by energy and power derived from running water, wind, electricity, and par-ticularly combustion of fossil fuels. When, at the beginning of this century, large-scale harnessing of energy from these sources had to a great extent replaced the labor of human beings and domestic animals, optimistic prophets could look forward to an age in which degrading toil would no longer be the inexorable fate of the masses of humanity.

Of course, the early stages of the energy revolution did not liberate humanity from degrading forced labor. On the contrary, it produced on a large scale the "dark, satanic mills" of Britain, continental Europe, and later the United States. Mid-nineteenth-century prophets like Marx and Engels saw no relief from this degradation except by violent revolution, which they thought could produce an age of reason, tolerance, and cooperation. Both these prophets and the industrial leaders who hoped to perpetuate their power and affluence by ruthless repression of the labor force were proved wrong. They were both frustrated by unforeseen events. The ballot box and the principle of representative government, which had been used previously by industrial and commercial leaders to break the control of the landed aristocracy, now became weapons turned against them by an increasingly organized labor force. The printing press and later the telephone, which were first used to establish and maintain the power of industrial leaders and were not available to previous generations of exploited laboring classes, became new tools that could be used to gain allies for the labor movement among dissident members of the exploiting class. Through collective activity, the laboring classes were able to gain an increasing degree of control over their own lives. Their success was greatly aided by conflicts between capitalists belonging to different nations, particularly World War I. The present state of society in countries that have not experienced communist revolutions is, therefore, to a large extent due to the opportunistic nature of social change.

The comparison given in this chapter between recorded history and biological evolution can be briefly summarized as follows. The facts of history do not point to any internal force, "manifest destiny," or divine guidance that has caused any nation or society to evolve in any particular direction. On the contrary, cultural evolution has been and still is determined to a great extent by interactions between societies and their environments. These interactions have been largely responsible for directional trends. They have also caused some societies to change rapidly and others to stagnate for millennia. Biological evolution has been basically opportunistic, being governed largely by short-term population–environment interactions, but is strongly influenced by epigenesis, particularly the nature of the gene pool, which has been determined largely by past evolutionary events. Similarly, the earlier course of cultural evolution was basically opportunistic and epigenetic, having been governed largely by interactions between societies and their physical, biological, and cultural environment as well as their cultural pool, the nature of which depended on their past history. As civilizations emerged, ruling classes and religious leaders increased their efforts to inject human purposes into the course of history, with varying degrees of success. In modern times, many optimists believe that human societies are increasing their power to chart and follow a purposeful course of change toward a better life for all. They are, however, balanced by an equal number of pessimists, who maintain that eventually the inexorable laws of nature and evolution will override human purpose and cause the human species to decline and disappear, as other animal species have done in the past. In the final chapter of this book, I present my evaluation of some of these conflicting claims and prophecies.

The author as a teacher of natural history. We can expect that humanity will respond to future challenges by increasing its knowledge and disseminating knowledge more widely. (Courtesy of Ansel Adams and the Regents of the University of California.)

Chapter Fifteen

PRESENT UNCERTAINTY
AND FUTURE HOPE

The second half of the twentieth century marks a major crisis not only in human history but in the entire course of the evolution of life since living organisms emerged billions of years ago. Older people like myself can recall the time when parts of the earth—much of the Amazon Basin, the forests of central Africa, and the ice cap of Greenland—were poorly known to civilized nations and still retained an aura of mystery. Now, thanks to human technology and exploration, even the most remote corners of the earth have become accessible and familiar to anyone who has the desire and the means to explore them. Practically the only regions that have escaped human disturbance are those that are being purposely conserved by far-sighted people, chiefly in the developed nations, who realize the future value of their preservation.

At the same time, the urge of humans to transform our surroundings into the tropical savanna environment in which our ancestors evolved has caused civilizations to construct enormously complex systems of capsules within capsules. Clothing, houses and apartments, elaborate office buildings, and public structures all insulate us two or three times from the wintry storms of New England, Minnesota, or the Russian steppes, from the icy fortress of Antarctica or the inhospitable climates of the American Southwest and the Arabian desert. Movable capsules—automobiles, railroad trains, ships, and airplanes—carry us over land with twice the speed of the swiftest cheetah, over the ocean many times faster than any fish can swim, and through the air faster than the speed of sound or the rotation of the earth. The first time I saw it, I marvelled at the sight of a sunset over Dulles airport in Washington, followed by sunrise in the west as our plane gained altitude, and a second sunset a few minutes later. This sight has now become so commonplace that I hardly ever think about it.

The most ominous aspect of this advance in human technology has been the ability to wipe out our civilization, if not life itself, in an atomic holocaust. If for no other reason than removing the shadow of this sword of Damocles, humanity must either curb its advances in artisanship and technology or deflect them from a course of fatal collisions between nations.

This explosive advance in human technology has been achieved at tremendous, perhaps unbearable cost. Technical progress has far outstripped the present capacity of human societies for solving social problems. The capacity of societies and their leaders for conscious time binding, particularly planning for the future on the basis of past experience, has increased far too little. Even greater has been the lag in the development of imaginal thinking—the awareness of the needs, both physical and emotional, of our fellow men. Arnold Toynbee has neatly characterized this unbalance by an aphorism: "The head has moved far beyond the heart." How can we restore a proper balance, or can we do so at all? This is the most urgent problem that modern humanity faces.

WILL HUMANITY BECOME EXTINCT?

Many modern thinkers have reached the conclusion that this imbalance will be fatal; either through atomic warfare or through irremediable ecological imbalance, humanity appears to be doomed to extinction. Three major arguments are advanced to support this view. First, whenever humans have invented a machine for a particular purpose, they have always managed to use it sooner or later, and the hydrogen bomb is not likely to be an exception. Second, even if some humans should survive an atomic war, the earth's environment would be so badly upset, particularly with respect to ozone and ultraviolet radiation in the atmosphere, that adjustment would be impossible. Third, the fossil record shows that all species, particularly mammals, become extinct sooner or later, so mankind must eventually suffer the same fate.

As a biologist, I shall deal first with the third of these arguments. There are excellent reasons for believing that humanity is exempt from all the factors that have caused mammalian species to become extinct in the past. Danger from predators was removed thousands of years ago; danger from plagues and pests will return only if, through our own stupidity, we remove the protective blanket of modern hygiene and medicine. To a species that can survive in virtually any climate, climatic change, even as drastic as that of a new Ice Age, does not pose a global threat. If New York, London, Paris, Moscow, and Beijing become ice-bound, the centers of culture will move to Los Angeles, Miami, Cairo, Delhi, Sydney, and Buenos Aires. Even over a time span of hundreds of thousands or millions of years, the extinction of humanity from external causes is inconceivable. If we become extinct, the cause will be species suicide, something that has never happened to any of the millions of species that have existed in the past. How likely is it to happen to us?

Species suicide through direct killing via a series of atomic blasts is virtually impossible. If an atomic war should break out between superpowers, their capacity for destruction will have become blunted long before bombs have fallen on such militarily unimportant regions as Africa, South America, and Malaysia. Global radioactivity from fallout, which twenty years ago was regarded as an imminent danger, is now regarded by few contemporary experts on atomic physics as a serious threat. The fallacy of the argument that "because enough nuclear bombs exist to wipe out the human species several times, they will certainly be used" has been pointed out by geneticist J. B. S. Haldane, who reminded us that for a century enough rifle bullets have existed to do the same, if the desire to do so existed. Moreover, even if a combination of blasts and fallout should destroy ninety percent of the global population of four and a half billion, the ten percent that survived, 450 million, would constitute a population larger than that which existed on the entire earth 1200 years ago.

The only reasonable question to ask is: Would the environmental imbalance resulting from an atomic holocaust increase the present maladjustment to such an extent that humanity could not survive, even at the level of a bare existence? To anyone familiar with the unspeakable conditions of degradation that conquered peoples and slaves have endured in the past, the answer should be clear. Human beings can survive adversity to an almost unbelievable extent. In this respect, we resemble such pests as rats and mice, or human commensals, such as feral dogs and cats. In some way or another, *Homo sapiens* will persist as long as the earth's environment is capable of supporting large animals.

WILL HUMANS EVOLVE INTO SUPERMEN?

We can also imagine that sometime in the future a part of the human population will become converted into a new species as different from ourselves as we are from *Homo erectus*. What are the chances that such a revolutionary change will take place?

I believe that such a new course of evolution is highly unlikely, for two reasons. First, in all examples of the origin of species among animals for which the facts are known or can be reasonably inferred, speciation is accompanied by isolation of a small population for many generations and exposure to entirely new selection pressures. The chances that any human population of small size will be isolated and faced with a new environmental challenge for as many as 100 generations, or 2000 to 2500 years, are vanishingly small. Second, for the past 15,000 years, humans have responded to environmental challenges by cultural rather than genetic changes, and there is every reason to believe that they will continue to do so in the future. Natural selection in humans is destined to be largely normalizing and thus to favor the status quo.

Some geneticists are afraid that natural selection will guide humanity not toward superior intelligence and knowledge but back toward the inferior intelligence of our remote ancestors. Their worry is based on the fact that, in many civilized countries, families of unskilled laborers and others who have a low intelligence quotient (IQ) are larger than families raised by people of higher intellect. Geneticist Sheldon Reed of the Dight Institute, University of Minnesota, has shown that this worry is largely unfounded. Since a high proportion of people having inferior intelligence don't marry at all, the actual number of offspring per *individual* (not per married couple) is greatest among those scoring IQ values near the mode of the population.

Still another possibility is that humans can breed a race of supermen by consciously practicing artificial selection, in the same way that animal breeders have produced vastly superior races of domestic animals. Is this a practical ideal?

Again, I believe not, for two reasons. First, animal breeders have worked with isolated individuals, flocks, or herds, which they carefully prevent from mating with outsiders. They can absolutely control the sex lives of their charges. Such control is so repugnant to humans that it would be possible only under an absolute dictatorship lasting hundreds or thousands of years, a situation that has never existed since the dawn of civilization.

The second reason this is impossible is that hardly any two people, and certainly never the majority in any nation, would agree on what characteristics a superior race of humans should possess. The leaders of Hitler's "thousand year Reich" were to have been bred to be tall and blond, with superior physique and intelligence but with merciless disdain for those who have hooked noses or dark skin. Some idealists might opt for a race having most highly developed the Christian virtues of generosity, meekness, and consideration for all mankind. More practical critics would maintain rightly that such a race could never survive times of turmoil and stress.

Many people, after bewailing the stupid actions of leaders and the masses throughout the world, ask in exasperation: Why can't we breed a more intelligent race of human beings? In addition to the general difficulties mentioned above, even more serious barriers prevent this from being done. First, how would we define intelligence? Surely, it is not equivalent to a high score on any IQ test that has yet been devised. One might opt for choosing people having the largest possible brains. Such a course would overlook the fact that, in modern humans, intellectual capacity is poorly correlated with brain size. For example, the writer Anatole France had one of the smallest brains on record for modern humans. Even if, by some miracle, a more intelligent race of humans could be bred, or could evolve by natural selection, would such a race be more capable than we are of solving our current problems? Many contemporary intellectuals tend to shy away from the problems of their less gifted fellow men, and if they do try to aid humanity, they are likely to make the mistake of assuming that all people are as rational and understanding as they are. Humanity will receive its greatest boost not from increased intelligence but by using more efficiently the brains we have and by becoming more aware of the needs, thoughts, and desires of others.

The great majority of geneticists believe, as I do, that tampering with the human gene pool will do more harm than good. As long as humanity continues to survive, we must tolerate the diversity of hereditary characteristics, both good and bad, that exists in contemporary gene pools.

By arguing that humans will not evolve into supermen, I do not mean to imply that the biological evolution of the human species has completely ceased. However, adjustments to the radically new environment that we ourselves have created will be largely invisible. Natural selection is most likely to favor those genetic changes that will ease the problem of mental and emotional adjustment to modern society. If social conditions become favorable to those societies that devote most of their collective efforts to saving our environment and reducing the extremes between riches and poverty, natural selection may favor genetic changes that reduce the number of people who are addicted to the pursuit of power.

In other words, the future biological evolution of humans will be intimately bound with cultural evolution and will affect chiefly the invisible characteristics of the human brain and hormonal system. For other characteristics, normalizing selection will prevail.

FUTURE EVOLUTION BASED ON PRESENT ADAPTATIONS

The gist of the argument presented in the last section is that the future evolution of humanity, in response to the awesome environmental and social challenges that now face us, must be based on cultural and social adjustment rather than genetic change. This conclusion is hardly surprising since, for at least 15,000 years, human responses to environmental challenges have consisted largely of cultural changes rather than genetic changes. Although this kind of cultural response is almost unique to *Homo sapiens* among animals, its presence in humans is entirely in accord with evolutionary trends. Whenever any line of evolution has embarked on a particular course, involving a special way of becoming adapted to new conditions, the most successful species have been those that continued along the course set by their predecessors. Starting with the earliest Australopithecines, and continuing until the evolution of *Homo sapiens*, the human evolutionary line has succeeded because of its emphasis on three characteristics defined in a previous chapter—artisanship, conscious time binding, and imaginal thinking. Since we are committed to this course, the most logical prediction would be that future humans will be most successful to the extent that they continue to rely on these characteristics. Which of the three should be emphasized the most in the foreseeable future? Should we be devising and making more complex and sophisticated machines? Should we learn how to make more accurate predictions on the basis of past experience? Or should we develop and perfect our ability to think, reason, and devise new ways of analyzing our situation?

Many people would argue that in the atomic age we have already invented too many complex machines for both creation and destruction. To repeat the insight of historian Arnold Toynbee mentioned previously in this chapter, humanity suffers because the head has gone too far beyond the heart. Surely, humans will continue to invent new machines and other artifacts for making society more efficient and life more comfortable, but I doubt that most people believe that this should be our major concern.

As to prophesying on the basis of past experience, during the past twenty years a number of prophesies have been made, some of them relying on one of the most sophisticated of modern machines, the digital computer. Will these be any more accurate than the prophesies made in ancient times by biblical prophets and others? Only time will tell. Many people complain of the short-sightedness of our leaders, and to some extent their complaints are justified. Society would certainly derive more benefit from a more far-sighted attitude on the part of leaders and followers alike. Nevertheless, even the most accurate predictions will fail unless social conditions enable them to succeed. Prophesies are helpful, but not decisive.

The final question, therefore, is: To what extent can humanity meet current challenges by using more efficiently and developing further our powers of thought and reason? I believe that this development should be a major concern. Nevertheless, thinking alone is not enough. Philosophers who have tried to imagine an ideal society, such as Plato's *Republic* and Thomas More's *Utopia*, have given us designs that are hardly more acceptable than their modern cynical caricature, Aldous Huxley's *Brave New World*. Philosophers who believe that social structure should be based entirely on thought and reason overlook the fact that, in its earliest stages, human social behavior was guided largely by emotions and that, right up to the present, emotion tends to dominate reason. Thinking and reasoning can help to guide our emotions in the right directions, but reason can never replace emotion. Consequently, some kind of religion, having as its purpose the balancing of emotional excesses and guidance toward a better understanding of our fellow man, is a necessary and integral part of human society. Paradoxically, reason, rather than emotion, may be the best tool for judging which of the numerous modern religions is best suited for this task. Readers may be wondering at this point whether I as a biologist haven't reasoned myself into a corner. Is an evolutionist better equipped than anyone else to recommend what is best for human society? Perhaps not. I would freely accept this criticism and will therefore end my chain of reasoning here.

As a final point, I would maintain that greater knowledge and understanding must be part of any design for the improvement of humanity in the future. Among the greatest achievements of our century is the invention of mass media by which knowledge may become more widely communicated. The unfortunate fact that these media are most often used for depicting violence and for diverting the public from consideration of more serious issues is not a fault of the media themselves but of the society that supports their use in this way. Moreover, these faults are not new; mass entertainment in the Roman Empire was far more violent and brutal than anything tolerated by modern society. As an optimist, I believe that the value of these media for instruction will be increasingly recognized.

One can hardly be so optimistic as to assume that society will not be faced with serious crises in the future, some of which may destroy much of what we value highly in our civilization. Nevertheless, in the past, even the most serious crises have been overcome eventually. I predict, therefore, that if human society is ready and willing to apply all the knowledge, understanding, and good will that we have at our disposal, present and future crises can also be overcome or avoided.

NOTES

PART ONE

Chapter One

p. 8   These examples are described by T. Dobzhansky (1951), (1970).

p. 10   Huxley's essay is published in several different volumes of his essays, notably Huxley (1909), a book well worth reading for its own sake.

p. 11   A thorough and readable account of the geological time scale, including radiometric dating, is Eicher (1968). See also Cloud (1977).

p. 11   Animal species that have remained stable over long periods of time are discussed by G. G. Simpson (1944); long-lived, stable species of forest trees, by Axelrod (1976) and Wolfe (1972); and constant primitive organisms, especially blue-green bacteria, by J. W. Schopf (1978).

p. 17   This problem, particularly the theory of punctuated equilibria, is carefully discussed by S. M. Stanley (1979).

p. 20   See Van Valen (1973).

p. 18   Simpson (1944) wrote the first carefully presented discussion of exceptionally slow and fast rates of evolution.

Chapter Two

p. 29   An authentic comparison of the Lamarckian and Darwinian theories appears in the essay by E. Boesiger in F. Ayala and T. Dobzhansky, eds. (1974).

p. 31   Most modern textbooks include good accounts of the nature and roles of DNA, RNA, and protein. See Hardin (1961) and Dobzhansky et al. (1977).

p. 32   An autoradiographic photo of the entire DNA molecule that extends the length of the chromosome of the fly *Drosophila* was published by biochemists R. Kavenoff, L. D. Klotz, and B. H. Zimm (1974).

p. 37   Darwin's *The Origin of Species* is a typical classic, always cited but almost never read except by specialists. With respect to the number of facts, both demonstrated and anecdotal, that are packed into almost every page, it has rarely been equalled. It is heavy reading, but anyone who really wants to understand evolutionary theory and its history should at least peruse it.

p. 46   This extraordinary story is discussed in detail in Mayr and Provine, eds. (1980).

p. 46   The lectures that I heard as an undergraduate were published by Professor Parker (1925).

p. 47   See Hardin (1961) or Ayala and Kiger (1980).

p. 50   In Mayr and Provine (1980), Alexander Weinstein, who spent his graduate student years in Morgan's laboratory, presents a most interesting account of the changes in Morgan's point of view. According to Weinstein, Morgan was sometimes greatly influenced by discussions with his colleagues, who believed in natural selection more strongly than he did, while at other times, particularly in his 1932 book, he expressed his intuitive opinion.

p. 50   During the 1920s, the geneticists who did the most to develop and promote the idea that quantitative characters are determined by genes at many different loci (multiple locus or polygenic inheritance) were Harvard Professors E. M. East and W. E. Castle and their students, including Sewall Wright, one of the founders of the quantitative study of evolutionary processes. Morgan was lukewarm or sometimes actually opposed to the idea; the geneticists of central Europe and Scandinavia wanted to resolve this kind of inheritance into identifiable genes, while those of Great Britain completely ignored it. R. A. Fisher, who did support it, was regarded by geneticists of the 1920s as a statistician and outsider.

p. 51   See Fisher (1930), Haldane (1932), and Wright (1931). Chetverikov's (1959) work had little influence because it was published in Russian and was not available to the geneticists of that time. By the time it was translated, it had only historical interest.

p. 51   The first readable synthesis of these ideas was *Genetics and the Origin of Species* by T. Dobzhansky (1937).

p. 52   The electrophoretic technique and the significance of the genetic variation that it reveals are discussed by F. J. Ayala (1978).

p. 54   For further details, see Stebbins (1950).

p. 60   See Stebbins and Lewontin (1971).

Chapter Three

p. 69   See Monod (1971), and Teilhard de Chardin (1959).

p. 74   See Ayala and Valentine (1979).

p. 77   See Ayala and Kiger (1980, chapter 16).

p. 82   In a 25-page article on Principles of Thermodynamics in the *New Encyclopedia Britannica*, Vol. 18, pp. 290–315, no mention is made of states of increasing order or disorder, even though several pages are devoted to a discussion of the Second Law. Entropy is defined as follows (p. 294): "Entropy will be defined as the extensive property S, the change of which DS, in a change of state is equal to a positive constant c_R times the difference between the corresponding changes of energy DE and available work D evaluated with respect to a standard reservoir." The reader is entitled to interpret this definition according to his or her preferences. Note that the emphasis is on changes of state with relation to the flow of energy. This is in accord with the discussion in the text.

p. 86   The founder principle is discussed by Ernst Mayr in several of his books, notably (1963), (1970).

p. 89   See Lorenz (1970, p. 224). The original observations were made by O. Heinroth.

p. 90   The genetical and ecological basis for the origin of species is discussed in several books. See Dobzhansky et al. (1977), Stebbins (1977), Grant (1971), and White (1978).

Chapter Four

p. 106   The great imperfections of the fossil record and the reasons for them are well discussed by G. G. Simpson (1949) and D. M. Raup and S. M. Stanley (1978).

p. 108   The relations between the study of evolution at the level of species and populations ("microevolution") and at higher levels in the taxonomic hierarchy ("macroevolution") are discussed by F. J. Ayala and J. W. Valentine (1979) and by Dobzhansky et al. (1977).

p. 114   The condition of the earth's atmosphere during the early years of life on our planet is well discussed by Preston Cloud (1977) and by W. J. Schopf (1978).

p. 118   The evolutionary tree of horses is presented and discussed by G. G. Simpson (1951).

p. 119   The extinction of the sabertooths (Machairodontinae) following narrow specialization and reduction in frequency of their only prey is
described by G. G. Simpson (1953, pp. 221–222).

p. 120   For the evolution of snakes, see Colbert (1955), Romer (1966), and
Stanley (1979, chapter 9).

p. 122   See S. J. Gould (1977).

p. 125   The topic of homology has an extensive literature from both the nineteenth and twentieth centuries. The leading ideas are reviewed in
Dobzhansky et al. (1977, chapter 9).

p. 126   The construction of phylogenies based on the amino acid sequences
of protein and the effect of this knowledge on the major groupings
or kingdoms of organisms is discussed in detail by Dobzhansky et
al. (1977, chapters 9 and 12).

p. 129   The chemical similarities and anatomical differences between humans
and chimpanzees are discussed by M. C. King and A. C. Wilson
(1975).

p. 131   This discussion is based on an article by R. Lande (1976) and a review
of it by S. M. Stanley (1979, pp. 56, 57, and 182).

p. 132   The fossil history of the rice grasses is reviewed by G. L. Stebbins
(1950, chapter 14), and has been updated by J. R. Thomasson (1979).

p. 134   The probable evolution of the Hawaiian honeycreepers is well presented by W. J. Bock (1970).

p. 136   See Axelrod (1975).

p. 140   See Simpson (1949, chapter 15) and Dobzhansky et al. (1977, chapter 16).

Chapter Five

p. 143   See Darwin (1972, chapter 1, 8th section, "Domestic pigeons . . .").

p. 144   See Goldschmidt (1940).

p. 145   See Mayr (1942, pp. 137–138).

p. 146   See Wright (1977).

p. 149   See Oparin (1961).

p. 150   The evolutionary history of maize has been a highly controversial
topic, but recent evidence points strongly to some species of teosinte,
a wild grass that in Mexico is often found around the edges of corn-
fields, as the most probable ancestor. In teosinte, pollen- and seed-
bearing structures are already quite distinct, but the large tough cob
has not been evolved, while the husks are much smaller and the silks
shorter than in corn. When teosinte is compared with various wild
species and genera of grasses, strong evidence exists for its ultimate
origin from a grass that bore kernels in structures similar to the pollen-bearing tassel of modern corn. See Mangelsdorf (1958), Galinat
(1975), and Beadle (1980).

p. 154   For the evolution of fish scales and teeth, see Romer (1970).

p. 154   For the evolution of globin genes, see Ayala and Kiger (1980, chapter 22).

p. 157   Zuckerkandl's hypothesis is presented in Petit and Zuckerkandl (1976, chapter X).

p. 158   The standard reference on the biota of the cow's rumen is Hungate (1966).

p. 161   Dobzhansky's original discussion of transcendence is in Dobzhansky (1969).

p. 167   See the discussion by J. T. Bonner (1980).

PART TWO

Chapter Six

p. 175   See Cloud (1977, chapters 1, 9, and 10). See also Orgel (1973) and Dickerson (1978).

p. 176   See Fox (1965).

p. 179   See Bernal (1967), and Cairns-Smith (1971).

p. 181   See Rao, Odom, and Oro (1980).

p. 182   This and several other fascinating sidelights on evolution are presented by Calder (1973).

p. 183   See Barghoorn (1971).

p. 188   See Woese, Mangrum, and Fox (1978).

p. 190   See Schopf (1978).

p. 195   See Himes and Beam (1975) and Beam and Himes (1977).

p. 195   See Margulis (1970), (1981).

p. 197   See Lazarides and Revel (1979).

p. 199   See Hardin (1961); see also any one of various textbooks of protozoology, such as Grell (1973).

p. 200   See Williams (1975) and Maynard-Smith (1978).

p. 201   See Ribbert (1972).

Chapter Seven

p. 205   Much of the the material in this chapter can be found in various textbooks of botany. One of the most modern and complete of these is Raven, Evert, and Curtis (1981).

p. 206   The relationship of size to longevity and death is well discussed by J. T. Bonner (1965).

p. 206   The nature of death with reference to cells, tissues, and the organism as a whole is discussed further in Stebbins (1969, pp. 20–24).

p. 212   Distinctive features of the origin of species in plants are discussed further by Grant (1971).

p. 213   See Oliver (1979).

p. 217   The position of the fungi as a separate kingdom is discussed by Whittaker (1969).

p. 222   The abundance and diversity of single-celled green "algae" that live on moist soil is well described by Bold (1970).

p. 222   See Stebbins and Hill (1980).

p. 225   Much of the material presented in the remainder of the chapter is discussed more fully in Foster and Gifford (1959).

p. 243   For various points of view on the controversial problem of the origin and evolution of the flowering plants, see Stebbins (1974), Beck (1976), and a collection of papers in *Annals of the Missouri Botanical Garden* 62, 3 (1975).

p. 244   For a full discussion of evolutionary trends in flowering plants, see Stebbins (1974).

Chapter Eight

p. 247   In addition to Hardin (1961, chapter 2), the following works give a more detailed account of the principal material of this chapter: Clark (1964), Clark and Panchen (1974), Gardiner (1972), and Valentine (1973), (1978).

p. 247   See Glaessner (1971).

p. 251   See Ward, Greenwald, and Greenwald (1980).

p. 254   See Moore, Lalicker, and Fisher (1952).

p. 265   See Beklemishev (1969) and Jägersten (1972); and for a different point of view, Clark (1964) and Valentine (1973).

Chapter Nine

p. 272   The pterobranchs are well described in Barrington (1965).

p. 272   *Balanoglossus*, the tunicates, and their relatives are described and illustrated in many textbooks of zoology and biology, such as Hardin (1961).

p. 273   A good account of *Amphioxus* and its significance is by Alexander (1975).

p. 275   For the evolution of fishes, see Colbert (1969), Romer (1966), and Carter (1967).

p. 277   P. F. Thomson (1969) has presented a detailed and modern account of lobe-finned fishes.

p. 277   The rediscovery and biology of modern *Latimeria* is described by Millot (1955).

p. 282   The suggestion that Rhipidistian fishes made sudden lunges after their prey was made by Schaeffer (1965).

p. 282   A dramatic account of the modern African lungfish and a bit of philosophy connected with it is "Kamongo" by Homer W. Smith (1949).

p. 286   See Romer (1966), and Colbert (1969). For a more detailed and recent account, see Carroll (1969).

p. 292   See Crompton and Jenkins (1968), Crompton and Parker (1978), and Romer (1969).

p. 293   McNab (1978) has made a good case for the early origin of temperature regulation in mammals.

Chapter Ten

p. 306   See Campbell (1976), Leakey and Lewin (1977), Pilbeam (1972), Poirier (1977), and Washburn and Moore (1974).

p. 314   See Simons (1972).

p. 319   See King and Wilson (1975).

p. 319   This material is well covered by Washburn and Moore (1974). See also Sarich and Cronin (1977).

p. 321   See Poirier (1972) and Lancaster (1975). The account by Wilson (1975, chapter 27) is strongly biased in favor of genetic control.

p. 323   See Goodall (1971).

p. 323   See Curtin and Dolhinow (1978).

p. 326   See Schaller (1964).

p. 326   For a full account of these fascinating experiments, see Gardner and Gardner (1969) and Premack and Premack (1972).

p. 328   The acquisition of potato washing and other cultural traits by the Japanese monkeys is well described by Wilson (1975, pp. 168-172).

p. 329   See Curtin and Dolhinow (1978).

PART THREE

Chapter Eleven

p. 324   The story of the Australopithecines and the significance of these creatures in human evolution are well told in Leakey and Lewin (1977).

p. 339   The monograph by Mary Leakey (1971) documents very well the gradual nature of the transition from Oldovan to Acheulian stone tools.

p. 341   The Terra Amata dwelling site is described and reconstructed by H. de Lumley in *Scientific American* (1969). See also Poirier (1977, pp. 285–290).

p. 342   A dramatic account of the finds in Shanidar Cave, by their discoverer, is Solecki (1971).

p. 347   The detective-type deduction that led to the hypothesis of primitive calendars carved on pieces of bone is related by Marshack (1972).

p. 350   The two contrasting theories for the origin of agriculture are presented by Whyte (1977) and Cohen (1977).

p. 352   See Jolly (1970).

p. 353   For the seed-eating hypothesis, see Jolly (1970). A survey of late Tertiary fossils of woody plants from the Sivalik region by Lakhanpal (1970) includes several species of nut-bearing trees.

p. 360   Wallace presented his theory in his book *Social Environment and Moral Progress* (1913).

p. 360   See Wilson (1975).

p. 362   A highly readable, intimate account of life among the chimpanzees is by Jane Goodall (1971). Gorilla societies are well described by Schaller (1964).

p. 362   See Baumgartel (1976).

p. 363   The term *imaginal thinking* was coined by Miller, Galanter, and Pribram (1960).

Chapter Twelve

p. 369   The book by Dubos (1974) is a sophisticated essay on the cultural flexibility of human nature and its significance to modern and future society.

p. 371   See Lancaster (1973), (1975).

p. 374   Kenny's definition of mind is quoted by Grene (1978).

p. 375   Bowlby (1969) has produced an array of facts that emphasize the attachments between parents and offspring as an important source of the early stages of learning.

p. 376   See Tiger (1969).

p. 377   An early stage in the use of foster children for increasing the size of the group exists among the Bushmen of South Africa, as described by Lee (1972).

p. 378   See Wilson (1975), (1978).

p. 380   Most of the evaluations and criticisms of sociobiology mentioned here are in the collection of papers edited by Gregory, Silvers, and Sutch (1978), particularly Barash (1978), Washburn (1978), Hull (1978), and Alper (1978). See also Sahlins (1976).

p. 380   Alexander (1975, 1979) has placed great emphasis on the importance of inclusive fitness.

Chapter Thirteen

p. 385   For a philosopher's opinion of sociobiology, see Ruse (1979).

p. 385   Fuller (1978) provides valuable information indicating how genetics and breeding can mold behavior patterns of domestic animals, at the same time making one realize how different the situation is in humans, where mating is not controlled by a superior intelligence.

p. 387   The length and complexity of the development pathway leading to the adult mind is emphasized by Anastasi (1968). Klopfer (1977) adds to the complexity by stressing the importance of feedback interactions.

p. 387   The importance of the early environment in determining human intellectual capacity and behavior patterns is brought out clearly by both Bloom (1964) and Fishbein (1976).

p. 387   Aguilar and Williamson (1968) as well as Conel (1939–1967) have described the increase in the number of connections between neurons that takes place in the brains of infants and young children.

p. 388   See Latter (1980).

p. 389   The descriptions by Freedman (1974), (1979) of differences between neonatal infants make fascinating reading.

p. 390   Hinde's (1974) account of the biological basis of human behavior is extensive and carefully worked out.

p. 391   See Barash (1977).

p. 392   The possible genetic basis of a feeling for grammatical structure in infants is discussed by Miller, Galanter, and Pribram (1960), Wilson (1978), and many other authors.

p. 393   See Sahlins (1976).

p. 394   See Medawar (1972) and Dawkins (1976).

p. 395   See Alexander and Borgia (1978, p. 471).

p. 397   See Harlow (1962), Harlow et al. (1966).

p. 398   See Neel (1970).

p. 398   See Alexander and Borgia (1978, p. 470).

p. 399   See Mead (1956).

p. 399   See Dobzhansky (1969).

p. 401   See Richerson and Boyd (1977).

p. 401   Evidence that the desire for power is an important factor controlling human behavior is presented by Durham (1976b). Observations by Bandura, Ross, and Ross (1968), (1971) indicate that this kind of motivation often appears in young children, and it may have a fairly strong genetic component.

p. 404   See Sipes (1973).

p. 404   See Durham (1976a).

p. 404   See Berkowitz (1969).

p. 405   The material on this and the next few pages is from Johnson (1972).

p. 409   See Devore (1971).

p. 409   See Durham (1976a).

p. 411   See Bonner (1980).

Chapter Fourteen

p. 415   See Spengler (1929), Toynbee (1935–1961), and Wells (1921).

p. 416   See Bigelow (1970), Cameron (1973), Coon (1962), Darlington (1969), Harris (1974), (1977), MacNeill (1970, 1979), Service (1975), and Wittfogel (1957).

p. 426   Oliver's study is reviewed by Service (1975).

p. 433   I refer here to a well-known poem from Blake's "Milton" that was used by British labor as a rallying hymn. See Blake (1970, p. 99).

Chapter Fifteen

p. 439   See Haldane (1963).

p. 441   See Reed (1965).

REFERENCES

PART ONE

Axelrod, D. I. 1975. Evolution and biogeography of Madro–Tethyan sclerophyll vegetation. *Annals of the Missouri Botanical Garden* 62:280–344.

Axelrod, D. I. 1976. History of the coniferous forests, California and Nevada. *University of California Publications in Botany* 70:1–62.

Ayala, F. J. 1978. The mechanisms of evolution. *Scientific American* 239(3):56–69.

Ayala, F. J., and Dotzhansky, T. (eds). 1974. *Studies in the Philosophy of Biology.* New York: Macmillan.

Ayala, F. J., and Kiger, J. A., Jr. 1980. *Modern Genetics.* Menlo Park, Calif.: Benjamin/Cummings.

Ayala, F. J., and Valentine, J. W. 1979. *Evolving.* Menlo Park, Calif.: Benjamin/Cummings.

Beadle, G. W. 1980. The ancestry of corn. *Scientific American* 242(1):112–119, 162.

Bock, W. J. 1970. Microevolutionary sequences as a fundamental concept in macroevolutionary models. *Evolution* 24:704–722.

Bonner, J. T. 1980. *The Evolution of Culture in Animals.* Princeton: Princeton University Press.

Chetverikov, S. S. 1959. On certain aspects of the evolutionary process from the standpoint of genetics. Translated from the Russian (1926) by M. Barker with commentary by I. M. Lerner. *Proceedings of the American Philosophical Society* 105(2):167–195.

Cloud, P. 1977. *Cosmos, Earth and Man.* New Haven: Yale University Press.

Colbert, E. 1955. *Evolution of the Vertebrates.* New York: Wiley.

Darwin, C. 1872. *The Origin of Species*. (The First Edition is not usually reprinted; most available copies are of the Sixth London Edition.)

Dobzhansky, T. 1937. *Genetics and the Origin of Species*. New York: Columbia University Press.

Dobzhansky, T. 1951. *Genetics and the Origin of Species*, 3rd ed. New York: Columbia University Press.

Dobzhansky, T. 1969. *The Biology of Ultimate Concern*. London: Rapp and Whiting.

Dobzhansky, T. 1970. *Genetics of the Evolutionary Process*. New York: Columbia University Press.

Dobzhansky, T., Ayala, F. J., Stebbins, G. L., and Valentine, J. W. 1977. *Evolution*. San Francisco: W. H. Freeman and Company.

Eicher, D. L. 1977. *Geologic Time*. Englewood Cliffs, N.J.: Prentice-Hall.

Fisher, R. A. 1930. *The Genetical Theory of Natural Selection*. Oxford: Clarendon Press.

Galinat, W. 1975. The evolutionary emergence of maize. *Bulletin of the Torrey Botanical Club* 102:313–324.

Goldschmidt, R. 1940. *The Material Basis of Evolution*. New Haven: Yale University Press.

Gould, S. J. 1977. *Ontogeny and Phylogeny*. Cambridge: Harvard University Press.

Grant, V. 1971. *Plant Speciation*. New York: Columbia University Press.

Haldane, J. B. S. 1932. *The Causes of Evolution*. New York: Harper & Row. (The most significant material in this book is contained in its Appendix.)

Hardin, G. 1961. *Biology, Its Principles and Implications*. San Francisco: W. H. Freeman and Company.

Hungate, R. 1966. *The Rumen and Its Microbes*. New York: Academic Press.

Huxley, T. H. 1909. *Autobiography and Selected Essays*, edited by A. L. F. Snell. Boston: Houghton Mifflin.

Kavenoff, R., Klotz, L. C., and Zimm, B. H. 1974. On the nature of chromosome-sized DNA molecules. *Cold Spring Harbor Symposium on Quantitative Biology* 38:1–8.

King, M. S., and Wilson, A. C. 1975. Evolution at two levels: Molecular similarities and biological differences between humans and chimpanzees. *Science* 188:107–116.

Lande, R. 1976. Natural selection and random genetic drift in phenotypic evolution. *Evolution* 30:314–334.

Lorenz, K. 1970. *Studies in Animal and Human Behavior*, Vol. 1. Translated by R. Martin. Cambridge: Harvard University Press.

Mangelsdorf, P. 1958. Reconstructing the ancestor of corn. *Proceedings of the American Philosophical Society* 102(5):454–463.

Mayr, E. 1942. *Systematics and the Origin of Species*. New York: Columbia University Press.

Mayr, E. 1963. *Animal Species and Evolution*. Cambridge: Harvard University Press.

Mayr, E. 1970. *Populations, Species, and Evolution.* Cambridge: Harvard University Press (This is a shorter version of the 1963 edition.)

Mayr, E., and Provine, W. (eds.). 1980. *The Evolutionary Synthesis.* Cambridge: Harvard University Press.

Monod, J. 1971. *Chance and Necessity: An Essay on the Natural Philosophy of Modern Biology.* New York: Knopf.

Neurath, H. 1964. Protein-digesting enzymes. *Scientific American* 211(6):68–79.

Oparin, S. 1961. *Life: Its Nature, Origin and Development.* Translated by Ann Synge. Edinburgh: Oliver and Boyd.

Parker, G. H. 1925. *What Evolution Is.* Cambridge: Harvard University Press.

Petit, C., and Zuckerkandl, E. 1976. *Évolution: Génétique des Populations, Évolution Moléculaire.* Paris: Hermann Cie.

Raup, D. M., and Stanley, S. M. 1978. *Principles of Paleontology.* San Francisco: W. H. Freeman and Company.

Romer, A. S. 1966. *Vertebrate Paleontology.* Chicago: University of Chicago Press.

Romer, A. S. 1970. *The Vertebrate Body,* 4th ed. Philadelphia: Saunders.

Schopf, J. W. 1978. The evolution of the earliest cells. *Scientific American* 239(3):110–138.

Simons, E. L. 1972. *Primate Evolution.* New York: Macmillan.

Simpson, G. G. 1944. *Tempo and Mode in Evolution.* New York: Columbia University Press.

Simpson, G. G. 1949. *The Meaning of Evolution.* New Haven: Yale University Press.

Simpson, G. G. 1951. *Horses: The Story of the Horse Family in the Modern World and Through Sixty Million Years of History.* New York: Oxford University Press.

Simpson, G. G. 1953. *The Major Features of Evolution.* New York: Columbia University Press.

Stanley, S. M. 1979. *Macroevolution: Pattern and Process.* San Francisco: W. H. Freeman and Company.

Stebbins, G. L. 1950. *Variation and Evolution in Plants.* New York: Columbia University Press.

Stebbins, G. L. 1977. *Processes of Organic Evolution,* 3rd ed. Englewood Cliffs, N. J.: Prentice-Hall.

Stebbins, G. L., and Lewontin, R. 1971. Comparative evolution at the levels of molecules, organisms, and populations. In *Darwinian, Neo-Darwinian and Non-Darwinian Evolution. Proceedings of the Sixth Symposium on Mathematics and Statistical Probability,* Berkeley 5:23–42.

Teilhard de Chardin, P. 1959. *The Phenomenon of Man.* Translated by Bernard Wall. London: Collins.

Thomasson, J. R. 1979. Late Cenozoic grasses and other angiosperms from Kansas, Nebraska, and Colorado: Biostratigraphy and relationships to living taxa. *University of Kansas Bulletin* 218.

Van Valen, L. 1973. A new evolutionary law. *Evolutionary Theory* 1:1–30.

White, M. J. D. 1978. *Animal Speciation*. San Francisco: W. H. Freeman and Company.

Wolfe, J. A. 1972. An interpretation of the Alaskan Tertiary floras. In *Floristics and Paleofloristics of Asia and Eastern North America*, edited by A. Graham. New York: Elsevier.

Wright, S. 1931. Evolution in Mendelian populations. *Genetics* 16:97–159.

Wright, S. 1977. *Evolution and the Genetics of Populations, Vol. 3: Experimental Results and Evolutionary Deductions*. Chicago: University of Chicago Press.

PART TWO

Alexander, R. M. 1975. *The Chordates*. Cambridge: Cambridge University Press.

Barghoorn, E. 1971. The oldest fossils. *Scientific American* 224(5):30-42.

Barrington, E. V. W. 1965. *The Biology of Hemichordata and Protochordata*. San Francisco: W. H. Freeman and Company.

Beam, C., and Himes, M. 1977. Sexual isolation and genetic diversification among some strains of *Crypthecodinium cohnii*-like flagellates: Evidence of speciation. *Journal of Protozoology* 24:532–539.

Beck, C. B. (ed.). 1976. *Origin and Early Evolution of Angiosperms*. New York: Columbia University Press.

Beklemishev, V. N. 1969. *Principles of Comparative Anatomy of Invertebrates*. Translated from the Russian by M. McLennon; edited by Z. Kabata. Edinburgh: Oliver and Boyd.

Bernal, J. D. 1967. *The Origin of Life*. Cleveland: World.

Bold, H. C. 1970. Some aspects of the taxonomy of soil algae. In *Phylogenesis and Morphogenesis in the Algae*, edited by J. F. Frederick and R. M. Klein. *Annals of the New York Academy of Science* 175:601–616.

Bonner, J. T. 1965. *Size and Cycle*. Princeton: Princeton University Press.

Cairns-Smith, A. G. 1971. *The Life Puzzle*. Toronto: University of Toronto Press.

Calder, N. 1973. *The Life Game: Evolution and the New Biology*. New York: Dell.

Campbell, B. G. 1976. *Humankind Emerging*. Boston: Little, Brown.

Carroll, R. L. 1969. Problems of the origin of reptiles. *Biological Reviews* 44:393–432.

Carter, G. S. 1967. *Structure and Habit in Vertebrate Evolution*. University of Washington Press Biology Series, 520 pp.

Clark, R. B. 1964. *Dynamics in Metazoan Evolution: The Origin of the Coelom and Segments*. New York: Oxford University Press.

Clark, R. B., and Panchen, A. L. 1974. *Synopsis of Animal Classification*. London: Chapman and Hall.

Cloud, P. 1977. *Cosmos, Earth and Man*. New Haven: Yale University Press.

Colbert, E. H. 1969. *Evolution of the Vertebrates*, 2nd ed. New York: Wiley.

Crompton, A. W., and Jenkins, F. A., Jr. 1968. Molar occlusion in later Triassic mammals. *Biological Reviews* 43:427–458.

Crompton, A. W., and Parker, P. 1978. Evolution of the mammalian masticatory apparatus. *American Scientist* 66:192–201.

Curtin, R., and Dolhinow, P. 1978. Primate social behavior in a changing world. *American Scientist* 66:468–475.

Dickerson, R. E. 1978. Chemical evolution and the origin of life. *Scientific American* 239(3):30–46.

Foster, A. S., and Gifford, E. R. 1959. *Morphology of Vascular Plants,* 2nd ed. San Francisco: W. H. Freeman and Company.

Fox, S. (ed.). 1965. *The Origins of Prebiological Systems.* New York: Academic Press.

Gardiner, M. S. 1972. *The Biology of Invertebrates.* New York: McGraw-Hill.

Gardner, R., and Gardner, B. 1969. Teaching language to a chimpanzee. *Science* 165:664–672.

Glaessner, M. F. 1971. Geographic distribution and time range of the Ediacara Precambrian fauna. *Bulletin of the Geological Society of America* 82:509–514.

Goodall, J. 1971. *In the Shadow of Man.* Glasgow: Fontana/Collins.

Grant, V. 1971. *Plant Speciation.* New York: Columbia University Press.

Grell, K. B. 1973. *Protozoology.* New York: Springer-Verlag.

Hardin, G. 1961. *Biology, Its Principles and Implications.* San Francisco: W. H. Freeman and Company.

Harlow, H. F., Harlow, M. K., Dodsworth, R. O., Arling, G. L. 1966. Maternal behavior of rhesus monkeys deprived of mothering and peer associations in infancy. *Proceedings of the American Philosophical Society* 110:58–66.

Himes, M., and Beam, C. 1975. Genetic analysis in the dinoflagellate *Crypthecodinium cohnii:* Evidence for unusual meiosis. *Proceedings of the National Academy of Sciences* 72:4546–4549.

Jägersten, G. 1972. *Evolution of the Metazoan Life Cycle: A Comprehensive Theory.* New York: Academic Press.

King, M. C., and Wilson, A. C. 1975. Evolution at two levels: Molecular similarities and biological differences between humans and chimpanzees. *Science* 188:107–116.

Lancaster, J. 1975. *Primate Behavior and the Emergence of Human Culture.* New York: Holt, Rinehart and Winston.

Lazarides, E., and Revel, J. P. 1979. The molecular basis of cell movement. *Scientific American* 240(5):110–113.

Leakey, R. L., and Lewin, R. 1977. *Origins.* New York: Dutton.

McNab, B. K. 1978. The evolution of endothermy in the phylogeny of mammals. *American Naturalist* 112:1–27.

Margulis, L. 1970. *Origin of Eukaryotic Cells.* New Haven: Yale University Press.

Margulis, L. 1981. *Symbiosis in Cell Evolution: Life and Its Environment on the Early Earth.* San Francisco: W. H. Freeman and Company.

Maynard-Smith, J. 1978. *The Evolution of Sex*. Cambridge: Cambridge University Press.

Millot, J. 1955. The coelacanth. *Scientific American* 193(6):34–39.

Moore, R. C., Lalicker, C. G., and Fischer, A. G. 1952. *Invertebrate Fossils*. New York: McGraw-Hill.

Oliver, C. G. 1979. Genetic differentiation and hybrid viability within and between some Lepidoptera species. *American Naturalist* 114:681–694.

Orgel, L. E. 1973. *The Origins of Life: Molecules and Natural Selection*. New York: Wiley.

Pilbeam, D. 1972. *The Ascent of Man*. New York: Macmillan.

Poirer, F. E. 1977. *In Search of Ourselves: An Introduction to Physical Anthropology*, 2nd ed. Minneapolis: Burgess.

Poirier, F. E. (ed.). *Primate Socialization*. New York: Random House.

Premack, A., and Premack, D. 1972. Teaching language to an ape. *Scientific American* 227(4):92–99.

Rao, M., Odom, D. G., and Oro, J. 1980. Clays in prebiological chemistry. *Journal of Molecular Evolution* 15:317–332.

Raven, P. H., Evert, R., and Curtis, H. 1981. *Biology of Plants*, 3rd ed.. New York: Worth.

Ribbert, D. 1972. Relation of puffing to bristle and footpad differentiation in *Calliphora* and *Sarcophaga*. In *Developmental Studies on Giant Chromosomes*, edited by W. Beermann. New York: Springer-Verlag.

Romer, A. S. 1966. *Vertebrate Paleontology*, 3rd ed. Chicago: University of Chicago Press.

Romer, A. S. 1969. Cynodont reptile with incipient mammalian jaw articulation. *Science* 166:881–882.

Schaeffer, B. 1965. The Rhipidistian–Amphibian transition. *American Zoologist* 5:267–276.

Schopf, J. W. 1978. The evolution of the earliest cells. *Scientific American* 239(3):110–138.

Simons, E. L. 1972. *Primate Evolution*. New York: Macmillan.

Smith, H. W. 1949. *Kamongo, or the Lungfish and the Padre*. New York: Viking Press.

Stebbins, G. L. 1969. *The Basis of Progressive Evolution*. Chapel Hill: University of North Carolina Press.

Stebbins, G. L. 1974. *Flowering Plants: Evolution Above the Species Level*. Cambridge: Harvard University Press.

Stebbins, G. L., and Hill, G. H. C. 1980. Did multicellular plants invade the land? *American Naturalist* 115:342–353.

Thomson, K. S. 1969. The biology of the lobe-finned fishes. *Biological Reviews* 44:91–154.

Valentine, J. W. 1973. *Evolutionary Paleoecology of the Marine Biosphere*. Englewood Cliffs, N.J.: Prentice-Hall.

Valentine, J. W. 1978. The evolution of multicellular plants and animals. *Scientific American* 239(3):140–152.

Ward, P. W., Greenwald, L., and Greenwald, O. E. 1980. The buoyancy of the chambered nautilus. *Scientific American* 243(4):190–203.

Washburn, S. L., and Moore, R. 1974. *Ape into Man: A Study of Human Evolution.* Boston: Little, Brown.

Whittaker, R. H. 1969. New concepts of kingdoms of organisms. *Science* 163:150–160.

Williams, G. C. 1975. *Sex and Evolution.* Princeton: Princeton University Press.

Wilson, E. O. 1975. *Sociobiology.* Cambridge: Harvard University Press.

Woese, C. R., Magrum, L. E., and Fox, G. E. 1978. Archaebacteria. *Journal of Molecular Evolution* 11:245–252.

PART THREE

Aguilar, M. J., and Williamson, M. L. 1968. Observations on growth and development of the brain. In *Human Growth,* edited by D. B. Cheek. Philadelphia: Lea and Febiger.

Alexander, R. D. 1975. The search for a general theory of behavior. *Behavioral Science* 20:77–100.

Alexander, R. D. 1979. *Darwinism and Human Affairs.* Seattle: University of Washington Press.

Alexander, R. D., and Borgia, G. 1978. Group selection and levels of organization. *Annual Reviews of Ecology and Systematics* 9:449–474.

Alper, J. S. 1978. Ethical and social implications. In *Sociobiology and Human Nature,* edited by M. S. Gregory, A. Silver, and D. Sutch. San Francisco: Jossey-Bass, 195–212.

Anastasi, A. 1968. Heredity, environment, and the question "How?". In *Contemporary Issues in Developmental Psychology,* edited by N. S. Endler, L. R. Boulter, and H. Osser. New York: Holt, Rinehart and Winston.

Bandura, A., Ross, D., and Ross, S. A. 1968. A comparative test of the status, envy, social power, and secondary reinforcement theories of identificatory learning. In *Contemporary Issues in Developmental Psychology,* edited by N. S. Endler, L. R. Boulter, and H. Osser. New York: Holt, Rinehart and Winston, 626–635.

Barash, D. P. 1977. *Sociobiology and Behavior.* New York: Elsevier.

Barash, D. P. 1978. Evolution as a paradigm for behavior. In *Sociobiology and Human Nature,* edited by M. S. Gregory, A. Silver, and D. Sutch. San Francisco: Jossey-Bass, 13–32.

Baumgartel, W. 1976. *Up Among the Mountain Gorillas.* New York: Hawthorn Books.

Bennett, J. W. 1976. *The Ecological Transition: Cultural Anthropology and Human Transition.* New York: Pergamon Press.

Berkowitz, L. 1969. *Roots of Aggression*. New York: Atherton.

Bigelow, R. 1970. *The Dawn Warriors: Man's Evolution Toward Peace*. Boston: Little, Brown.

Blake, W. 1970. *A Choice of Blake's Verse*. Selected and with an Introduction by K. Raine. London: Faber and Faber.

Bloom, B. S. 1964. *Stability and Change in Human Characteristics*. New York: Wiley.

Bonner, J. T. 1980. *The Evolution of Culture in Animals*. Princeton: Princeton University Press.

Bowlby, J. 1969. *Attachment*. New York: Basic Books.

Butzer, K. W. 1977. Environment, culture, and human evolution. *American Scientist* 65:572–584.

Cameron, K. N. 1973. *Humanity and Society: A World History*. Bloomington: University of Indiana Press.

Cohen, M. N. 1977. *The Food Crisis in Prehistory: Overpopulation and the Origins of Agriculture*. New Haven: Yale University Press.

Conel, J. 1939–1967. *The Postnatal Development of the Human Cerebral Cortex*, Vol. 8. Cambridge: Harvard University Press.

Coon, C. R. 1962. *The Story of Man*. New York: Knopf.

Darlington, C. D. 1969. *The Evolution of Man and Society*. London: Allen and Unwin.

Dawkins, R. 1976. *The Selfish Gene*. New York: Oxford University Press.

de Lumley, H. 1969. A Paleolithic camp at Nice. *Scientific American* 220(5):42–50.

Devore, I. 1971. The evolution of human society. In *Man and Beast: Comparative Social Behavior*, edited by J. F. Eisenberg. Washington, D. C.: Smithsonian Institution Press, 299–311.

Dobzhansky, T. 1969. *The Biology of Ultimate Concern*. London: Rapp and Whiting.

Dubos, R. 1974. *Beast or Angel? Choices That Make Us Human*. New York: Scribners.

Durham, W. 1976a. Resource competition and human aggression, I: A review of primitive war. *Quarterly Review of Biology* 51:385–415.

Durham, W. 1976b. The adaptive significance of cultural behavior. *Human Ecology* 4:89–121.

Fishbein, H. D. 1976. *Evolution, Development, and Children's Learning*. Pacific Palisades: Goodyear.

Freedman, D. G. 1974. *Human Infancy: An Evolutionary Perspective*. Hillsdale, N.J.: Lawrence Erlbaum.

Freedman, D. G. 1979. *Human Sociobiology: A Holistic Approach*. New York: Free Press.

Fuller, J. L. 1978. Genes, brains and behavior. In *Sociobiology and Human Nature*, edited by M. S. Gregory, A. Silver, and D. Sutch. San Francisco: Jossey-Bass, 98–115.

Goodall, J. 1971. *In the Shadow of Man*. Glasgow: Fontana/Collins.

Gregory, M. S., Silver, A., and Sutch, D. (eds.) 1978. *Sociobiology and Human Nature*. San Francisco: Jossey-Bass.

Grene, M. 1978. Sociobiology and the human mind. In *Sociobiology and Human Nature*, edited by M. S. Gregory, A. Silver, and D. Sutch. San Francisco: Jossey-Bass, 183–194.

Haldane, J. B. S. Biological possibilities in the next ten thousand years. In *Man and His Future*, edited by G. Wolstenholme. Boston: Little, Brown, 337–361.

Harlow, H. F. 1962. The heterosexual affectional system in monkeys. *American Psychology* 17:1–9. Reprinted in *Readings in Animal Behavior*, 3rd ed., edited by T. McGill. New York: Holt, Rinehart and Winston, 1977, 304–313.

Harlow, H. F., Harlow, M. K., Dodsworth, R. O., Arlington, G. L. 1966. Maternal behavior of rhesus monkeys deprived of mothering and peer associations in infancy. *Proceedings of the American Philosophical Society* 110:58–66.

Harris, M. 1974. *Cows, Pigs, Wars, and Witches*. New York: Random House.

Harris, M. 1977. *Cannibals and Kings: The Origin of Culture*. New York: Random House.

Hinde, R. A. 1974. *Biological Bases of Human Social Behavior*. New York: McGraw-Hill.

Hull, D. L. 1978. Scientific bandwagon or travelling medicine show. In *Sociobiology and Human Nature*, edited by M. S. Gregory, A. Silver, and D. Sutch. San Francisco: Jossey-Bass.

Johnson, R. N. 1972. *Aggression in Man and Animals*. Philadelphia: Saunders.

Jolly, C. L. 1970. The seed-eaters: A new model of hominid differentiation based on a baboon analogy. *Man* n.s. 5:5–26.

Klopfer, P. H. 1977. Social Darwinism lives! Should it? *Yale Journal of Biological Medicine* 50(1):77–84.

Lakhanpal, R. N. 1970. Tertiary floras of India and their bearing on the historical geology of the region. *Taxon* 19:675–694.

Lancaster, J. B. 1973. On the evolution of tool-using behavior. In *Man in Evolutionary Perspective*, edited by C. L. Bruce and J. Metress. New York: Wiley, 79–90.

Latter, B. D. H. 1980. Genetic differences within and between populations of the major human subgroups. *American Naturalist* 116:220–234.

Leakey, M. 1971. *Olduvai Gorge, Vol. 3: Excavations in Beds I and II, 1960–1963*. Cambridge: Cambridge University Press.

Lee, R. B. 1972. The !Kung bushmen of Botswana. In *Hunters and Gatherers Today*, edited by M. G. Bicehieri. New York: Holt, Rinehart and Winston, 327–368.

MacNeill, W. H. 1970. *The Rise of the West*. Chicago: University of Chicago Press.

MacNeill, W. H. 1979. *A World History*, 3rd ed. New York: Oxford University Press.

Marshack, A. 1972. *The Roots of Civilization.* New York: McGraw-Hill.

Mead, M. 1956. *New Lives for Old: Cultural Transformation—Manus 1928–1953.* New York: Morrow.

Medawar, P. B. 1972. *The Hope of Progress.* London: Methuen.

Miller, G., Galanter, F., and Pribram, K., 1960. *Plans and the Structure of Behavior.* New York: Holt, Rinehart and Winston.

Neel, J. V. 1970. Lessons from a "primitive" people. *Science* 170:815–822.

Reed, S. C. 1965. The evolution of human intelligence. *American Scientist* 53:317–326.

Richerson, P. J., and Boyd, R. 1978. A dual inheritance model of the human evolutionary process, I: Basic postulates and a simple model. *Journal of Sociological Structure* 1:127–154.

Ruse, M. 1979. *Sociobiology: Sense or Nonsense?* Dordrecht: D. Reidel.

Sahlins, M. 1976. *The Use and Abuse of Biology: An Anthropological Critique of Sociobiology.* Ann Arbor: University of Michigan Press.

Sarich, V. M., and Cronin, J. E. 1977. Molecular systematics of the primates. In *Molecular Anthropology,* edited by M. Goodman and R. E. Tashian. New York: Plenum Press, 141–170.

Schaller, G. 1964. *The Year of the Gorilla.* Chicago: University of Chicago Press.

Service, E. R. 1975. *Origins of the State and Civilization.* New York: Norton.

Sipes, R. G. 1973. War, sports, and aggression: An empirical test of two rival theories. *American Anthropologist* 75:64–86.

Solecki, R. S. 1971. *Shanidar, the First Flower People.* New York: Knopf.

Spengler, O. 1929. *The Decline of the West.* New York: Knopf.

Tiger, L. 1969. *Men in Groups.* New York: Random House.

Toynbee, A. 1953. *A Study of History.* 5 Vols. New York: Oxford University Press.

Wallace, A. R. 1913. *Social Environment and Moral Progress.* New York: Cassell.

Washburn, S. L. 1970. Comment on a possible evolutionary basis for aesthetic appreciation in men and apes. *Evolution* 24:824–825.

Washburn, S. L. 1978. Animal behavior and social anthropology. In *Sociobiology and Human Nature,* edited by M. S. Gregory, A. Silver, and D. Sutch. San Francisco: Jossey-Bass, 53–74.

Washburn, S. L., and Harding, R. S. 1970. Evolution of primate behavior. In *The Neurosciences: Second Study Program,* edited by F. O. Schmitt. New York: Rockefeller University Press, 39–47.

Washburn, S. L., and Lancaster, C. S. 1968. The evolution of hunting. In *Man the Hunter,* edited by R. B. Lee and I. DeVore. Chicago: Aldine, 293–303. Boston: Little, Brown.

Wells, H. G. 1921. *The Outline of History.* New York: Macmillan.

Whyte, R. O. 1977. The botanical Neolithic revolution. *Human Ecology* 5:209–222.

Wilson, E. O. 1975. *Sociobiology*. Cambridge: Harvard University Press.

Wilson, E. O. 1978. *On Human Nature*. Cambridge: Harvard University Press.

Wittfogel, K. A. 1957. *Oriental Despotism*. New Haven: Yale University Press.

GLOSSARY

adenosine triphosphate (ATP) An organic molecule that consists of the purine base adenine linked to a 5-carbon ribose sugar and three phosphate (PO_4) groups. By losing a phosphate group it delivers chemical energy that is used in a great variety of cell reactions.

aerobic respiration The breakdown of complex molecules by means of a sequence of biochemical reactions that requires oxygen, releases carbon dioxide, and generates energy for biological activity.

alga (pl. algae) Any of a large number of simply constructed green plants, most of which live in water. Many are seaweeds; others form pond scum.

allele One of two or more closely similar but not identical genes located at a particular position or locus on a chromosome.

amino acid An organic or carbon-containing acid molecule that also contains an amino (NH_2) group. Of the numerous different kinds of amino acids that have been identified by chemists, twenty are units of the polypeptide chains that make up proteins.

analogy, analogous Similarity in structure, function, or both, due to convergent evolution rather than to genetic or evolutionary relationship.

arthropod Any of a large group of animals, including crabs, shrimps, scorpions, spiders, centipedes, and insects, that have jointed legs and external hard parts or skeletons.

artificial selection Improvement of breeds of domestic animals or varieties of cultivated plants by conscious human selection of desired individuals.

autocatalysis The stimulation of a reaction by one or more of its products. Applied to evolution, the stimulation of evolutionary change by an environmental challenge generated by the evolution of a particular kind of organism.

balanced polymorphism   The existence in a population of two alternative morphological, physiological, or biochemical conditions that are maintained in a more or less stable equilibrium by opposing selective pressures.

bilateral symmetry   A type of symmetry in which an organ or organism has a front and hind end and two similar sides.

biotype   A general term for any genotype that is adapted to a particular habitat.

bottleneck   A figure of speech used to denote the reduction of a population to a small size followed by its later expansion.

bradytelic   Very slow rates of evolution, approaching constancy over long periods of time.

canine teeth   In mammals, the four teeth, usually longer and sharper than the rest, found in the front of the mouth between the incisors and the molars.

carotene   An orange-yellow pigment found in chloroplasts and in some plastids of yellow flowers.

cell   A living structure bounded by a cell wall, containing various organelles and a nucleus or a naked double helix of DNA.

cellular skeleton   A system of protein fibrils and microtubules found in the cytoplasm of eukaryotic cells. It requires photos taken with the electron microscope for its detection.

chitin (ký-tin)   A nitrogen-containing substance that forms the body wall of many animals, particularly insects and most fungi.

chlorophyll   The green pigment that performs photosynthesis in plants, made up of large organic molecules with a central core of magnesium.

chloroplast   A green plastid containing chlorophyll, carotene, and xanthophyll, through which light energy is converted into sugars and starch in green plants and eukaryotic microorganisms.

chromosome   An organelle found in the nuclei of eukaryotic organisms that contains an extremely long double helix of DNA plus associated proteins.

ciliates   A group of protozoa having unusually large cells that are partly or entirely covered by large numbers of cilia.

cilium (pl. **cilia**)   An organelle of eukaryotic organisms, similar to flagella but usually smaller and more numerous, that performs various functions by active motility.

class (of organisms)   The category of the taxonomic hierarchy above the order and below the phylum.

cleavage   In animals, the process by which the fertilized egg forms several cells of the early embryo.

coacervate   A mixture consisting of droplets containing large organic macromolecules suspended in a watery medium.

coenocyte   A large cell containing many nuclei, found in some algae and fungi.

coevolution   The simultaneous evolution of two genetically independent but ecologically interdependent evolutionary lines, such as predator–prey, stimulated by selection pressures caused by feedback interactions between populations belonging to the two lines.

colloid   A jellylike substance made up of large suspended organic molecules.

corolla   In flowering plants, the structure formed by the congenital union of petals.

cultural diffusion   The spread of cultural traits from one society to another by various means, such as trade, oral communication, or written communication.

cultural fitness   The capacity of an individual to contribute to the cultural advancement of a society.

cultural pool   The sum total of the cultural traits of a particular society.

cultural template   Any manmade object that aids in the transmission of a cultural trait from one generation to the next, for example, books, diagrams, mathematical formulas.

cytochrome   A type of protein that plays an important role in electron transfer in living cells.

deoxyribonucleic acid (DNA)   A long linear molecule in the form of a double helix, which consists of a particular sequence of nucleotides containing the purine bases adenine and guanine and the pyrimidines cytosine and thymine, plus a 5-carbon sugar and a phosphate group. It differs from ribonucleic acid (RNA) in that it lacks a hydroxyl (OH) group at the second position of the 5-carbon sugar and possesses thymidine rather than uracil.

differential survival   Differences in longevity or the capacity to persist over long periods of time; can be applied to individuals, populations, or species.

dominant   As applied to properties of genes, the ability of a gene to express its phenotype when associated with its opposite allele in the heterozygous condition.

drift   The tendency of the frequency of a gene or allele in a population to fluctuate about a mean value, due to errors of chance sampling. In small populations, drift can lead to fixation of one allele and loss of the opposite allele, thereby reducing genetic variation.

ecological niche   The position of a species in a community of organisms, based on its adaptive requirements and functions.

ecosystem   An interacting group of organisms that occupies a particular kind of territory.

electrophoresis   The separation of proteins on an electrically polarized grid, based on the electric charges of their molecules.

entropy   A measure of the unavailable energy in a closed thermodynamic system.

enzyme   A large complex organic molecule, usually a protein, that speeds up the chemical reactions taking place in living systems.

epigenesis   The appearance of new, distinctive forms based on drastic modification of pre-existing forms; applied usually to stages in the development of an embryo or fetus, but also applied to successive stages in the evolution of form.

epoch   A division of geological time, included within a period and era.

era   One of the major divisions of geological time.

eukaryotes A term applied to all organisms that have cells containing a nucleus surrounded by a membrane and that perform cell division by mitosis or meiosis.

exosomatic transmission Transmission of a trait by nongenetic means, that is, outside of the body.

family A subdivision of the taxonomic hierarchy that includes one or more genera and is included within an order.

fertilization The union of a sperm or pollen nucleus with an egg nucleus to form a zygote.

flagellates A class of protozoa having relatively small cells that bear one or a few flagella.

flagellum (pl. **flagella**) A whiplike process attached to the cells of many organisms, particularly bacteria and smaller unicellular eukaryotes. Flagella of eukaryotes have the same ultrastructure as cilia.

founder principle The concept that in very small, newly founded populations, often consisting of a single gravid female, entirely new adaptive complexes are likely to become established, often associated with new, adaptively neutral traits.

gamete One of the two cells that unite to form a zygote, usually an egg or sperm cell.

gametophyte In mosses or ferns, the kind of plants that produce gametes rather than spores (see **sporophyte**).

gene A part of the DNA molecule that provides information for a particular morphological or chemical characteristic of the organism, usually because it codes for a protein or for one of its chains.

gene pool The total amount of variation in the genes present among the individuals of a population.

genotype The total amount of genetic information contained in an organism; the genetic constitution of an organism with respect to one or a few gene loci under consideration.

genus A category in the hierarchy of biological classification, above the species and below the family.

hemoglobin The protein molecule in red blood cells that carries oxygen through the body.

heterozygous An organism containing two different alleles at a particular gene locus.

holism A methodology of evolutionary study that emphasizes synthesis and systems relationships rather than reductional analysis.

homologous Similar with respect to both structure and function resulting from common descent.

homozygous Containing two similar alleles at a particular gene locus.

hydrostatic Relating to pressure and equilibrium of liquids, especially water.

incisor teeth The front teeth of mammals and humans.

inclusive fitness   The capacity of an individual animal or human to produce a maximum of vigorous, well-adapted adult offspring.

ingest   To take in solid food by surrounding or engulfing it rather than by absorption.

ion   An electrified particle formed when a neutral atom or group of atoms loses or gains one or more electrons.

kingdom   The highest category in the taxonomic hierarchy of biology.

larva (pl. larvae)   The young stage of an animal that differs greatly from the adult form.

locus (of a gene)   The particular position on the DNA molecule of a chromosome at which a gene is located.

macromolecule   A large organic molecule found in living systems or their derivatives made up of many smaller units, usually in a linear order. Examples: proteins, nucleic acids.

marsupial   One of a subclass of mammals that nurture their young in a pouch.

meiosis   The pair of successive mitotic divisions that take place in reproductive organs and bring about reduction of chromosome number for the formation of gametes.

metaphase   The stage of mitosis or meiosis at which chromosomes are centrally aligned.

microorganism   Any of various one-celled organisms such as bacteria and protozoa.

microtubule   An organized, tubelike aggregate of protein molecules that forms part of the cytoplasmic ultrastructure or submicroscopic skeleton of a living cell.

mitochondrion (pl. mitochondria)   Tiny organelles found in the cytoplasm of eukaryotic cells that generate chemical energy for cellular activity.

mitosis   The succession of stages passed through by chromosomes during nuclear division. Mitosis ensures transmission of the organism's genes to every new nucleus.

modifying factors   Genes having slight effects on the expression of other genes.

molar teeth   In mammals, back teeth that are differentiated for chewing and grinding food.

molluscs   Members of a phylum of animals that incudes snails, slugs, oysters, clams, squids, and octopi.

mosaic evolution   The tendency of different characteristics of the organism to evolve at different rates in the same evolutionary line.

multiple factors   Genes situated at different loci that have similar effects on the phenotype and contribute to the inheritance of a quantitative character. Sometimes called *polygenes*.

mutation pressure   The effect of many mutations having similar effects in altering the characteristics of a population over successive generations.

mutations   Changes in the structure of DNA that usually alter the organism's appearance or behavior.

natural selection   The process—based on overreproduction, genetic variability, and survival of best-adapted organisms in a particular environment—that maintains or alters gene frequencies in populations.

neoteny   The shifting of reproductive maturity to early stages of development.

notochord   An elastic rod of cells that, in the lowest vertebrates, forms the supporting axis of the body.

nucleotide   A single unit of a nucleic acid molecule, consisting of a base (purine or pyrimidine), a 5-carbon sugar, and a phosphate group.

nucleus   The part of a eukaryotic cell that contains the chromosomes and other organelles.

organelle   Any of several kinds of organized cellular structures consisting chiefly of protein and performing specific functions.

organic   Containing carbon and usually derived from the activity of living organisms.

orthogenesis   The supposed tendency for evolutionary lines to progress steadily in a particular direction, regardless of environmental influences or natural selection. The theory of orthogenesis was maintained by many paleontologists between 50 and 100 years ago, but is now obsolete.

ovary   The egg-containing organ of an animal. Applied also to the ovule-containing structure of plants that usually forms the seed pod or capsule.

paedomorphosis   Retention of juvenile characteristics in later stages of life.

paleontology   The branch of science devoted to uncovering and analyzing the fossil record.

period   A span of geological time consisting of several million years, longer than an epoch and shorter than an era.

phenotype   The observable characteristics of an individual, resulting from the interaction between the genotype and the environment.

phosphorylation   Adding a phosphate (PO_4) group to an organic compound.

photosynthesis   In green plants and other organisms containing chlorophyll, the use of light energy to synthesize sugar, starch, and other organic compounds.

phylum (pl. phyla)   The taxonomic category below the kingdom and above the class.

placenta   In most mammals, the blood-filled structure by which the fetus is nourished in the uterus.

plankton   Various forms of life that inhabit the surface layers of oceans and lakes.

plastid   An organelle of photosynthetic eukaryotes, usually green, containing chlorophyll and performing photosynthesis.

polygenes   See **multiple factors.**

polymer   A macromolecule that consists of many similar units.

polymorphism   In populations, the condition of containing, usually in a balanced condition, two or more alleles or chromosomal types.

polypeptide   A macromolecule consisting of a chain of amino acid residues in a definite linear sequence. It forms part of a protein molecule.

preadaptation   Having a combination of genes that can be converted relatively easily into an adaptive response to a new environmental challenge.

prebiotic   Events related to the origin and nature of life that occurred in the primeval earth, before the origin of living cells.

prokaryote   Organisms like bacteria that lack clearly defined nuclei, chromosomes, or any other kind of organelles.

prophase   The first stage of mitotic division, consisting of the contraction of chromosomes while still enclosed in the nuclear membrane.

protein   Any of numerous different kinds of organic macromolecules that contain one or more polypeptide chains consisting of amino acids in a definite sequence.

protozoan (pl. protozoa)   The class of single-celled animals or microorganisms in which the entire body consists of a single cell.

punctuated equilibria   A theory that postulates evolution as having occurred not as a series of continuous sequences of change but as a succession of rapid quantum changes interrupting the essentially static continuation of established species.

pure line   A population containing genetically homozygous and genetically similar individuals, produced by self-fertilization over many generations.

quantum evolution   The theory that evolution often includes periods of rapid change, usually in a small population exposed to an unstable environment.

radial symmetry   A form of symmetry of an organism or organ in which similar parts are arranged in a circle around a central point.

radiometric dating   The dating of geological strata or the fossils contained in them on the basis of proportions of isotopes of elements that are subject to atomic decay at a regular rate, for example, uranium–lead, potassium–argon.

recessive allele   An allele that does not express its effects when present with its alternative dominant allele.

reductionism   A philosophy of biology or evolution that emphasizes analysis of a process or phenomenon in terms of its component parts.

revertant   An individual that has reverted from one state to another previous state.

ribonucleic acid (RNA)   A form of nucleic acid produced by transcription from deoxyribonucleic acid (DNA) in which the second carbon position contains a hydroxyl (OH) group, and uracil exists in place of thymidine. It forms the intermediary between the information stored in DNA and the translation of this information to produce a definite amino acid sequence of proteins.

ribosome   A submicroscopic cellular organelle consisting of RNA and protein that forms part of the translation mechanism in protein synthesis.

segregation (Mendelian)   The regular distribution to different gametes of alternative alleles at a gene locus, as a result of meiosis.

sessile   Animals that always remain in the same spot, such as oysters and corals.

spacer DNA   Stretches of the DNA molecule that exist between the coding genes of eukaryotic organisms that may be transcribed but do not code for proteins. Equivalent to middle repetitive DNA; sometimes called "nonsense," "junk," or "parasitic" DNA.

species   A category of the taxonomic hierarchy above the race or subspecies and below the genus. Species are usually delimited by barriers of reproductive isolation.

spindle (mitotic)   A structure consisting of protein macromolecules joined together to form parallel or slightly divergent microfibrils. It is formed during mitosis and provides the framework around which daughter chromosomes become separated from each other.

sporangium (pl. sporangia)   Any of various kinds of sacs or containers found in lower plants that contain single-celled dispersal units, or spores.

sporophyte   In mosses and ferns, the plant or "generation" that has the unreduced chromosome number and produces sporangia and spores.

sport   A word used by pre-Mendelian animal breeders, including Darwin, for an individual that differs greatly from the rest of the flock or herd, usually the result of a mutation.

stamen   The pollen-bearing organ of a flowering plant.

stromatolite   Any of various mineralized bodies formed by excretion from large colonies of bacteria or algae, usually along the seashores. Some of them are contemporary; others are among the oldest known fossils.

symbiont   A species that has a particular symbiotic relationship with another species.

symbiosis   Expressing a mutual interaction between two unrelated species from which both symbiotic partners derive benefit.

template   A mold or model that can be copied and that serves as a guide to the formation of a particular structure.

testis (pl. testes)   The male or sperm-producing organs of an animal.

tissue   In many-celled organisms, an aggregate of cells specialized for performing a particular function.

transcendence   Applied to evolution, a term invented by T. Dobzhansky to characterize evolutionary steps that originate a completely new way of life, with unpredictable consequences.

transcription   In organisms, the formation of a molecule of ribonucleic acid (RNA) from a template of deoxyribonucleic acid (DNA).

translation   In organisms, the formation of a protein (polypeptide) chain from an RNA template with the aid of specific molecules of transfer RNA that can translate the nucleotide sequence of the nucleic acid into the amino acid sequence of the protein, following the genetic code.

tubulin   In most cells, a kind of protein in which the chains of amino acids are joined together in a linear fashion into ultramicroscopic tubular fibrils that form the framework or skeleton.

vascular   As applied to plants, the tissue or system that contains tubular walls of dead cells that transport water, and tough, thick-walled, woody cells that aid in support.

vertebra (pl. vertebrae)   The individual bones that together form the backbone of fishes, reptiles, amphibians, birds, and mammals.

vertebrate   A member of a subphylum of animals possessing vertebrae.

viviparous   In higher animals, the ability to bring forth young alive, rather than by laying eggs.

xanthophyll   In plants, a yellow pigment associated with chlorophyll in the chloroplast.

zoospore   In aquatic plants, chiefly algae, reproductive cells produced by mitosis that swim through the water with the aid of flagella.

zygote   The cell produced by the union of gametes—usually a fertilized egg.

INDEX

(Italic page numbers refer to illustrations.)